Microtechnology and MEMS

The series *Microtechnology and MEMS* comprises textbooks, monographs, and state-of-the-art reports in the very active field of microsystems and microtechnology. Written by leading physicists and engineers, the books describe the basic science, device design, and applications. They will appeal to researchers, engineers, and advanced students.

More information about this series at http://www.springer.com/series/4526

Martin Dienwiebel
Maria-Isabel De Barros Bouchet
Editors

Advanced Analytical Methods in Tribology

Springer

Editors
Martin Dienwiebel
Fraunhofer Institute for Mechanics of Materials
Freiburg, Baden-Württemberg, Germany

and

Institute for Applied Materials
Karlsruhe Institute of Technology KIT
Karlsruhe, Germany

Maria-Isabel De Barros Bouchet
Laboratoire de Tribologie et Dynamique des Systèmes, CNRS-UMR5513, Ecole Centrale de Lyon
Université de Lyon I
Écully, France

ISSN 1615-8326
Microtechnology and MEMS
ISBN 978-3-030-07641-2 ISBN 978-3-319-99897-8 (eBook)
https://doi.org/10.1007/978-3-319-99897-8

Softcover re-print of the Hardcover 1st edition 2018

This Springer imprint is published by the registered company Springer Nature Switzerland AG
The registered company address is: Gewerbestrasse 11, 6330 Cham, Switzerland

Preface

Friction- and wear-related phenomena are experienced by almost everybody in daily life. Whether it is by walking on slippery ground, commuting to work, or when travelling by train, airplane or on a cruise ship, the science known as tribology is omnipresent but mostly hidden to the eye. However, although we appreciate low friction, for example, when we go inline skating or skiing during the weekend or high friction when we want to get our car to stop in front of a red traffic light, we usually do not know how to control the coefficient of friction using established theories. Scientists strive to unravel the mysteries of friction and wear processes since the very first systematic studies more than 500 years ago yet universal guidelines for the selection of the best materials, coatings and chemicals are not easily at hand. The reason for this is not simple as frictional properties are governed by mechanisms that can appear on many different length scales from several metres to the molecular scale and can be related to the topography of sliding surfaces, to the composition and distribution of phases or to the surface chemistry directly. Because of this complexity, an experimentalist needs to use, besides a well-designed tribological apparatus, also other instruments in order to characterize the surfaces, lubricants and materials.

A century ago, the optical microscope was probably the main and only tool for the analysis of sliding surfaces. Many important properties of tribological systems were still hidden at that time. Since then, many new scientific instruments have been invented that can and need to be utilized to improve our understanding of what causes high or low friction in various environments. These analytical instruments undergo constant improvements in resolution, data acquisition and analysis speed. The constantly filling toolbox of analytical methods available to tribologists enabled novel technologies such as coatings and lubricants that provide ultra-low friction and very small wear rates or that allow changing the friction by applying a potential. It was possible to understand the fundamental steps that lead to these improvements by a combination of suitable analytical techniques and also by a combination of experiments and computer simulations. In many recent developments, those techniques have proven to be particularly important that could be applied *in situ* during the experiment or shortly afterwards.

The aim of this book is to provide a broad overview of the possibilities that modern state-of-the-art analytical and simulation techniques give in combination with tribological experiments. In 11 chapters, leading experts introduce the analytical methods and present examples from the recent research on how the method can be successfully applied.

Chapters 1 and 2 deal with microstructural characterization of sliding interfaces. The first chapter presents the method of focused ion beam analysis and related techniques in order to obtain information on shear-induced changes in the microstructure. The second chapter demonstrates the use of the transmission electron microscope (TEM) for the analysis of ultra-low wear tribocouples. The following six chapters are devoted to various techniques for the investigation of tribochemical mechanisms. The application of Near Edge X-Ray Absorption Fine Structure Spectroscopy (NEXAFS) and related Synchrotron-based spectroscopic techniques for the investigation lubricant chemistry and surface termination are presented in Chap. 3. Chapter 4 is giving an overview over the possibilities of gas-phase lubrication experiments in combination with in situ X-ray photoelectron spectroscopy (XPS). Chapter 5 demonstrates the possibilities of the in situ analysis of tribochemical reactions using Auger electron spectroscopy and low-energy electron diffraction (LEED). The analysis of anti-wear tribofilms by X-ray absorption near-edge spectroscopy (XANES) is covered in Chap. 6. Chapter 7 demonstrates the possibilities of micro-FTIR for the in situ analysis of lubricants and greases. Chapter 8 is devoted to the analysis of the use of synchrotron radiation in order to measure real-time wear dynamics. The following three chapters deal with in situ microscopy techniques. Electrochemical atomic force microscopy and atomic-scale friction measurements in an electrolytical environment are introduced in Chap. 9. Examples for friction measurements inside an electron microscope are presented in Chap. 10, and Chap. 11 then reviews instrumentation and the application of in situ topography measurements. The final chapter of this book is dedicated to simulations as a numerical analytical tool.

Obviously, this book is not intended to be an exhaustive review of all analytical techniques available nor does it cover all research in tribology as this would go beyond the bounds of the content of a single book. At this point, the editors are extremely thankful to all authors who contributed to the chapters of the book and their time spent in carefully selecting interesting portions of their work. The editors gratefully acknowledge support from their funding agencies (Centre National de la Recherche Scientifique de France and Deutsche Forschungsgemeinschaft).

Freiburg/Karlsruhe, Germany Martin Dienwiebel
Écully, France Maria-Isabel De Barros Bouchet

Contents

Part I Microstructural Characterization

1 **Electron Microscopy and Microanalysis for Wear Surface Characterization** 3
Joseph R. Michael, Paul G. Kotula and Somuri V. Prasad
1.1 Introduction 3
1.2 Focused Ion Beam Preparation and Scanning Electron Microscopy 5
1.2.1 Focused Ion Beam (FIB) Preparation of Samples 5
1.2.2 Electron Backscatter Diffraction and Transmission Kikuchi Diffraction 10
1.3 Transmission Electron Microscopy and Microanalysis 12
1.3.1 Transmission Electron Microscopy 14
1.3.2 Scanning Transmission Electron Microscopy 15
1.3.3 Microanalysis 16
1.3.4 Spectral Image Analysis 17
1.3.5 Multivariate Statistical Analysis 18
1.4 Tribology Case Studies 19
1.4.1 Diamond-Like Carbon Coating Characterization 20
1.4.2 Wear-Induced Recrystallization 20
1.4.3 Tribology of Nanocrystalline Metals 23
1.4.4 Transition Metal Dichalcogenides 25
References 27

2 **Analyzing Mild- and Ultra-Mild Sliding Wear of Metallic Materials by Transmission Electron Microscopy** 29
Alfons Fischer, Wlodzimierz Dudzinski, Birgit Gleising and Priska Stemmer
2.1 Introduction 30
2.1.1 TEM Analytics of the Cyclic Strain Gradient 33

2.1.2 TEM Analyses of the Tribomaterial 38
2.1.3 Wear Particle Analyses by TEM 47
2.2 Conclusions and Outlook 51
References 56

Part II Tribochemical Characterization Techniques

3 Near Edge X-Ray Absorption Fine Structure Spectroscopy: A Powerful Tool for Investigating the Surface Structure and Chemistry of Solid Lubricants 63
Filippo Mangolini and J. Brandon McClimon
3.1 Introduction 64
3.2 Principles of Near Edge X-Ray Absorption Fine Structure (NEXAFS) Spectroscopy 65
3.3 Experimental Methods in NEXAFS Spectroscopy 70
3.3.1 Data Acquisition 70
3.4 Applications of NEXAFS Spectroscopy 76
3.4.1 NEXAFS Spectroscopic Investigation of the Surface Structure and Chemistry of Thin Films for Tribological Applications 76
3.4.2 NEXAFS Spectroscopic Investigation of the Thermally-Induced Structural Evolution of Solid Lubricating Films 83
3.4.3 NEXAFS Spectroscopic Investigation of the Tribochemistry of Solid Lubricants 87
3.5 Conclusions 96
References 96

4 Tribochemistry of n-Alkane Thiols Examined by Gas-Phase Lubrication (GPL) 107
Maria-Isabel De Barros Bouchet and Jean-Michel Martin
4.1 Introduction 108
4.2 Experimental 109
4.2.1 Gas-Phase Lubrication (GPL) Coupled with In Situ XPS/AES Surface Analyses 109
4.2.2 Materials 111
4.3 Adsorption of Hexanethiol on Steel 113
4.3.1 Procedure for Adsorption Study 113
4.3.2 Adsorption of Hexanethiol on the Native Oxide Layer of Steel 114
4.3.3 Adsorption of Hexanethiol on Metallic Iron (Etched Steel Surface) 115
4.4 Model GPL Experiment for Understanding the Formation Mechanism of Hexanethiol-Based Tribofilm 116

4.4.1 Tribochemical Reactions of Hexanethiol on Oxidized Steel and Metallic Iron 117
4.4.2 Study of the Role of Nascent Metallic Surfaces on the Tribochemistry of Hexanethiol 118
4.5 Ex Situ Analyses of the Iron Sulfide Tribofilm Formed by Thiol. 120
4.5.1 Formation of a 1-Hexanethiol Tribofilm by GPL 121
4.5.2 ToF-SIMS Chemical Analysis of the 1-Hexanethiol Tribofilm Formed by GPL 122
4.5.3 TEM Observations and EDS Chemical Analyses of the 1-Hexanethiol Tribofilm Cross Section. 123
4.6 Conclusions 127
References 127

5 In-Situ Measurement of Tribochemical Processes in Ultrahigh Vacuum 129
Wilfred T. Tysoe
5.1 Introduction 129
5.2 In-Situ Analyses of Sliding Interfaces 131
5.2.1 In-Situ Analyses with Opaque Interfaces 131
5.2.2 In-Situ Analyses of Transparent Interfaces 135
5.2.3 Pseudo In-Situ Analyses 137
5.3 Elementary Steps in Tribochemical Reactions. 144
5.3.1 Gas-Phase Lubrication of Copper by DMDS 146
5.3.2 Thermal Chemistry of DMDS on Copper 147
5.3.3 Identifying Elementary Steps During Sliding 149
5.3.4 Microkinetic Modelling of Tribofilm Formation for DMDS on Copper. 152
5.4 Conclusions 155
References 155

6 Tribochemistry and Morphology of P-Based Antiwear Films 159
A. Dorgham, A. Neville and A. Morina
6.1 Introduction 159
6.2 Chemical Nature of ZDDP and DDP 161
6.3 Reactions of ZDDP and DDP 163
6.3.1 Ligand Exchange 163
6.3.2 Decomposition of Peroxides and Peroxy-Radicals 163
6.3.3 Formation of Tribo- and Thermal Films. 164
6.3.4 Hydrolytic Decomposition 169
6.4 Composition of Antiwear Films. 170
6.4.1 Phosphorus, Zinc and Iron Species 171
6.4.2 Sulphur Species 172
6.4.3 Factors Affecting the Tribofilms' Composition and Formation 175

6.5 Mechanical Properties of Antiwear Films 187
6.5.1 Structure 187
6.5.2 Hardness and Elastic Modulus 192
6.5.3 Tenacity and Durability 196
6.6 Rheological Properties of Antiwear Films 197
6.7 Tribological Properties of Antiwear Films 198
6.7.1 Friction Mechanism 198
6.7.2 Antiwear Mechanism 201
6.8 Antiwear Films in Humid Environments 203
6.8.1 Effect of Water on the Composition of Tribofilms 204
6.8.2 Effect of Water on the Tribological Properties of Tribofilms 204
6.9 Concluding Remarks 205
References 206

7 In Situ Observation of Lubricating Films by Micro-FTIR 215
Shigeyuki Mori
7.1 Background 215
7.1.1 Tribological Contact 215
7.1.2 In Situ Observation 216
7.2 Principle and Method 217
7.2.1 Infrared Spectroscopy 217
7.2.2 Lubrication Tester 217
7.2.3 Materials 218
7.3 Structure of EHL Film 219
7.3.1 Film Thickness and Pressure 219
7.3.2 Component of Lubricant 220
7.3.3 Chemical Information 222
7.4 Molecular Interaction 223
7.4.1 Interaction of Components 223
7.4.2 Effect of Load 224
7.5 Molecular Alignment 225
7.5.1 Effect of Shear 225
7.5.2 Effect of Molecular Structure 225
7.6 Boundary Lubrication 227
7.6.1 Adsorbed Layer 227
7.6.2 Reaction Layer 227
7.7 Applications 228
7.7.1 Water-Based Lubricant 228
7.7.2 Semi-Solid Lubricant 229
7.7.3 Paper Friction Material (Wet Clutch) 231

7.7.4 Wire Drawing 231
7.8 Future Work 233
References 234

8 Micro-scale Real-Time Wear Dynamics Investigated by Synchroton Radiation 237
M. Belin, Jean-Michel Martin, J. Schou, I. L. Rasmussen, R. Feidenhans'l, T. Straasø and N. J. Mikkelsen
References 243

Part III In-Situ Microscopy and Topography Measurements

9 Electrochemical Friction Force Microscopy 247
Florian Hausen
9.1 Introduction 247
9.2 Electrochemical Foundations 249
9.2.1 Electrochemical Cell Design 253
9.3 Historical Development 255
9.4 Friction Force Microscopy Under Electrochemical Conditions 256
9.4.1 Aqueous Electrolytes 256
9.4.2 Ionic Liquids as Electrolytes 263
9.5 Summary 267
References 268

10 In Situ Friction Tests in a Transmission Electron Microscope 273
Fabrice Dassenoy
10.1 Introduction 273
10.2 Example 1: In Situ Compression Test on a Single MoS_2 Nanoparticle 274
10.3 Example 2: In Situ Compression Test on a Single WS_2 Nanoparticle 279
10.4 Example 3: In Situ Sliding Test on a Single MoS_2 Nanoparticle 279
10.5 Example 4: Friction Coefficient Measured on a Single WS_2 Nanoparticle 282
10.6 Conclusion 286
References 287

11 In Situ Digital Holography for 3D Topography Analysis of Tribological Experiments 289
Martin Dienwiebel and Pantcho Stoyanov
11.1 Introduction 289
11.1.1 Optical Roughness Measurements 290
11.1.2 Digital Holographic Microscopy (DHM) 291

11.2 Integration of DHM into a Tribometer 292
11.2.1 Analysis of Plowing Friction 293
11.3 Topography and Wear Evolution in Metal/Ceramic Tribosystems 298
11.3.1 Tungsten Sliding Against Tungsten Carbide 298
11.3.2 Tungsten Sliding Against Diamond-like Carbon 299
11.3.3 Aluminum Sliding Against Sapphire 302
11.4 Conclusion 304
References 304

Part IV Computer Simulations

12 Understanding the Tribochemistry of Lubricant Additives by *Ab initio* Calculations: The Case of Phosphites 309
M. Clelia Righi
12.1 Introduction 309
12.2 Combining First Principles Calculations and XPS Experiments to Study the Adsorption of Lubricant Additives 312
12.3 Effects of Mechanical Stresses on Additive Decomposition at the Tribological Interface 315
12.4 Functionality of Phosphorous in Reducing Adhesion and Friction 317
12.5 Conclusions 321
References 322

Index 325

Contributors

Maria-Isabel De Barros Bouchet Laboratoire de Tribologie et Dynamique des Systèmes, Ecole Centrale de Lyon, CNRS, UMR5513, Université de Lyon I, Lyon, Ecully, France

M. Belin Laboratoire de Tribologie et Dynamique des Systèmes, CNRS UMR 5512, Ecole Centrale de Lyon, Ecully, France

M. Clelia Righi Department of Physics, Informatics and Mathematics, University of Modena and Reggio Emilia, Modena, Italy

Fabrice Dassenoy Laboratoire de Tribologie et Dynamique des Systèmes, CNRS, UMR 5512, Ecole Centrale de Lyon, Ecully, France

Martin Dienwiebel Institute for Applied Materials, Microtribology Center µTC, Karlsruhe Institute of Technology KIT, Karlsruhe, Germany; Fraunhofer-Institute for Mechanics of Materials, Freiburg, Germany

A. Dorgham School of Mechanical Engineering, Institute of Functional Surfaces, University of Leeds, Leeds, UK

Wlodzimierz Dudzinski Faculty of Technology and Engineering, Wroclaw University of Science and Technology, Wroclaw, Poland

R. Feidenhans'l Niels Bohr Institute, University of Copenhagen, Copenhagen, Denmark

Alfons Fischer Materials Science and Engineering, University of Duisburg-Essen, Duisburg, Germany; CeNIDE-ICAN, Interdisciplinary Center for Analytics on the Nanoscale, University of Duisburg-Essen, Duisburg, Germany

Birgit Gleising CeNIDE-ICAN, Interdisciplinary Center for Analytics on the Nanoscale, University of Duisburg-Essen, Duisburg, Germany

Florian Hausen Forschungszentrum Jülich, Institute of Energy and Climate Research, Jülich, Germany; Institute of Physical Chemistry, RWTH Aachen University, Aachen, Germany; Jülich-Aachen Research Alliance, Section JARA-Energy, Jülich, Germany

Paul G. Kotula Sandia National Laboratories, Albuquerque, NM, USA

Filippo Mangolini Department of Mechanical Engineering, Materials Science and Engineering Program, The University of Texas at Austin, Austin, TX, USA

Jean-Michel Martin Laboratoire de Tribologie et Dynamique des Systèmes, Ecole Centrale de Lyon, CNRS, UMR5513, Université de Lyon I, Lyon, Ecully, France

J. Brandon McClimon Department of Materials Science and Engineering, University of Pennsylvania, Philadelphia, PA, USA

Joseph R. Michael Sandia National Laboratories, Albuquerque, NM, USA

N. J. Mikkelsen CemeCon AS, Åbyhøj, Denmark

Shigeyuki Mori Department of Chemical Engineering, Faculty of Engineering, Iwate University, Morioka, Iwate, Japan

A. Morina School of Mechanical Engineering, Institute of Functional Surfaces, University of Leeds, Leeds, UK

A. Neville School of Mechanical Engineering, Institute of Functional Surfaces, University of Leeds, Leeds, UK

Somuri V. Prasad Sandia National Laboratories, Albuquerque, NM, USA

I. L. Rasmussen DTU Fotonik, Technical University of Denmark, Roskilde, Denmark

J. Schou DTU Fotonik, Technical University of Denmark, Roskilde, Denmark

Priska Stemmer Materials Science and Engineering, University of Duisburg-Essen, Duisburg, Germany; CeNIDE-ICAN, Interdisciplinary Center for Analytics on the Nanoscale, University of Duisburg-Essen, Duisburg, Germany

Pantcho Stoyanov Institute for Applied Materials, Microtribology Center µTC, Karlsruhe Institute of Technology KIT, Karlsruhe, Germany; Fraunhofer-Institute for Mechanics of Materials, Freiburg, Germany

T. Straasø DTU Fotonik, Technical University of Denmark, Roskilde, Denmark; Niels Bohr Institute, University of Copenhagen, Copenhagen, Denmark

Wilfred T. Tysoe Department of Chemistry and Biochemistry, University of Wisconsin Milwaukee, Milwaukee, WI, USA

Part I
Microstructural Characterization

Chapter 1
Electron Microscopy and Microanalysis for Wear Surface Characterization

Focused Ion-Beam, Electron Backscattered Diffraction and Transmission Electron Microscopy

Joseph R. Michael, Paul G. Kotula and Somuri V. Prasad

Abstract Wear surface analysis, specifically in the subsurface regions, is essential for understanding the fundamental mechanisms of friction and material removal during wear. The advent of focused ion-beam systems (combined scanning electron and focused ion-beam microscopes) has made it possible to prepare site-specific cross sections of wear surfaces for scanning electron microscopy, electron backscattered diffraction, transmission electron microscopy, and X-ray microanalysis. During the past decade, these state-of-the-art electron microscopy techniques are being increasingly used to characterize wear-induced changes to microstructures and crystallographic textures.

1.1 Introduction

Wear damage is typically initiated in the subsusurface regions of the contacting sliding surfaces before it is fully manifested on the top surface, see Fig. 1.1. In metallic alloys, sliding contact results in plastic deformation, which often leads to recrystallization and development of crystallographic texture [1]. In the case of brittle materials, microscopic fracture events in the subsurface regions are the principal source

J. R. Michael · P. G. Kotula · S. V. Prasad (✉)
Sandia National Laboratories, Albuquerque, NM 87185-0886, USA
e-mail: svprasa@sandia.gov

J. R. Michael
e-mail: jrmicha@sandia.gov

P. G. Kotula
e-mail: pgkotul@sandia.gov

M. Dienwiebel and M.-I. De Barros Bouchet (eds.), *Advanced Analytical Methods in Tribology*, Microtechnology and MEMS,
https://doi.org/10.1007/978-3-319-99897-8_1

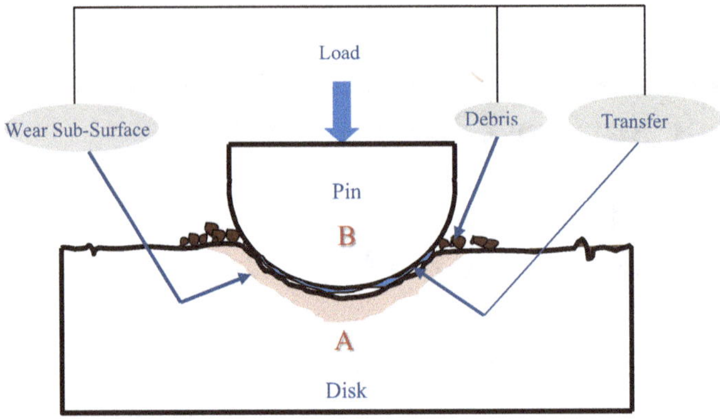

Fig. 1.1 Schematic illustration of the cross-section of a typical pin-on-disk wear system showing the subsurfaces

for wear damage [2]. Amorphous to crystalline transitions in the subsurface regions of transition metal dichalcogenide solid lubricating thin films (e.g., MoS_2, WS_2), with basal planes aligning in the sliding direction, and formation of transfer films on the counterface have been well documented [3]. Now it is widely acknowledged that a fundamental understanding of the friction-induced microstructural changes in the subsurface regions is essential for analyzing the wear mechanisms.

Up until the 1960s, wear surface analysis has primarily relied on visualization of wear damage from optical microscopy of wear surfaces. Metallographic cross-sections were sometimes prepared by cutting the sample obliquely to enhance the vertical magnification relative to the horizontal one. Like in the case of focused ion beam milling, which will be described in the later sections, it is necessary to protect the surface from damage from metallographic sample preparation. This is typically done by electroplating the surface with a metal of similar hardness. Using this technique, Moore [4] resolved the features indicative of plastic deformation and work hardening underneath the finely scratched, machined and lapped surfaces of metals (see also Bowden and Tabor [5]).

The advent of scanning and transmission electron microscopes in the 1960s, provided a powerful tool to the tribologists with high resolution imaging for visualizing the surface features, orders of magnitude higher than the optical microscopy [6]. Simultaneously, elemental and chemical analyses of the wear surfaces were made possible by X-ray and electron spectroscopy [7], while electron diffraction in the transmission electron microcopy enabled the determination of the crystal structures. However, specimen preparation for cross-sectional analyses remained a challenging task. SEM examination of subsurfaces continued to rely on dicing followed by conventional metallographic sample preparation techniques. Earlier research on cross-sectional transmission electron microscopy (TEM) of wear surfaces did not gain widespread application due to the difficulty in preparing the TEM samples by conventional techniques, which involved core drilling or dicing by diamond saw, grinding, electropolishing and dimpling [8, 9]. These steps are not only cumbersome and time consuming, but also are inadequate in locating the substructures in specific locations on the wear surface. This chapter will first highlight the application of modern focused ion beam (FIB) microscopy techniques for visualization of friction-induced deformation in the subsurfaces as well as the preparation of TEM samples in site-specific locations on a wear surface. This will be followed by specific examples from our recent research on the application of electron backscatter diffraction (EBSD) and high-resolution TEM for analyzing the friction and wear mechanisms in nanocrystalline metals and transition metal dichalcogenides.

1.2 Focused Ion Beam Preparation and Scanning Electron Microscopy

1.2.1 Focused Ion Beam (FIB) Preparation of Samples

The development of the combined focused ion beam/scanning electron microscope platforms (referred to as FIB/SEM) is possibly one of the most important developments that has enabled an improved understanding of sub-surface microstructural changes due to wear [11, 12]. The FIB/SEM at the simplest level is the combination of a high-resolution imaging tool (SEM) with a precision machining tool (FIB). This combination has turned out to be extremely powerful and has enabled the production of site specific samples for SEM imaging and other analytical techniques, such as transmission electron microscopy (TEM), surface analysis, micromechanical testing and many other analytical methods.

There are now two main ion sources for FIB tools, the liquid metal ion source (LMIS) and the plasma source. The most common source is a LMIS that typically produces a $Ga+$ ion beam that is used to micromachine via sputtering of selected regions of the sample. The LMIS equipped FIB allows small ion beams to be formed that are very useful for the preparation of samples that are on the order of 20 μm by about 10 μm in size. The plasma source ion column can deliver much higher beam currents to the sample at the expense of the size of the area of the beam. Thus, plasma sourced FIB columns are quite useful for preparing much larger samples (as large as 1 mm in width) than the LMIS FIB. The beam size of the plasma FIB is not as small as those that can be produced by LMIS sourced FIB tools. Selection of the plasma FIB over the LMIS sourced FIB depends on the end use. If larger cross sections of surfaces are to be routinely made then the plasma FIB may be the best choice. If TEM samples are more important, then the LMIS FIB may be the better choice. One also needs to keep in mind that the LMIS produces a beam of $Ga+$ ions that may react chemically with the sample. Plasma sources generally utilize inert gases (most commonly Xe) so chemical reactivity is not a concern. FIB instrumentation has been covered in detail elsewhere [11, 12].

To properly utilize any FIB tool, it is important for the user to have some understanding of the interactions that occur when an energetic ion interacts with the atoms in a sample. There are many events that occur when an energetic ion interacts with the atoms in a solid, but for the case of SEM and TEM sample preparation and ion imaging, we are mainly interested in sputtering (the removal of atoms from the surface of the sample), secondary electron production and damage to the sample in terms of ion implantation and loss of crystalline structure. Sputtering is the process that removes atoms from the target. Secondary electron production is important as images formed with secondary electrons induced by ions have some important advantages over secondary electron imaging. In principle, it is impossible to have an ion beam interact with a sample without causing some form of surface damage during ion irradiation.

Understanding the interaction of ions with the target material is helpful in reducing the amount of damage to acceptable or tolerable levels through appropriate sample preparation techniques. Energetic ions interact with a target in many ways. Once the ion enters the sample its path can be deflected by interactions with the atomic nuclei and the electron charges. As the ion moves through the sample it has sufficient energy to knock other atoms off their respective lattice positions. The target atoms that are knocked off their atomic positions can have enough energy to knock other target atoms off their atomic positions. This series of moving atoms and ions within the sample is referred to as a collision cascade. Some of the atoms that have been knocked from their atomic positions may reoccupy a lattice position or may end up in interstitial sites. There can also be lattice sites that are not reoccupied by target atoms and these are vacancies. Both interstitials and vacancies are considered damage to the crystalline structure of the sample. Most of the time, the original beam ion will end up coming to rest within the sample. This is termed ion implantation. Ion implantation results in the detection of the ion beam species in the sample and thus it is desired to minimize this as much as possible. Many of the collision cascades will eventually reach the surface of the sample. When the ion leaves the sample, it may have sufficient energy to knock an atom from the surface into the vacuum. This process is called sputtering and results in a net loss of material from the sample, usually from areas very close to the beam impact point. At the same time when the ion is either entering or leaving the sample, secondary electrons are generated that are useful for producing images of the sample surface scanned by the ion beam. It is important to remember that scanning an energetic ion beam over the surface of the sample will always result in some damage to the sample. The details of this process have been covered in other texts [10, 12].

It is most common to use the FIB to remove material to produce site-specific specimens via sputtering, but ions also produce secondary electrons that can be collected and imaged just like secondary electrons in the SEM. Ions generate many more secondary electrons than do electrons, thus ion induced secondary electron imaging can provide quality low-noise images. In addition, ions tend to channel more strongly than electrons and thus can produce striking grain contrast images. It is important to remember that during ion imaging of a sample, the continuous process of ion damage is occurring that will eventually (given sufficient time and ion dose) alter the sample to the point that it may no longer represent the original material.

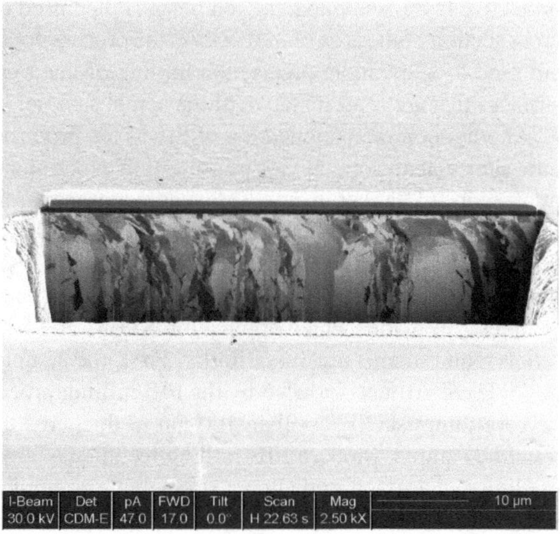

Fig. 1.2 Ion induced secondary electron image of a FIB prepared cross section of a wear scar (100 gm load for 1000 cycles) in Ni collected at 30 kV with an ion beam current of 47 pA. The dark layer on top of the Ni is a solid lubricant thin film

Figure 1.2 is a FIB prepared cross section of a wear scar produced by unidirectional sliding on electrodeposited Ni. This is an ion induced secondary electron (ISE) image obtained with a Ga+ ion beam using an accelerating voltage of 30 kV and a beam current of 47 pA. Note the high grain contrast produced by the ion beam and it is apparent that the stylus had moved from the right to the left-hand side of the image. The excellent grain contrast provided by ISE imaging provides useful insights into the wear process.

One common application of FIB is the preparation of cross sections of specific features. An example, as shown in Fig. 1.2, is that of a cross section normal to a wear surface that exposes sub-surface microstructural details. Cross sections prepared by FIB can vary from a few 10's of μm wide by a maximum of about 10 μm deep due to the time required to prepare the cross section. However, new developments in the use of plasma ion sources that can deliver much higher beam currents and thus faster milling rates have allowed sections to be milled as large as 1 mm wide by 10's of μm deep.

This text is not intended to describe in detail all of the steps required to produce quality cross sections as there are many texts already available that do this [11–13]. Cross sections of surfaces are accomplished in the FIB by careful control of the ion beam current. The first step of the process is to deposit a protective layer over the area of interest using electron or ion assisted deposition. Electron or ion assisted deposition is a process where a precursor gas is injected into the FIB through a needle that is very close to the sample surface. The electrons or ions in the beam disassociate the precursor resulting in a very impure Pt layer or a C layer to be formed only where the ion or electron beams are scanned during gas injections. The purpose of this protective layer is to prevent the energetic ions from interacting with the area of interest on the sample and causing damage as discussed previously. Once the protective layer is applied, the ion beam is then used to remove material to expose a cross section of the area of interest. As the process continues lower ion beam currents are used to allow more precise ion milling of the area and to produce very smooth surfaces that are representative of the actual sample.

Another common application of FIB is the production of thin samples from specific sites that are suitable for transmission experiments in the transmission electron microscope (TEM) or in the SEM (scanning transmission electron microscopy in the SEM also called STEM in SEM or just STEM.). One other application that requires thin samples is transmission Kikuchi diffraction (TKD) [14, 15]. This technique is done in the SEM and provides high resolution (2–10 nm) orientation mapping of thin samples. No matter which application STEM or TKD is to be used, sample preparation requirements are very similar. First, the sample must be free (or as close as possible) of artifacts induced by the FIB milling process. Second, the samples must be very thin, usually less than 100 nm in thickness to allow the electrons to penetrate the sample. Modern FIB milling techniques have been developed that reduce or eliminate artifacts and allow extremely thin samples to be prepared and mounted so that the sample can be introduced into the STEM or SEM for imaging. This text is not intended to discuss all the methods of sample preparation as there are many complete texts on this subject [11, 12]. A brief discussion of the steps to produce thin samples with the FIB will be given.

Just as with cross sections made for SEM imaging the first step is to locate the area of interest and then protect this area with a layer of Pt or C deposited by either ion beam assisted or electron beam assisted deposition. Once this is done then trenches are placed on either side of the area of interest as the final thin sample will come from the material left in between these trenches. The sample thickness is carefully reduced by using decreasing beam current ion beams in the FIB until the desired thickness has been reached. At some point during the process (usually before the sample is at the final thickness) the ion beam is used to cut the sample free from the surrounding material so that it can be lifted out from the trench and placed on a suitable support for subsequent imaging. It is often advantageous to use a lower energy ion beam as a final step to remove the ion beam damage caused during the sample milling at higher energies. This step is extremely important when high resolution imaging in the TEM or STEM is to be done. Figure 1.3 is an example of a completed thin sample prepared by FIB milling. Here the sample has been thinned to the final dimensions and is ready to be manipulated to a suitable support for imaging. The sample is manipulated outside the FIB tool to the support and this is called ex situ lift-out [16]. Figure 1.4 shows the completed TEM specimen after manipulation to a carbon-coated TEM grid.

Thin samples may also be produced by in situ lift-out in the FIB system using an internal manipulator [17]. The general procedure adds a couple of steps to those described above. The first step is to excise a chunk of material from the bulk sample using the ion beam and the in situ manipulator. The chunk of material is then attached to a suitable grid or support structure and held in place with ion beam assisted deposition of Pt. The sample is then thinned by the removal of material from both sides of the sample until the desired thickness is reached. The advantage to in situ

Fig. 1.3 Thin sample prepared by FIB that is ready for manipulation onto a suitable support for transmission imaging

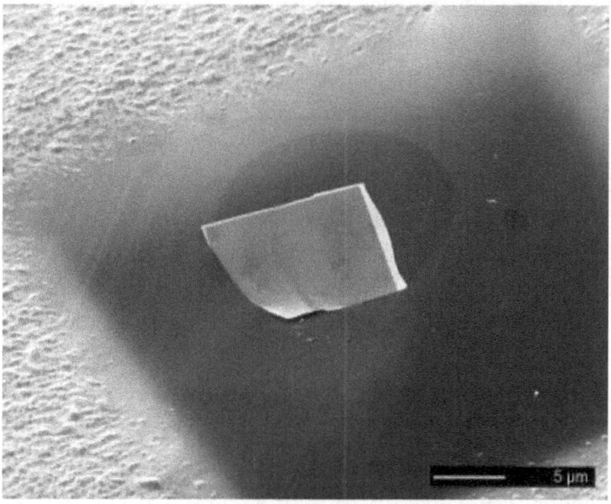

Fig. 1.4 FIB prepared thin sample manipulated onto a thin carbon film support

lift-out samples is that there is no support film to interfere with subsequent analysis and the sample is now firmly attached to the support structure and may be handled more easily.

The study of the evolution of sub-surface microstructures during wear requires that site-specific samples be produced. FIB tools can provide site-specific samples that are suitable for SEM imaging and analysis, TEM and STEM analysis. Many different types of samples can be produced in this manner that enhance the study of wear.

1.2.2 *Electron Backscatter Diffraction and Transmission Kikuchi Diffraction*

An understanding of the microstructure and crystallography of wear scars is required to fully describe the structure property relationships that control the tribological response of crystalline materials. By linking the microstructure to the crystallography of the sample, a full picture of the sample can be developed. Electron backscattering diffraction (EBSD) and the related technique of transmission Kikuchi diffraction (TKD) provide new insights into the microstructural changes that occur beneath a wear scar.

EBSD is accomplished in the SEM by the addition of a suitable camera to image the diffraction patterns and software to analyze the patterns and determine the crystallographic orientation of the sample. Most often, EBSD is conducted on polished bulk samples or bulk samples prepared in the FIB tool. Improved resolution over EBSD of bulk samples has been achieved by utilizing thin samples and TKD. TKD allows many of the fine details in wear microstructures to be observed. Both EBSD and TKD will be introduced in this section, and the reader is directed to many of the excellent texts available for more details [12, 18, 19].

EBSD patterns are obtained by illuminating a tilted sample surface with a stationary electron beam. The beam electrons interact with the sample and are initially inelastically scattered. The scattered electrons are diffracted by the crystalline planes in the sample to form the EBSD pattern that is then observed using a suitable camera. In the case of TKD, the thin sample is held normal or near-normal to the electron beam and the electrons transmit through the sample and are diffracted to form the TKD pattern. Note that EBSD and TKD patterns appear visually to be quite similar and can be indexed in the same manner (Fig. 1.5).

The EBSD pattern consists of sets of parallel lines. These are Kikuchi bands and the distance between the parallel pairs of lines is inversely related to the lattice spacing of the specific direction and the angles between the parallel sets of lines are the angles between the reflecting planes in the crystal. Thus, the angles between the planes can be compared to a table of angles prepared from known crystal structure data to determine the orientation of the patterns obtained from each pixel in an image.

EBSD or TKD orientation imaging is conducted by arranging the EBSD or TKD camera and the sample geometry in the SEM. The sample must be prepared so that the surface is free of preparation induced artifacts. Typically with bulk samples metallographic sample preparation is adequate and for TKD either FIB prepared surfaces or electropolished surfaces must be used. Figure 1.6 shows the arrangement of the sample and the EBSD or TKD cameras in the SEM. Note that a standard EBSD camera may also be used for TKD. The electron beam is scanned pixel-by-pixel over the areas of interest and patterns are collected at each pixel. The patterns are indexed and the orientation of the crystal is determined for each pixel. This procedure is repeated at each pixel within the image. Modern EBSD systems can now do this reliably at over 3000 patterns/sec allowing large areas to be efficiently imaged. Figure 1.7 shows inverse pole figure maps (IPF) of a wear scar made with a 10 g load

Fig. 1.5 EBSD pattern obtained from Mo at 20 kV

on a <111> single crystal of Ni. The wear scar was made in the <211> direction. The sample for these maps was made by FIB prepared thick lift-outs from the center of the wear scar. Figure 1.7a is the IPF map with respect to the surface normal and this is the <111> direction as can be seen by comparing the color of the substrate (blue) with the color key shown in Fig. 1.7c where the <111> directions on the stereogram is colored blue. Figure 1.7b is colored with respect to the wear direction and one can see that the <211> direction on the stereogram is between the blue of the <111> and the red color of the <100>. Thus by inspection it is relatively easy to understand the sample crystallography. In these images it is also apparent that the surface of the sample has been transformed from a single crystal to a layer of finer grained Ni with more random orientations.

An example of TKD of a wear scar in Ta is shown in Fig. 1.8. In this case a Co pin was worn against a Ta counterface. Note that this was obtained from a thin sample that was prepared for STEM imaging. The TKD maps were obtained using a 30 kV electron beam energy and a step size of 10 nm. Figure 1.8a is formed by a measure of the pattern's sharpness at each pixel and then plotting this as a grey scale image. The result accurately represents the microstructure of the sample. Figure 1.8b is an orientation map with respect to the wear direction (horizontal). Note that the upper part of Fig. 1.8b represents very fine grained wear debris that contains some very small grains that are just resolved.

EBSD and TKD are important characterization tools for wear phenomena in crystalline materials. TKD has the advantage of higher resolution while standard EBSD can provide data over larger areas at a reduced resolution.

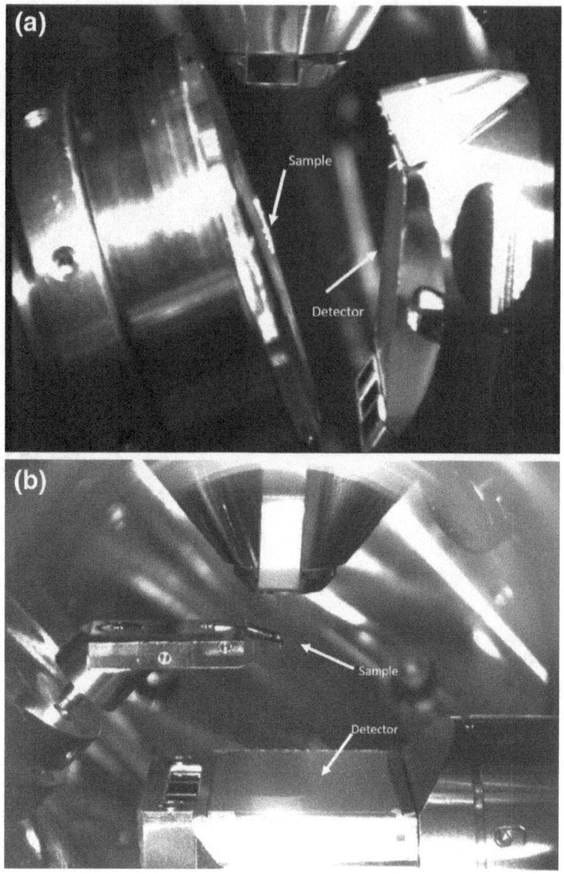

Fig. 1.6 Arrangement of the sample and the EBSD or TKD detectors in the SEM. **a** EBSD **b** TKD

1.3 Transmission Electron Microscopy and Microanalysis

Transmission electron microscopy (TEM) encompasses a few powerful techniques for characterizing wear surfaces at length scales from 100 pm to tens of μm. Since electrons interact strongly with matter we need to use high-energy electrons (100–300 kV) and create thin specimens (~100 nm in transmission). Before the advent of the FIB, the challenge was making a sample in the desired location and an appropriate thickness in projection. As discussed above the FIB now allows for routine creation of site-specific TEM samples which in many cases would be difficult or impossible to do by many using conventional sample preparation methods. In this section we will discuss the basic imaging methods of TEM, diffraction, and microanalysis and related data analysis methods.

In this section we give a general overview of TEM and associated techniques. The reader is directed to the text by Williams and Carter [7] for more detail on TEM and to Carter and Williams [20] for advanced discussion of TEM techniques including this

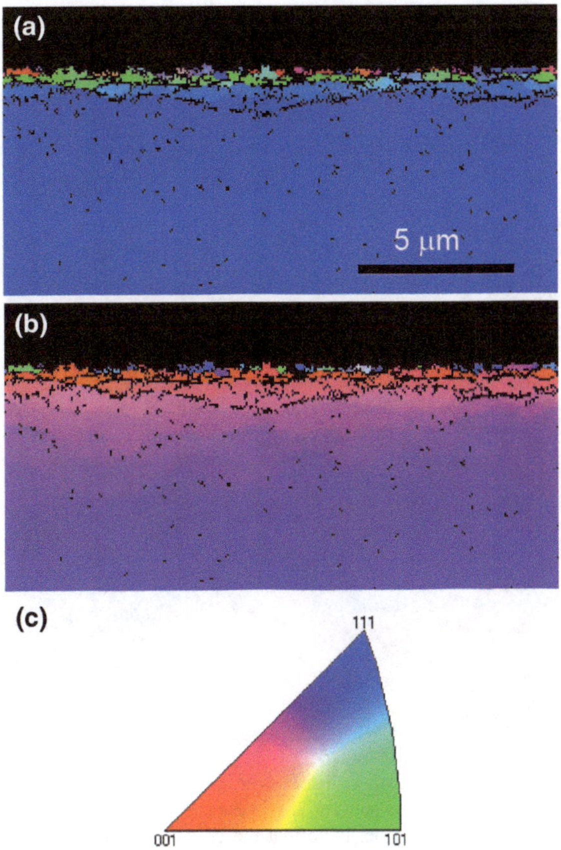

Fig. 1.7 Inverse pole figure maps obtained from wear scars on <111> single crystal Ni substrates. **a** IPF with respect to sample normal, **b** IPF with respect to wear direction, **c** color key for IPF maps

section's discussion of microanalysis (spectral imaging and data analysis methods). The TEM is analogous to a light microscope. We have an electron instead of light source, electromagnetic instead of glass lenses, condenser lenses which structure the electron illumination on the specimen, an objective lens, projection lenses and detectors to record the images. Since electrons interact strongly with matter we need to use both high-energy electrons, typically 100–300 kV, and thin specimens typically 100 nm thick or less in the projection direction. As with light microscopes we can think of the resolving power in terms of the wavelength of the illuminating light or in the case of TEMs the electrons which are 3.7 pm at 100 kV and 2 pm at 300 kV (at these energies the electrons are relativistic). One of these requirements, electron energy, leads to additional characterization capabilities to TEMs allowing us to analyze crystallography of our materials. In particular, the wavelength of the electrons (which are relativistic and thus act as both particles and waves) is smaller than the crystallographic plane spacings in materials which are typically 1 nm and

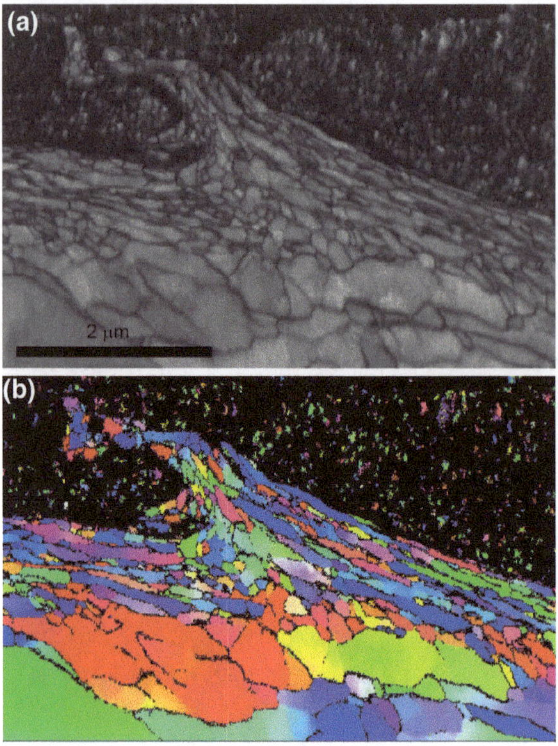

Fig. 1.8 TKD of Ta worn with a Co pin collected with a 10 nm step size. **a** A band contrast image formed by a measure of the collected patterns sharpness at each point in the image. **b** Orientation map with respect to the wear direction. Note that the colors represent orientations as shown in Fig. 1.7c

smaller, so that we get diffraction. Additionally, given the samples must be thin and analyzed in transmission we can improve the resolution of microanalysis versus a SEM. This will be discussed in the following sections.

1.3.1 Transmission Electron Microscopy

TEM is both a general term describing an instrument and technique, namely passing electrons through a specimen, and a specific term referring to the use of broad electron illumination on a specimen. In this sense, it is analogous to the basic visible-light microscope. However, because electrons interact strongly with matter, the electrons can be scattered or diffracted multiple times while passing through a specimen. This can be used to great advantage for imaging defects like dislocations and stacking faults. Additionally, twins, precipitates, cracks, and amorphous regions can be imaged in the conventional TEM modes. The types of data typically acquired for this include bright-field (BF) and dark-field (DF) images which are distinguished by the use of a small objective aperture (placed in the back-focal plane of the objective lens where a diffraction pattern is located) centered on the direct or a diffracted beam respectively. In TEM (specific definition) we have broad illumination on the sample (created by the condenser lenses). The sample sits within the objective lens which has two parts, an upper lens and a lower lens. The upper objective lens is used for focusing the illumination to a point, and this is important for scanning-TEM (STEM). The lower objective lens is the image-forming lens. Depending on how the projector lenses are set we can form an image of our specimen, or a diffraction pattern on the detectors. The former setting is used to collect a BF, DF or other image. The latter setting allows us to collect a diffraction pattern. Furthermore, we can place an aperture at a conjugate image plane within the projector lens system. This aperture can be moved to capture a small image region from the specimen. By then switching to diffraction mode, we can see a selected-area diffraction (SAD) pattern. BF, DF and SAD are powerful methods for characterizing and surveying the coarse-scale structure of a specimen.

In order to take advantage of the resolving power of the TEM we must operate in a different mode, high-resolution TEM (HR-TEM). In this operational mode, we use a larger objective aperture which allows the phase of several diffracted beams to interfere. In this way can resolve lattice planes from a specimen provided they are oriented parallel to the electron beam and of a sufficient spacing to be resolved by the microscope. Recent advances in electron optics, namely the commercial availability of spherical aberration correctors, have allowed for routine resolution of sub-100 pm lattice spacings [21]. HR-TEM is useful for imaging basal planes in graphite and MoS_2 especially when aligned parallel to the electron beam as in looking at the cross-section of a wear surface with those solid lubricants.

1.3.2 Scanning Transmission Electron Microscopy

STEM is analogous to SEM in that we focus the electrons to a point on the specimen and then scan it across our specimen. We then get contrast from the specimen with one of several possible detectors. These detectors sit below the projector lenses at a conjugate diffraction plane and the projector optics can be changed to magnify or de-magnify the diffraction pattern on the various detectors. By reciprocity [7] we can form images equivalent to conventional TEM BF and DF images in STEM mode. We can also form images with electrons scattered out to higher angles. These are called high-angle annular dark-field (HAADF) images and contrast from the specimen (presuming constant sample thickness in projection) is proportional to average atomic number. STEM is a versatile imaging mode for looking at both coarse structure (tens of microns) and atomic or near-atomic resolution 0.3 nm (lattice in some materials is resolvable) in a conventional STEM. As with TEM, spherical aberration correctors [22] have been developed for STEMs and have pushed direct resolution to 0.08 nm or better. It's important to provide a caveat for such resolution numbers, namely that these numbers are for a crystalline specimen oriented with lattice planes parallel to the electron beam and are dependent on electron channeling. For a specimen that is amorphous or not in a special orientation, the beam will broaden as it passes through lowering the actual resolving power.

1.3.3 Microanalysis

When a high-energy electron hits a specimen, it's possible for electrons to be knocked out of atoms in the specimen. As a result of the subsequent electronic relaxation, one possibility is that X-rays will be emitted from the excited atoms. The energy of the emitted X-rays is characteristic of the elements present in the specimen. Therefore we can use our electron probe to perform X-ray microanalysis to understand the local chemistry of our specimen with sensitivity (varies by element) to the entire periodic table above Li. As with SEMs, we can add an X-ray detector to detect the X-rays emitted from the specimen. In the SEM, the resolution of this signal is complicated and dependent upon both the elements in the specimen and the energy of the incident electrons. The volume from which X-rays are emitted is known as the interaction volume. In some cases, this can be a micron or more. This provides a distinct advantage in resolution for TEM specimens where the bulk of material has been removed, leaving only a thin lamella of material from which X-rays can be generated and subsequently detected. The result of this is that fewer X-rays are generated compared with a typical SEM specimen. The benefit is that the X-rays that are generated come from a region comparable to the electron probe plus the amount that the beam broadens as it passes through the specimen. The electron beam will broaden more in a high atomic number material than in a lower one of similar thickness. Thinning the specimen can improve resolution but at the expense of less material which can generate an X-ray signal. Modern analytical electron microscopes (AEM) are STEMs with X-ray and other chemically sensitive detectors (e.g., electron energy-loss spectrometers). To make up for the smaller number of X-rays generated by thin samples, X-ray detectors with larger solid-angles of collection have been developed so that 0.5 steradians to 1 steradian systems are commercially available [23]. For comparison, older AEMs typically had solid-angles of collection of 0.06 steradians.

There are a number of ways we can interrogate our specimen with X-ray microanalysis. The simplest is to manually move the electron beam to a feature and collect a single high-signal spectrum. This can be repeated on additional features but entails a significant amount of subjectivity with respect to appropriate sampling. Did we sample enough regions? And what if regions with important chemical differences show the same image contrast? Will we sample those? X-ray microanalysis has undergone a number of advances in the past 25 or so years. The most important was the advent of spectral imaging systems for commercial X-ray detectors on SEMs [24] and later STEMs. A spectral image combines imaging with full spectrum acquisition. A spectral image consists of a full X-ray spectrum at each pixel in an image. We then are statistically sampling many different points in a regular array. The next most important development was the development of the silicon-drift detector (SDD) [25]. In its modern form [26, 27] the SDD has completely replaced the older liquid nitrogen cooled Si–Li technology [28]. SDDs are cooled to −20 °C allowing for windowless operation improving light-element sensitivity. They are also able to operate at higher X-ray fluxes (over 1 MHz is possible but practical only for an SEM geometry) at near theoretical resolution. Since the cooling requirements are less stringent compared with older detectors, novel geometries have been implemented [29, 30] improving solid-angle of collection in the AEM by more than ten times [23].

Spectral imaging solved the problem of sampling but created a new one, namely how to analyze all the data comprehensively. In the next section, we discuss conventional and advanced data analysis methods.

1.3.4 Spectral Image Analysis

In order to understand the structure and chemistry of our specimen with an AEM, we collect electron images and spectral images. We can make direct measurements from the images (e.g., film thickness, crack length, phase fraction, etc.) but the spectral images are more complex, consisting of tens of thousands to hundreds of thousands of spectra (pixels) each with 2000 energy channels typically. The challenge is how to comprehensively analyze the sheer quantity of data we can quickly generate on a modern AEM. The spectral image lends itself to retrospective analysis in that we can collect the data on the AEM and then analyze and re-analyze it offline.

Conventional analysis of a spectral image is an iterative process [20, 31] starting with looking at the sum of all the spectra in the spectral image. We can then form region-of-interest maps of the elements identified from that spectrum. From the maps, we can identify regions in the image and look at sum spectra from identified regions. This forms the basis to confirm the analytical results. In X-ray spectroscopy, there are a number of artifacts in maps. These include Bremsstrahlung or breaking radiation generated as electrons in the specimen are decelerated. The result is a continuum X-ray background which is stronger for heavy elements than for light elements. Looking only at a map for a light element like oxygen, in a specimen with a heavy element like gold will illustrate the artifact. Regions with gold will produce a large background signal. In the oxygen map, regions with gold but not oxygen will show elevated intensity. This can be visualized by looking at spectra from the gold regions and verifying that no oxygen is present. More importantly we can see that oxygen is present by looking at spectra from the other distinct regions in the oxygen map to verify the presence of oxygen. An additional artifact in X-ray spectroscopy comes from the fact that different elements and have X-ray lines at the same energy. For example, the silicon K line is at 1.739 kV while the tantalum M-α line is at 1.709 kV. The result is that the silicon map will show intensity everywhere there is Ta. Also, depending on which tantalum line we use to form the Ta map we might see silicon where it is not. Using the iterative analysis approach, we can help solve such pathological overlap problem cases.

Conventional analysis can be time consuming and subjective. We can also miss important features due to the artifacts mentioned above as well as just miss a weak but potentially important feature. For these reasons, we discuss the use of statistical analysis methods to quickly and comprehensively analyze X-ray spectral images with no preconceptions of what is or is not present.

1.3.5 Multivariate Statistical Analysis

Multivariate statistical analysis (MSA) encompasses a family of data analysis methods that are used to visualized data sets subject to various constraints. For X-ray spectral images the number of variables is typically the number of spectral channels, perhaps 1000–2000. Understanding how collections of these co-vary can be used to understand the chemistry of a specimen. One can think of the channels containing a particular spectral peak or families of peaks as co-varying when the element responsible for the peaks is present in regions in the specimen. Similarly, elements found together in the same amounts in regions of the specimen would produce channels that co-vary.

Before we discuss different MSA methods, it's important to discuss scaling the raw X-ray spectral image data for non-uniform noise [32, 33] and estimating the rank of the data. When we discuss X-ray spectroscopy we have X-ray counts where the noise is proportional to the signal. If we fail to account for the noise in our MSA methods then we risk fitting noise when it is greater in magnitude than weaker but chemically relevant signals in the spectral image. If we have a channel in our data where the intensity is 100 counts, the associated noise for that datum is the square root of 100 or ± 10. Variance-based MSA methods would preferentially fit such a variation over a chemical signal smaller than that. Noise scaling, for the purpose of the examples described in this chapter, amounts to dividing the spectral image by the square root of the mean spectrum in the spectral domain and the square root of the mean image in the image domain [32, 33]. Once we have performed the noise normalization, we can then calculate the eigenvalues from the scaled spectral image data. The sorted eigenvalues on a semi-log plot are referred to as a scree plot, an example of which is given in Fig. 1.9. Here we examine just the 64 largest eigenvalues. Sandia's Automated eXpert Spectral Image Analysis (AXSIA) software [34–36] performs an automated rank estimation by fitting a straight line to eigenvalues indexed 25 through 64, a region assumed to represent the noise baseline. When an eigenvalue rises above the fit to the line by greater than 25 σ it is deemed non-noise and usually represents chemical information. The index of this eigenvalue is the estimated rank, or number of components needed to describe the data minus the noise.

Following noise normalization and rank estimation we then factor the data into a form we can readily interpret. The reader may be familiar with one MSA variant called Principal Components Analysis (PCA) [37]. PCA seeks to serially maximize the

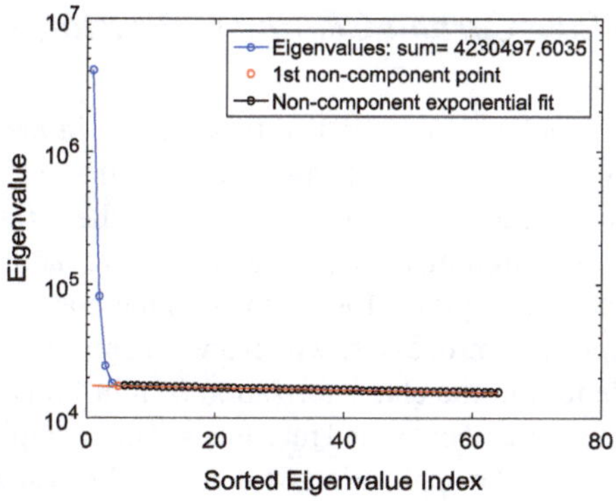

Fig. 1.9 Scree plot showing the results of an Eigenanalysis of a spectral image weighted for non-uniform noise. The automatically determined rank for the model of the data is four meaning that only four factors are needed to describe the chemical information in the spectral image consisting of hundreds of thousands of pixels and thousands of spectral channels

variance in components which are mutually orthogonal. The first principal component is the mean. The second principal component describes the second most variance in the data while being orthogonal to the first and so on. The problem with PCA is that the constraints (serial maximization of variance and orthogonality) result in non-physical components consisting of a component image and corresponding spectral shape which both consist of positive and negative intensities. For this reason we add an additional assumption to the factor analysis that none of the component images or spectral shapes has negative (non-physical) intensities. The non-negativity constraint relaxes the orthogonality imposed by PCA such that the resulting factors now are directly interpretable. AXSIA achieves this via application of a MSA technique called multivariate curve resolution (MCR) implemented by an alternating least squares (ALS) approach [36]. The rotational ambiguity, the fact that there are an infinite number of solutions to fit the data equally well, is addressed by performing a Varimax rotation [38] on the data to make it simple in either the spectral [34–36] or spatial [39] domain. Simple implies high contrast or zeros in many parts of the components. The end result for both spectral and spatially simple representations is a compact and physically interpretable model for the data. The factors making up this model contain spatial components and corresponding spectral shapes. The component images tell us where and how much of the elements are present, and the spectral shapes tell us the identities of the elements.

1.4 Tribology Case Studies

This section describes a few specific examples on the application of these novel electron microscopy techniques in tribology and tribological coatings.

1.4.1 Diamond-Like Carbon Coating Characterization

The most widespread application of FIB in tribology is in the characterization of the coatings on miniature parts, specifically those on the sidewalls of microelectromechanical systems (MEMS). With miniaturization, the surface interactions play a predominant role in determining the performance and reliability, necessitating the need for solid lubricating coatings. Most of the commercial coatings are deposited by physical vapor deposition processes, which are inherently line-of-sight. FIB has become an invaluable tool in the characterization of thin films and coatings on the sidewalls of miniature parts and on buried interfaces. The example shown in Fig. 1.10 is from our earlier study on the application on diamond-like carbon on Ni Microsystems part fabricated by LIGA (German acronym for lithography, electroforming, and molding). The DLC coating was applied by a commercial plasma enhanced chemical vapor deposition technique. TEM analysis was used to assure that the DLC coating coverage on the sidewalls, where it is most needed, is dense, conformal and without any interfacial defects [40].

1.4.2 Wear-Induced Recrystallization

The first reported application of EBSD for the study of wear-induced recrystallization in metals was on electroplated Ni surfaces [41]. Figure 1.11a shows the location of the FIB cut on the wear surface created by a Si3N4 ball at an initial Hertzian contact stress of 315 MPa for 1000 cycles of sliding in unidirectional mode. The direction of sliding was from left to right, as shown by the arrow. The electroplated Ni had a predominantly <001> fiber texture. There also appears to be a fine-grained region near the top, i.e. right underneath the wear track. Figure 1.11b is the orientation map of the substructure underneath the wear track. The colors represent the orientations normal to the sample surface based on the color key shown in the inset. This is an interesting area as there is a region that has a <110> fiber texture (designated by green) in the predominantly <001> fiber textured material. The thick and thin black lines represent high- and low-angle grain boundaries respectively. Figure 1.11b clearly reveals two characteristic zones, each with its own unique features which differ significantly from the microstructure in the bulk undisturbed material. A few microns below the wear track, the bending of columnar grains in the direction of sliding is observed, which is referred to as "Zone 1". As we approach the wear surface, a lineup of thin black lines appears in the microstructure indicating the formation of substructures within the deformed zone. Right underneath the wear track, the columnar structure broke down into more equiaxed submicron-size grain structure, which is referred to as "Zone 2". The depth of this zone extends to 1–2 μm. Zone 1 and Zone 2 are also referred to as "plastically deformed" and "highly deformed" zones respectively. It is also interesting to note that Zone 2 is thicker in the <110> fiber textured region than in the <001> fiber textured region. The pole figures corresponding to <001>

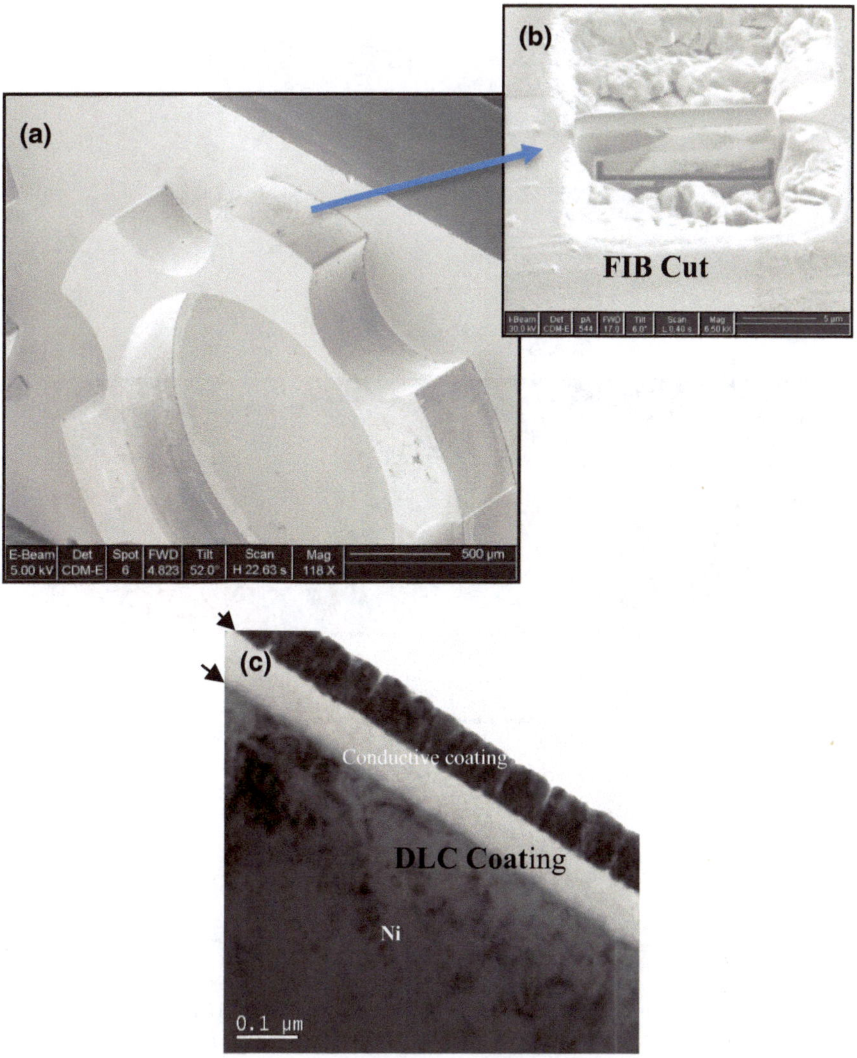

Fig. 1.10 **a** SEM micrograph of a Diamond-like carbon coated Ni-based microsystems part fabricated by LIGA, **b** FIB cut on the sidewall to prepare a TEM sample, **c** Bright-filed TEM image of the FIB cut showing the DLC coating on the sidewall. Reproduced by permission of IEEE Xplore [41, Figs. 3 and 8]

and <110> textured grains are shown in Fig. 1.11c. The spread in orientation of pole figures (Fig. 1.11c) can be used to judge the extent of wear-induced deformation in the subsurfaces. The <110> textured grains have wider orientation spread in the sliding direction (18°–44°) than the <001> textured grains (10°–17°), which is in agreement with the microstructural findings in Fig. 1.11b.

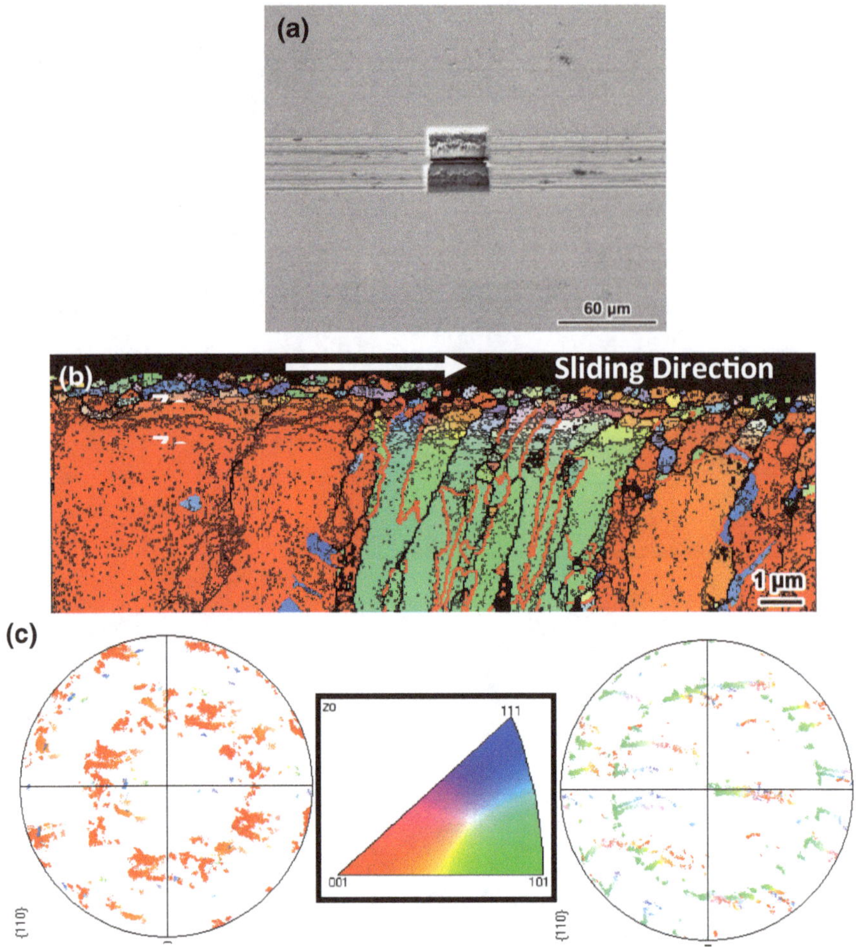

Fig. 1.11 EBSD analysis on the subsurface of a wear scar on electroplated Ni. **a** Location of the FIB cut on the wear scar; **b** orientation map with respect to the surface normal (the arrow represents the sliding direction); the heavy black lines represent orientation changes $>10^\circ$ and thin lines represent orientation changes of 1° or less, **c** pole figures of the region underneath the wear scar showing <001> and <110> fiber textured material (sliding direction is Y0). Inset at the center of Figure **c** is the stereographic triangle with color key for (**b**). Reproduced by permission of Pergamon Press [40, Fig. 3]

This very first study on the application of EBSD for friction-induced wear phenomena has demonstrated the unique role of focused ion beam techniques in preparing cross sections of narrow wear tracks generated under very light loads for electron backscatter diffraction studies. By suitably thinning the samples further, this technique can be easily extended to prepare cross sections of wear tracks for TEM analysis. Unlike in conventional specimen preparation techniques, the FIB enabled specimens are free of artifacts introduced during dicing, grinding, electropolishing, dimpling, etc., and sections can be precisely cut at a specific location on the wear track.

1.4.3 Tribology of Nanocrystalline Metals

While the deformation behavior in metals with relatively large grain sizes is predominantly dictated by dislocation dynamics, gain boundary processes dominate in the nanocrystalline regimes, when the grain sizes are below 20 nm [42]. Analogous to the grain size dependency on the mechanical behavior, recent studies seem to confirm that friction behavior of nanocrystalline metals could be significantly altered when the friction-induced deformation results in the formation of stable ultrananocrystalline structures in the subsurfaces. This was first highlighted by Prasad, Battaile and Kotula [42] from their study on friction behavior of nanocrstalline nickel, which exhibited two distinct friction behaviors, $\mu \sim 0.30$–0.35 or $\mu \sim 0.6$–0.7, depending upon the contact stress and sliding speed (Fig. 1.12a).

Figure 1.12b is an annular dark field STEM image showing detailed features of the subsurface for the case which exhibited low friction behavior (green circles in Fig. 1.12a). The arrow indicates the direction of sliding. The image shows three zones (labeled in Fig. 1.12b). The zones observed include an ultrafine grained region directly at the wear surface (Zone 1), a region with large, textured, flattened grains (Zone 2), and a gradual transition of grain size approaching that of the bulk material (Zone 3). The bulk has an average grain size of 20–100 nm. During frictional contact at stresses and sliding speeds corresponding to the green circles in Fig. 1.12a, ultrafine nanocrystalline gains with a size of 2–10 nm formed right underneath the wear surface (Zone 1). The authors attributed the experimentally observed transition from high friction to low friction regime to the formation of this ultrafine nanocrystlline structure underneath the wear surface elucidated by FIB-TEM. The authors also reported that in Zone 1 microstructure was absent in cases that did not exhibit a friction transition. Zone 2 consisted of platelet shaped grains larger than 100 nm, elongated in the sliding direction, in between Zone 1 and the bulk microstructure of Zone 3. The grains in Zone 2 represent a layer approximately 150 nm thick that showed significant grain growth and the development of crystallographic texture. Additionally, the transition between Zones 2 and 3 is not sharp but rather graded as the grain size returns to bulk dimensions over a distance of a few hundred nanometers. The study is a clear demonstration of the application of modern electron microscopy techniques for the fundamental understanding of the mechanisms of friction in nanocrystalline metals.

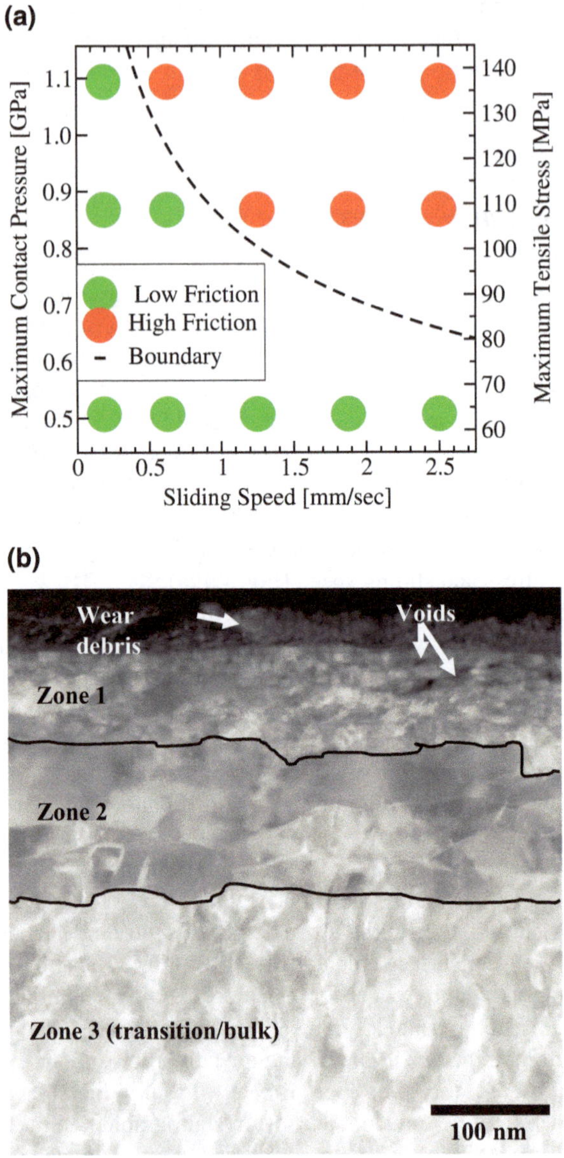

Fig. 1.12 **a** Friction transitions in nanocrystalline Ni [42]. The data was generated over a wide range of contact pressures and sliding speeds. The green dots represent the low friction regime while the red dots correspond to the high friction regime. The dotted line depicts the friction transition. **b** Annular dark field STEM image of the subsurface frim the low friction regime, showing ultrananocrystlline grain structure at the top. Reproduced by permission of Pergamon Press [42, Figs. 2 and 3b]

1.4.4 Transition Metal Dichalcogenides

Transition Metal Dichalcogenides (TMD), most notably, molybdenum disulfide (MoS_2) and tungsten disulfide (WS_2) are well known for their solid lubricating behavior and are widely used in several applications. MoS_2 crystallizes in the hexagonal structure where a sheet of molybdenum atoms is sandwiched between two hexagonally packed sulfur layers with a high *c*/*a* ratio (c = 12.29 Å, a = 3.16 Å). The bonding within the S–Mo–S sandwich is covalent, while weak Van der Waals forces hold the sandwich together resulting in interlamellar mechanical weakness. It has been widely reported [43] that under a shearing force the basal planes slide over one another by intracrystalline slip, similar to a 'deck-of-cards', resulting in the formation of a transfer film on the rubbing counterface (e.g., the ball surface in a ball-on-disk friction and wear test), shown schematically in Fig. 1.1. According to the widely accepted hypotheses, the main mechanisms for imparting low interfacial shear in TMD are: creation of (0002) basal planes and subsequent (re)orientation parallel to the sliding direction, and the development of a transfer film on the counterface to accommodate interfacial sliding. Until the advent of FIB, preparing TEM samples of these thin layers transferred onto the ball surfaces from ball-on-disk tribology tests of MoS_2 films that are suitable for high resolution lattice imaging has not been very successful. A recent study by Scharf, Kotula and Prasad [43] demonstrated the application of FIB to prepare TEM samples of transfer films form a molybdenum disulfide film doped with antimony trioxide and gold (MoS_2–Sb_2O_3–Au). This film, which has been successfully deployed in several space mechanisms, lacks any long-range crystallinity in the as-deposited condition. Using FIB-TEM, these authors have reported the presence of a thin layer at the top of the cross-section of the wear surface that has crystalline 2H–MoS_2 basal (0002) planes aligned parallel to the sliding direction (Fig. 1.13b). The arrow indicates the sliding direction, while the region above the arrow corresponds to the carbon film that was applied as part of the TEM FIB sample preparation. Just below this crystalline 2H–MoS_2 layer, the TEM image in Fig. 1.13b shows the presence of ~4–6 nm Au particles that are larger than the 2–3 nm size nanoparticles in the as-deposited coating seen in the bulk. This Au nanoparticle coarsening is clearly a result of the sliding process. The AXSIA map in Fig. 1.13d (where red is MoS_2/Sb_2O_3/Au and green is Au) also confirms Au particle coarsening due to frictional contact. A typical HRTEM image of a FIB cross-section of the transfer film (Fig. 1.13a) showed that it was comprised of predominantly crystalline MoS_2 through its entire thickness of ~1 μm. However, in the top layer, the basal planes are aligned parallel to the sliding direction. This layer is about 6 nm thick, which approximates to nine basal planes of MoS_2. A SAD pattern obtained from this top most region of the cross-section (Fig. 1.13c) showed that the interlamellar

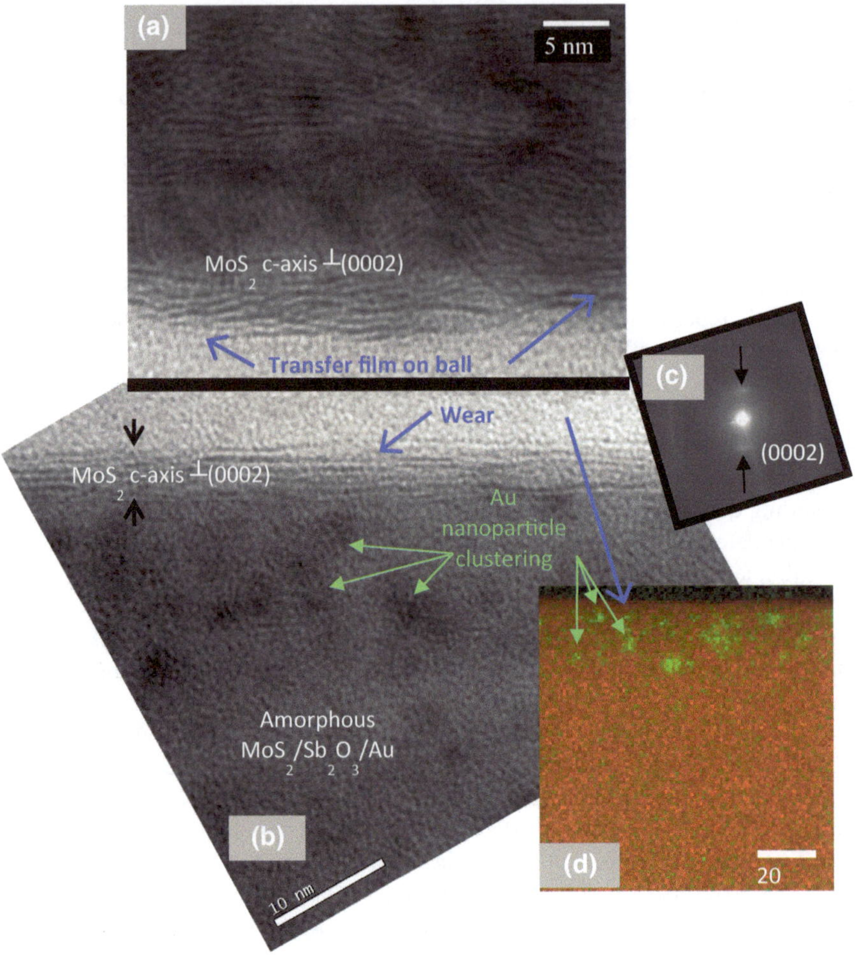

Fig. 1.13 **a** and **b** Cross-sectional HRTEM montage images of wear track on $MoS_2/Sb_2O_3/Au$ solid lubricant film with its transfer film on the ball surface. **c** SAED pattern of transfer film with interlamellar d-spacing between the (0002) $2H_2$–MoS_2 planes measured at ~6 Å, which compares well with indexed 6.1 Å lattice spacing of MoS2. **d** AXSIA results of the wear surface where red is $MoS_2/Sb_2O_3/Au$ and green is Au (with some Sb/O). Reproduced by permission of Pergamon press [43, Fig. 9]

d-spacing between the (0002) 2H–MoS_2 planes was ~6 Å, which compared well with indexed (PDF#01-087-2416) 6.1 Å lattice spacing for MoS_2 lattice. Once again, the application of modern electron microscopy and microanalysis techniques has been able to experimentally validate the widely held concept of self-mated 'basal plane-on-basal plane' sliding as the fundamental mechanism of lubrication in transition metal dichalcogenide thin films.

Acknowledgements Sandia National Laboratories is a multi-mission laboratory managed and operated by National Technology and Engineering Solutions of Sandia, LLC., a wholly owned subsidiary of Honeywell International, Inc., for the U.S. Department of Energy's National Nuclear Security Administration under contract DE-NA0003525. The authors gratefully acknowledge Professor Thomas W. Scharf, University of North Texas, for many years of collaboration with the authors on this subject, and Dr. Corbett Battaile for critically reviewing the manuscript. Special thanks to Michael Rye for the FIB work and to Bonnie McKenzie for the SEM, EBSD and TKD studies.

References

1. D.A. Rigney, J.P. Hirth, Plastic-deformation and sliding friction of metals. Wear **53**, 345–370 (1979)
2. S.V. Prasad, Wear, in *Concise Encyclopedia of Advanced Ceramic Materials*, ed. by R.J. Brook (Pergamon Press, Oxford, 1991), pp. 511–519
3. S.V. Prasad, T.W. Scharf, J. Mater. Sci. **48**, 511–531 (2013)
4. A.J.W. Moore, Proc. Roy. Soc. A **195**, 231 (1948)
5. F.P. Bowden, D. Tabor D, *The Friction and Lubrication of Solids* (Clarendon Press, Oxford, 1986), pp. 112, 120
6. R.F. Pease, W.C. Nixon, J. Sci. Instrum. **42**, 81–85 (1965)
7. D.B. Williams, C.B. Carter, *Transmission electron microscopy: A Textbook for Materials Science*, 2nd edn. (Springer, 2013) ISBN-13: 978–0387765006
8. P. Heilmann, D.A. Rigney, Metall. Trans. **12A**, 686 (1981)
9. W.M. Rainforth, Wear **245**, 162 (2000)
10. M. Nastasi, J. Mayer, J.K. Hirvonen, *Ion-Solid Interactions: Fundamentals and Applications* (Cambridge University Press, New York, 1996)
11. L. Giannuzzi, *Introduction to Focused Ion Beams: Instrumentation, Theory, Techniques and Practice* (Springer, New York, 2006)
12. J.I. Goldstein, D.E. Newbury, J.R. Michael, N.W.M. Ritchie, J.H. Scott, D.C. Joy, *Scanning Electron Microscopy and X-Ray Microanalysis* (Springer, New York, 2017)
13. J. Mayer, L.A. Giannuzzi, T. Kamino, J. Michael, TEM sample preparation and FIB-induced damage. MRS Bull. **32**(5), 400–407 (2007)
14. P.W. Trimby, Orientation mapping of nanostructured materials using transmission Kikuchi diffraction in the scanning electron microscope. Ultramicroscopy 16–24 (2012)
15. R.R. Keller, R.H. Geiss, Transmission EBSD from 10 nm domains in a scanning electron microscope. J. Microsc. 245–251 (2012)
16. L.A. Giannuzzi, Z. Yu, D. Yin, M.P. Harmer, Q. Xu, N.S. Smith, L. Chan, J. Hiller, D. Hess, T. Clark, Theory and new applications of ex situ lift out. Microsc. Microanal. **21**(4), 1034–1048 (2015)
17. N. Bassim, K. Scott, L.A. Giannuzzi, Recent advances in focused ion beam technology and applications. MRS Bull. **39**(4), 317–325 (2014)
18. V. Randle, *Microtexture Determination and its Applications* (Maney Publishing, London, 2003)
19. A.J. Schwartz, M. Kumar, B.L. Adams, D.P. Field (eds.), *Electron Backscatter Diffraction in Materials Science* (Springer, New York, 2009)
20. C.B. Carter, D.B. Williams (eds.), *Transmission Electron Microscopy* (Springer International Publishing Switzerland, 2016), Chapter 16, https://doi.org/10.1007/978-3-319-26651-0_16
21. M. Haider, S. Uhlemann, E. Schwan, H. Rose, B. Kabius, K. Urban, Electron microscopy image enhanced. Nature **392**, 768–769 (1998). https://doi.org/10.1038/33823
22. P.E. Batson, N. Dellby, O.L. Krivanek, Sub-ångstrom resolution using aberration corrected electron optics. Nature **418**, 617–620 (2002). https://doi.org/10.1038/nature00972

23. H.S. von Harrach, P. Dona, B. Freitag, H. Soltau, A. Niculae, M. Rohde, J. Phys.: Conf. Ser. **241**, 012015 (2010)
24. R.B. Mott, C.G. Waldman, R. Batcheler, J.J. Friel, Position tagged spectrometry: A new approach for EDS spectrum imaging. in *Proceeding of Microscopy Microanalysis*, ed. by G.W. Bailey, M.H. Ellisman, R.A. Hennigar, N.J. Zaluzec. Jones and Begell Publishing, New York, pp. 592–593
25. E. Gatti, P. Rehak, Nuc. Inst. Meth. Phys. Res. **225**, 608–614 (1984)
26. L. Strüder, N. Meidinger, D. Stotter, J. Kemmer, P. Lechner, P. Leutenegger, H. Soltau, F. Eggert, M. Rohde, T. Schulein, High-resolution X-ray spectroscopy at close to room temperature. Microsc. Microanal. **4**, 622–631 (1998)
27. A. Niculae, P. Lechner, H. Soltau, G. Lutz, L. Strüder, C. Fiorini, A. Longoni, Optimized readout methods of silicon drift detectors for high-resolution X-ray spectroscopy. Nucl. Instrum. Methods. Phys. Res. A **568**, 336–342 (2006)
28. R. Fitzgerald, K. Keil, K.F.J. Heinrich, Science **159**, 528–530 (1968)
29. B.L. Doyle, D.S. Walsh, P.G. Kotula, P. Rossi, T. Schulein, M. Rohde, X-Ray Spectrom. **34**, 279–284 (2005)
30. P.G. Kotula, J.R. Michael, M. Rohde, Microsc. Microanal. Suppl. **2**(14), 116–117 (2008)
31. P.G. Kotula, M.R. Keenan, Application of multivariate statistical analysis to STEM X-ray spectral images: interfacial analysis in microelectronics. Microsc. Microanal. **12**(6), 538–544 (2006)
32. M.R. Keenan, P.G. Kotula, Accounting noise in the multivariate statistical analysis of TOF-SIMS data. Surf. Interf. Anal. **36**, 203–212 (2004)
33. M.R. Keenan, P.G. Kotula, Optimal scaling of ToF-SIMS spectrum-images prior to multivariate statistical analysis. Appl. Surf. Sci. **231–232**, 240–244 (2004)
34. M.R. Keenan, P.G. Kotula, Apparatus and System for Multivariate Spectral Analysis. US Patent # 6,584,413, 2003
35. M.R. Keenan, P.G. Kotula, Method of Multivariate Spectral Analysis. US Patent # 6,675,106, 2004
36. M.R. Keenan, Multivariate analysis of spectral images composed of count data. in *Techniques and Applications of Hyperspectral Image Analysis*, ed. by H. Grahn, P. Geladi (Wiley & Sons, Chinchester, 2007)
37. I.T. Jollife, *Principal Component Analysis*, 2nd edn. (Springer, New York, 2002)
38. H.F. Kaiser, The varimax criterion for analytic rotation in factor analysis. Psychometrika **23**(3), 187–200 (1958)
39. M.R. Keenan, Exploiting spatial-domain simplicity in spectral image analysis. Surf. Int. Anal. **41**, 79–87 (2009)
40. S.V. Prasad, J.R. Michael, T.R. Christenson, EBSD studies on wear-induced subsurface regions in LIGA nickel. Scripta Mater. **48**, 255–260 (2003)
41. S.V. Prasad, T.W. Scharf, P.G. Kotula, J.R. Michael, T.R. Christenson, J Microelectromech. Syst. **18**, 695 (2009)
42. S.V. Prasad, C.C. Battaile, P.G. Kotula, Scripta Mater. **64**, 729–732 (2011)
43. T.W. Scharf, P.G. Kotula, S.V. Prasad, Acta Materiala **58**, 4100–4109 (2010)

Chapter 2
Analyzing Mild- and Ultra-Mild Sliding Wear of Metallic Materials by Transmission Electron Microscopy

Alfons Fischer, Wlodzimierz Dudzinski, Birgit Gleising and Priska Stemmer

Abstract Any understanding of tribological behavior is connected to sound analyses of the wear appearances, which render insight into the acting wear mechanisms and their sub-mechanisms. Today one important method to analyze wear appearances at high resolution is transmission electron microscopy (TEM) being invented by Knoll and Ruska in the early 1930th in Berlin (Knoll in Z fuer Phys 78:318–339, 1932 [1]).

Abbreviations and Terms

AES	Auger atom emission spectroscopy
AFM	Atomic force microscopy
ECC	Electron channeling contrast
EDS	Energy dispersive X-ray spectroscopy
EELS	Electron energy loss spectroscopy
EFTEM	Energy filtered transmission electron microscopy
FIB	Focussed ion beam
KAM	Kernel average missorientation

A. Fischer (✉) · P. Stemmer
Materials Science and Engineering, University of Duisburg-Essen, Duisburg, Germany
e-mail: alfons.fischer@uni-due.de

P. Stemmer
e-mail: priska.stemmer@uni-due.de

W. Dudzinski
Faculty of Technology and Engineering, Wroclaw University of Science and Technology, Wroclaw, Poland
e-mail: wlodzimierz.dudzinski@pwr.edu.pl

A. Fischer · B. Gleising · P. Stemmer
CeNIDE-ICAN, Interdisciplinary Center for Analytics on the Nanoscale, University of Duisburg-Essen, Duisburg, Germany
e-mail: birgit.gleising@uni-due.de

M. Dienwiebel and M.-I. De Barros Bouchet (eds.), *Advanced Analytical Methods in Tribology*, Microtechnology and MEMS,
https://doi.org/10.1007/978-3-319-99897-8_2

MD	Molecular dynamics computer simulation
RS	Raman-spectroscopy
SEM	Scanning electron microscopy
TEM	Transmission electron microscopy
BF	Bright field image
DF	Dark field image
DP	Diffraction pattern
XPS	X-ray photoelectron spectroscopy
nc	Nano-crystalline <100 nm
ufc	Ultrafine-crystalline 100–500 nm
µc	Micro-crystalline 500 nm

2.1 Introduction

Any understanding of tribological behavior is connected to sound analyses of the wear appearances, which render insight into the acting wear mechanisms and their sub-mechanisms. Today one important method to analyze wear appearances at high resolution is transmission electron microscopy (TEM) being invented by Knoll and Ruska in the early 1930th in Berlin [1]. The history of TEM and tribology started about in the 1950 by work about solid lubrication [2]. Certainly wear and its mechanisms were a crucial part of research already at that time [3], therefore, TEM was introduced into it in the 1960s by Bill Glaeser et al. [4] and wear mechanisms, sub-mechanisms and the underlying elementary processes have been investigated for a long time since then. Investigating metals at the desired small wear rates required high resolution analyses at and just below the worn surfaces either because of the materials properties of interest and their microstructural alterations under tribological loading [5–8] or under certain processing parameters [9]. Last-but-not least it is crucial to understand the influence of the interfacial media like lubricants [10] and wear debris [11] as well as any causal materials science aspects or elementary processes [12–14]. Thus the number of papers on TEM and wear related phenomena as well as conference proceedings, letters and short notices increased steadily (Fig. 2.1).

On the basis of the early investigations friction, wear, and lubrication were seen as interconnected parts of an area being named tribology in 1966. It became clear that any property is not just related to a certain material but it is connected to the entire tribological system and its specific structure and acting wear mechanisms (Fig. 2.2) [15, 16]. This somewhat rough scheme—and even though not all sub-mechanisms are known or have yet been sufficiently investigated—has been proven to allow for targeted countermeasures beyond just trial-and-error (or the common hardness vs. wear) approaches [17]. Due to the fact that in the current English literature the mechanisms and sub-mechanisms are described or designated by similar terms we will use quotes for the mechanisms and sub-mechanisms. In general we distinguish between

Table 2.1 Main wear mechanisms and their submechanisms

Main wear mechanisms	Tribochemical reactions	Surface fatigue	Abrasion [18]	Adhesion [19]
Submechanisms	Tribooxidation [20]	Delamination (by predominantly elastic interaction in solid contacts [21])	Microploughing	Material transfer
			Microcutting	
	Tribocorrosion [22]		Microcracking	
	Mechanical mixing [24]	Indentation (by predominantly plastic interaction in solid contacts [23])	Microfatigue	

four main wear mechanisms: "Abrasion", "Adhesion", "Surface Fatigue" and "Tribochemical Reactions" [18] which are related to and can be defined by certain wear appearances. Still depending on the structure of the tribosystem and the properties of the involved materials there are alterations of the surfaces as to topography, chemistry, and microstructure as well as changes of the interfacial and surrounding media. This did lead to the definition of so-called sub-mechanisms. Table 2.1 sums about up the current state of sub-mechanisms known to the authors from references of the last 50 years. The references given in Table 2.1 are far from being complete and should just point on some substantial contributions to this approach.

It also has been shown by [18] and others that "Adhesion" might lead to wear rates that are nine or more orders of magnitude larger than those caused by "Tribochemical Reactions". At the desired very small wear rates in the range of less

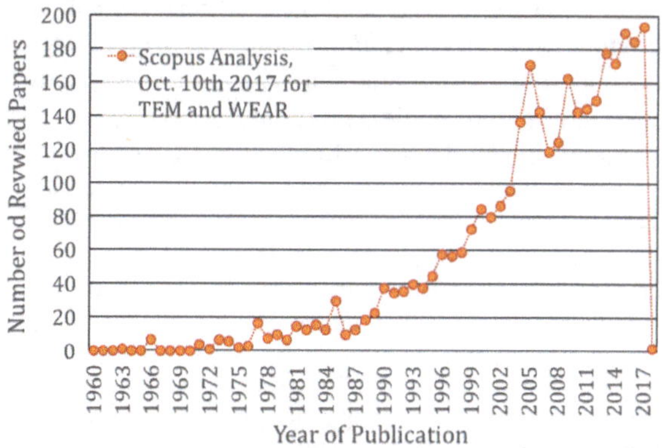

Fig. 2.1 Number of reviewed papers on transmission electron microscopy and wear (Scopus Search on Oct. 10th 2017)

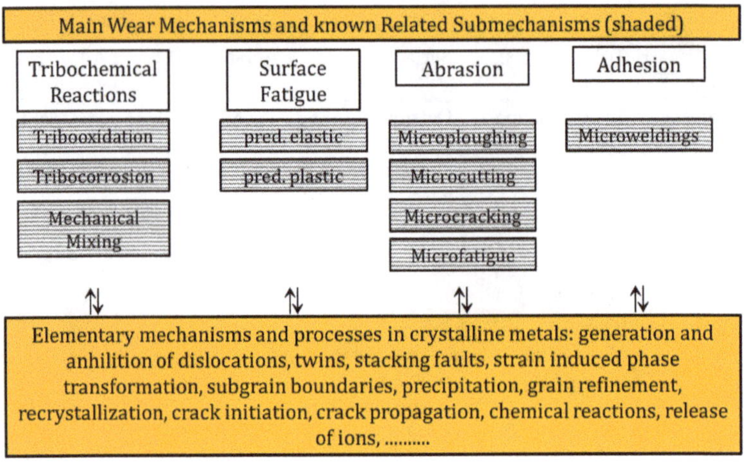

Fig. 2.2 The main wear mechanisms, known related submechanisms and examples for primary elementary mechanisms and processes

than e.g. <3 nm/h[1] one would in general expect sub-mechanisms of "Tribochemical Reactions" and "Surface Fatigue". The authors exemplified that a combination of "Mechanical Mixing" and "Indentation" a sub-mechanisms of "Surface Fatigue" characterized by predominantly plastic interaction generated by rotating sub-μm size wear particles may lead to relatively small wear rates [17, 23]. This is the more pronounced the better the mechanically mixed (or in other systems tribooxidized) layer at the surface is supported by the subsurface microstructure [25]. Another prerequisite is that such layer must not be overloaded by too large contact stresses [26]. Thus if "Indentation" as well as frictional contact shear forces trigger "Mechanical Mixing" and, therefore, generate a tribofilm and nano-size wear particles the total wear rates might become very small and even reach the so-called ultra-mild wear regime [26–28].

In order to properly analyze such near- or sub-surface structures one has to distinguish between either amorphous, nanocrystalline and/or chaotic surface layer and the underlying strain gradient. The latter is also called "transition zone" (here: subsurface or strain gradient) in many papers and is generated by the accumulation of cyclic plastic strains [25]. In contrast the uppermost layer (here: near-surface or tri-

[1]This number is a criterion for piston ring wear that would allow for a minimum of 200,000 km of a passenger car engine without substantial wear. It is chosen absolutely voluntarily in order to distinguish between ultra-mild wear and mild wear. While most laboratory tribometers for time reasons run under mild-wear conditions. Still most parts in application would require ultra-mild wear for a sufficient life time. In sliding wear mild and severe can be distinguished by the characteristics of contact mechanics; e.g. the plasticity index (s.a. Johnson (1985) Contact Mechanics, Cambridge University Press, Cambridge, England). If the contact is predominantly elastic it might lead to mild wear, while a predominantly plastic interaction brings about severe wear. Still all these terms are not standardized and the regimes might overlap.

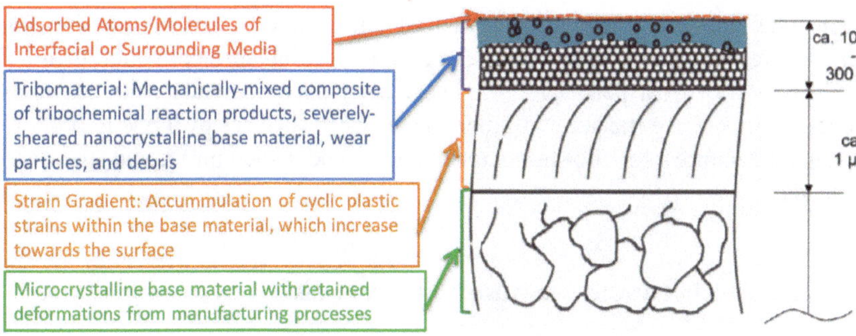

Fig. 2.3 Scheme of surface and subsurface structures generated under tribological stresses based on [16, 30, 31]

bomaterial) appears like a shear band [13] caused by monotonic stresses and strains similar to what is known from severe plastic deformation [29] (Fig. 2.3).

Despite all of the research of the last 70 years any direct relation or even quantitative connection to the so-called elementary processes (Fig. 2.2) is still missing. Hence further research is necessary comprising low and high resolution analyses in combination with the appropriate computer simulation method.

In this section we cannot give an overview over all the work that has been done by TEM in tribology. We would just like to show how one can get information on tribological characteristics of the elementary processes and mechanisms by TEM analyses and—at least—qualitatively relate it to the tribological behavior of exemplary selected metallic materials.

2.1.1 TEM Analytics of the Cyclic Strain Gradient

Accumulation of cyclic strains are known from fatigue of metals (also called cyclic-creep or ratchetting), but here the stress state is multiaxial. Hence no direct quantitative relation between classical fatigue and wear is possible, but elementary mechanisms and processes are similar like e.g. phase transformations or the influence of long- and short-range order effects. Finnie et al. pointed already in 1984 on the fact that the strain gradient is generated by accumulated cyclic strains [32] they found the conflicting requirement of high cyclic strength in combination with high ductility (ability to accumulate cyclic strains without crack nucleation) for wear resistant metals. In fatigue and, therefore, as in wear this is governed by the characteristics of either planar or wavy cyclic slip of metals. On the basis of TEM analyses of MgO and pure Ni, Rigney and Glaeser did also show earlier that such differences in slip behavior might govern sliding wear [5]. Glaeser pointed on such mechanisms already in 1977 by investigating wear particles detached from copper showing dislocation cells [33]. He hypothesized that crack nucleation must have taken place at such cell walls. In 1983 Saleski and Ritchie proposed a detailed model for plain steel showing that cracks are initiated at dislocation cells below a sliding surface, which then led to the formation of wear particles [34]. These early papers already depict towards the necessity to analyze the strain gradient under worn surfaces by TEM in order to understand the nature of cyclic sliding mechanisms under such multiaxial stress and strain fields.

2.1.1.1 The Strain Gradient in Austenitic Materials

By comparing Austenitic Materials with known planar, wavy, and mixed sliding characteristics Fischer et al. were able to show, that planar sliding improves the wear behavior distinctly (Table 2.2) [25].

Due to the fact that under planar sliding the lattice defects (dislocations, twins, stacking faults, ε-martensite, Fig. 2.4a) stay mobile larger cyclic strains can be accumulated over a bigger number of cycles. Thus crack initiation as well as propagation are hindered. By wavy slip cells are formed blocking dislocations (cell walls, Fig. 2.4b) which bring about subsurface crack initiation. It is also known that wavy slip promotes faster crack propagation [35]. If wear particles are generated by crack initiation and propagation inside the strain gradient their size is likely within the μm-range. Such elementary process can be attributed to "surface fatigue" and its submechanism "delamination". As a results μm-large wear particles deteriorate the contact situation e.g. by bringing about "microploughing" and destroying also any existing nanocrystalline tribomaterial (Fig. 2.5) or by "microcutting" generating instantly further μm-size chips and increase the wear rate (Table 2.2).

Table 2.2 also shows by the wear rates—as has been reported also by others [11, 36, 37] earlier—that the stacking fault energy as well as strain induced phase transformations play an important role. From fatigue tests it is known that short-range order effects are as important for the ability to accumulate cyclic strains as a so-called low damage accumulation rate [38].

Table 2.2 Sliding characteristics of austenitic materials and their wear rates

Material	Sliding characteristics	Wear rate in m/m under dry sliding after 90,000 cycles
X5CrNiMo17-13-2, ISO5832-1	Wavy	2×10^{-8}
X6CrNiMoN22-10-4-3, ISO5832-9	Mixed	1.8×10^{-8}
X13CrMnMoN18-14-3, ASTM F2229-02	Planar	7×10^{-10}
CoCr29Mo6, ISO5832-6	Planar, very low stacking fault energy	2×10^{-10}

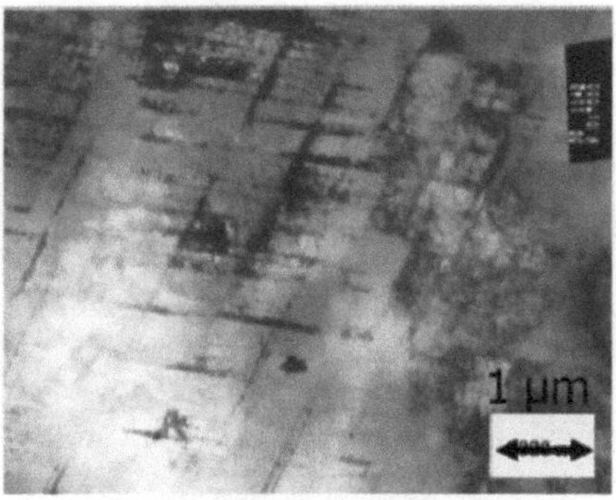

(a) Planar Sliding Structure of an Austenitic High Nitrogen CrMnMoN-Steel with Dislocations, Twins, Stacking Faults and ε-Martensite

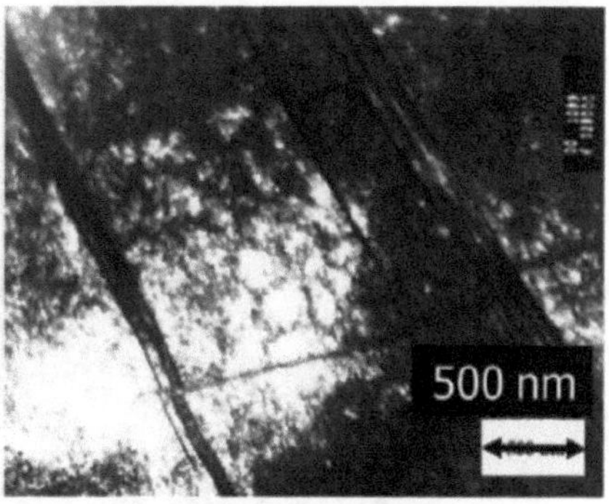

(b) Wavy Sliding Structure of an Austenitic CrNiMoN-Steel with Dislocations Cells and Twins

Fig. 2.4 TEM micrographs of defect structures within the subsurface strain gradient after dry sliding wear. **a** Planar and **b** Wavy sliding austenitic steels from [25]

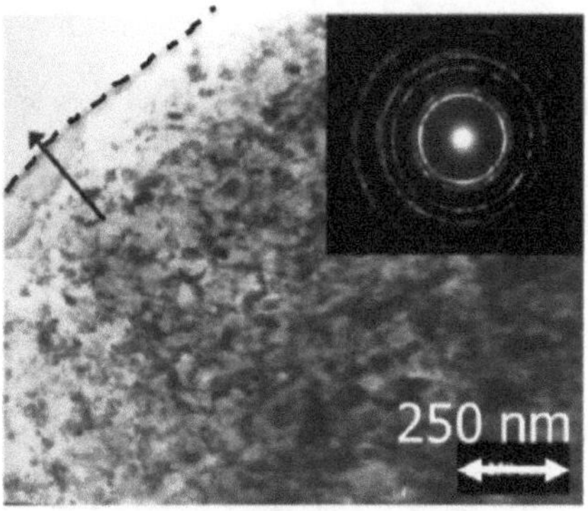

(a) Austenitic High Nitrogen CrMnMoN-Steel

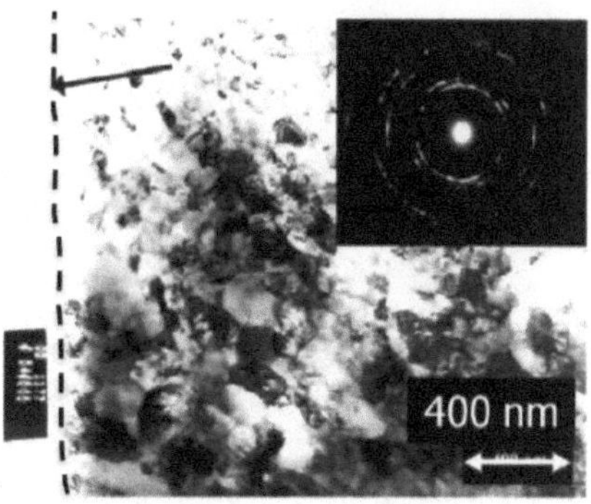

(b) Austenitic CrNiMoN-Steel

Fig. 2.5 TEM micrographs of nanocrystalline tribomaterial at the worn surfaces of **a** planar and **b** wavy sliding austenitic steels from [25]. The arrows are perpendicular to the contact surfaces marked by the dashed lines. It should be mentioned here that even though such tribomaterial is nanocrystalline it may show traces of a directional texture as can be seen by the oriented brightness of the diffraction pattern in this figure

2.1.1.2 Strain Gradient in Martensitic Materials

Now most steels used in mechanical engineering are not austenitic but ferritic or martensitic and are known for wavy slip as well. But in contrast to austenitic metals their strength is much higher, while for martensites the ductility is much smaller. Thus it is interesting to understand whether high strength martensitic steels would generate a strain gradient by cyclic multiaxial stresses as well. Figure 2.6 shows

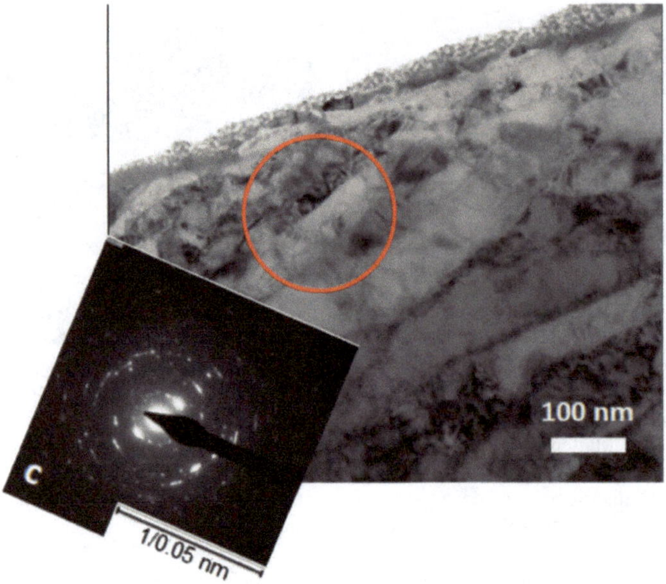

Fig. 2.6 TEM bright field image and diffraction pattern of a strain gradient of a cold work tool steel 56NiCrMoV7 (AISI L6) under boundary lubricated sliding wear at a normal load of 150 N from [39]. The diffraction pattern was rotated in order to parallelize its directional brightness to the lattice orientation of the crystals inside the red circle

an overview of the subsurface region of a worn cold work tool steel depicting a zone of distinct subsurface grain refinement parallel to the surface. The red circle focusses into such refined zone with a grain size of about 200 nm, while the bulk depicts martensite lathes of some μm. It is worth mentioning that within the strain gradient the grain size stays above 100 nm also known as the ultra-fine (100–500 nm) crystalline range.

Similar outcomes can also be found with carburized martensitic steels after 2×10^6 cycles of ultra-mild sliding wear under boundary lubrication in gear oil at 80 °C [40]. The bulk within the carburized zone shows μm-large martensite lathes while the strain gradient depicts again smaller grains as can be derived from the streaks around the diffraction reflexes (Fig. 2.7).

Now it is extremely important to notice that such strain gradient can already exist before the wear tests have started. Figure 2.8 shows a SEM cross section of a carburized and ground 18CrNiMo7-6. Both render similar diffraction patterns that are characteristic for an ultrafine grain size, while the obvious tendency to become circular can qualitatively be seen for about 100–200 nm grain size.

Even though the loading during grinding is completely different from those under ultra-mild wear the internal structure of the about 1 μm thick strain gradient (emphasized by the arrows) is very similar to that after wear (compare Figs. 2.7b and 2.9) even though the grains are smaller directly after grinding.

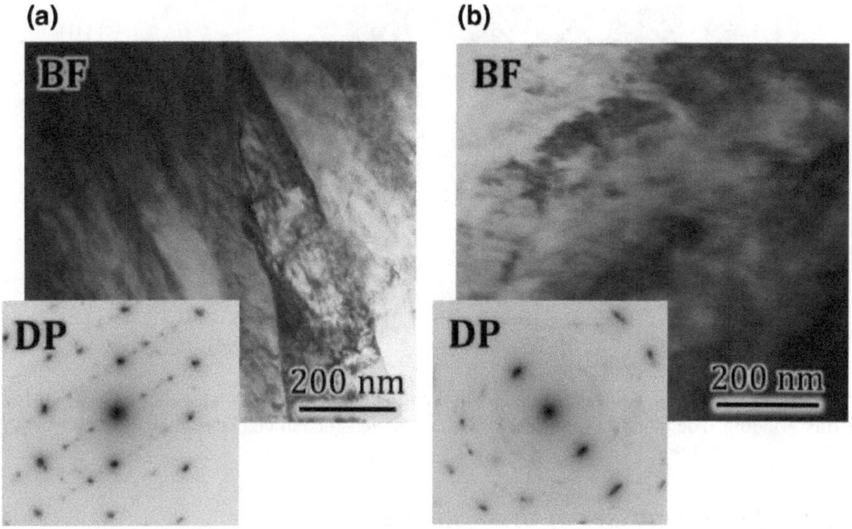

Fig. 2.7 TEM bright field micrographs and diffraction patterns of **a** the bulk and **b** the strain gradient of a carburized steel 18CrNiMo7-6 (SAE 18NCD6) after 2 × 10^6 cycles under boundary lubricated sliding wear at 80 °C in gear oil

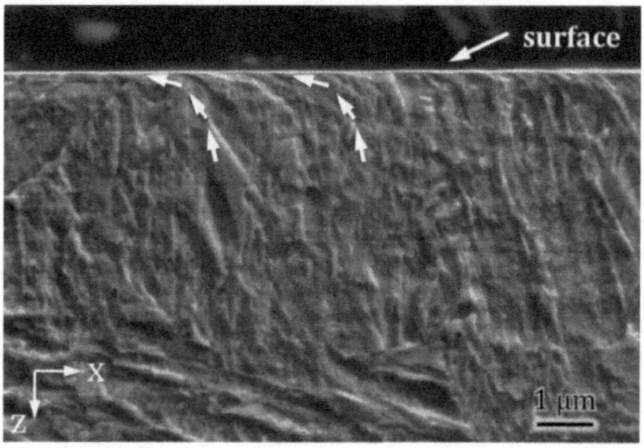

Fig. 2.8 SEM cross section of a carburized steel 18CrNiMo7-6 after grinding. The strain gradient is emphasized by white arrows

Hence it is absolutely mandatory, that any TEM analyses of near-surface and subsurface structures must be put into relation to those, that were there before the wear test (e.g. after machining, grinding, polishing, etc.). Moreover, any TEM analysis requires the knowledge of the absolute position of the final samples inside and underneath the wear track in relation to the incipient surface before wear. Otherwise neither the predeformation characterized by the lattice defect state (and properties) of that specific area is known nor their development into the investigated state after the wear test.

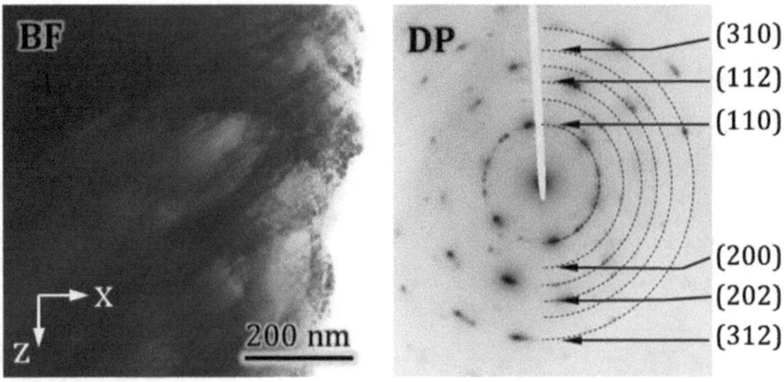

Fig. 2.9 TEM figure and diffraction pattern of the strain gradient of a carburized steel 18CrNiMo7-6 after grinding. The numbers represent indices of martensitic lattice planes

2.1.2 TEM Analyses of the Tribomaterial

The investigation of the uppermost layer by TEM and based on replicas started in the 1950s under the headline of transfer layer under mild-wear conditions [41] and was steadily improved. In the beginning the main target was to understand "material transfer" a submechanism of "adhesion" under more severe-wear conditions. But it could also be shown quite early that under mild-wear such layers consist in parts or in total of oxidized wear particles of a few nm in size. It was found later by SEM-EDS that such layers could be a mixture of materials from the body, the counterbody, the interfacial media, and the environment leading to an alteration of the internal structure but also of the chemical composition. In 1983 Heilmann et al. concluded on the basis of TEM analyses of near-surface and subsurface areas that such layer is a common appearance of wear, bares some nanostructured, nanocrystalline composite that represents the surface in contact (s.a. 3rd bodies [42]), is generated very early in the wear process (s.a. [43, 44]), mostly matches the wear particles, and is influenced by body, counter-body, interfacial medium (e.g. lubricant) as well as the environment [7]. Today these early findings can be related to different mechanisms and submechanisms of ultra-mild wear e.g. by "microploughing" (of "abrasion"), "indentations" (of "surface fatigue"), as well as "tribooxidation", "tribocorrosion", and "mechanical mixing" (of "tribochemical reactions").

2.1.2.1 The Tribomaterial in Austenitic Material

The generation of such nanocrystalline metal-base composite—from a former microcrystalline bulk that has been predeformed inside the ultrafine-crystalline strain gradient—appears like a shear band (Fig. 2.10) [45]. It is important to notice that within such shear bands the chemical composition does not alter but the grain size does over three orders of magnitude from 20 μm to about 20 nm. The elementary mechanisms

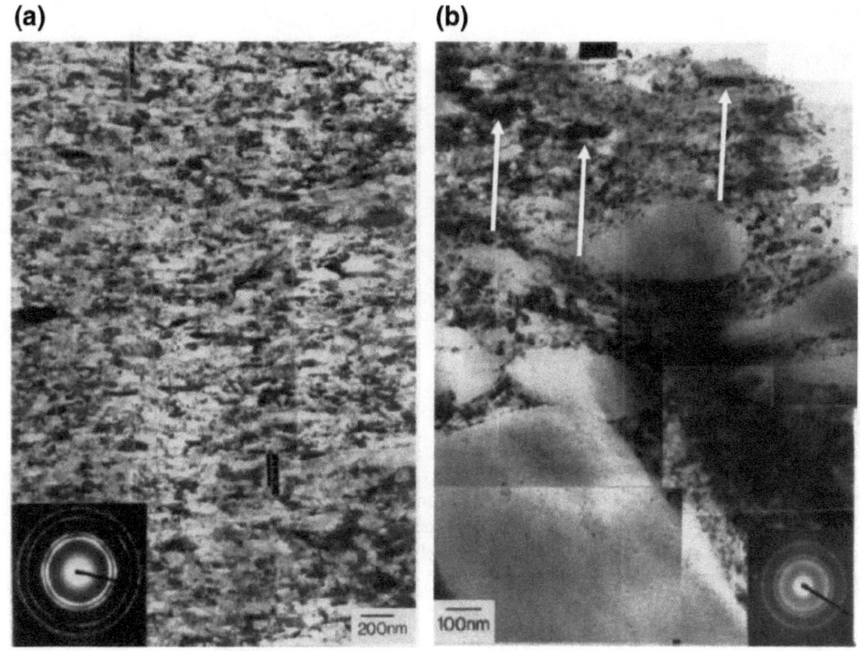

Fig. 2.10 TEM micrographs of **a** a shear-banded region and **b** nanocrystalline tribomaterial with incorporated metal-oxides marked by White Arrows from [45]

of such formation can be related to those known from severe plastic deformation [29].

Due to the fact that tribomaterial differs chemically from the bulk material Rigney proposed a model, that allows for such chemical changes just by the rotation of atoms or clusters of atoms and called it "mechanical mixing" [46]. As for atoms in a MD model this can also take place for nanocrystals which bring about plastic deformation by the rotation of grains [29, 44]. The marked difference between SPD and tribology is, that the stress field in tribology is not uniaxial. The shear band is supposingly brought about e.g. by rotating compact wear particles (or any other debris) within the interface generating "indentations". Besides shallow particles that slide would lead to "microploughing". Both submechanisms generate mainly a very localized plastic deformation while "indentations" is of cyclic nature and, therefore, a submechanism of "surface fatigue" whereas "microploughing" is more of monotonic kind and a submechanism of "abrasion". Yet such rotation of nanocrystals at the surface incorporates any interfacial media consisting e.g. of lubricant, wear particles, contaminants, reaction products, and further debris into such layer and a metal-based nano-composite of some sub-μm thickness is formed by "mechanical mixing[2]" a submechanism of "tribochemical reactions". At mild and ultra-mild wear this tribo-

[2]The term tribomaterial was proposed by David Rigney on a tribochemistry meeting in Hagi, Japan in 2011 in order to summarize the large number of already existing terms like Beilby layer,

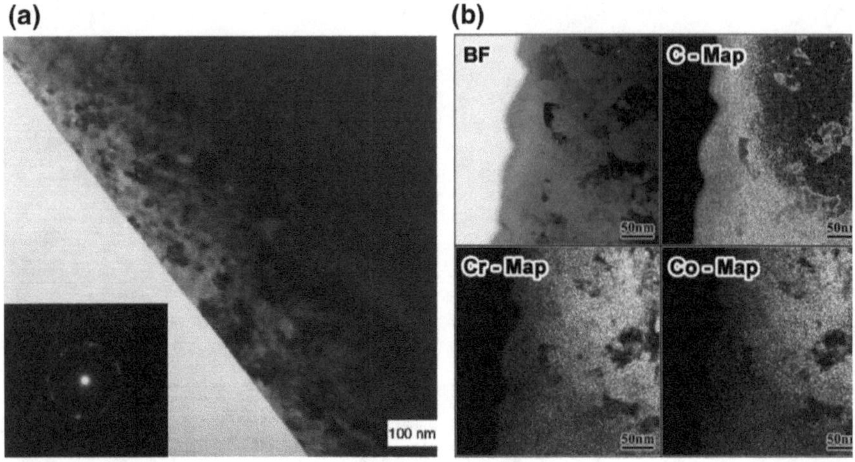

Fig. 2.11 BF micrographs of nanocrystalline tribomaterial of **a** a CoCrMo-retrieved cup after eight years in vivo from [47] and **b** a simulator tested head after 10×10^6 cycles from [48]. The EELS-maps of **b** show an extensive amount of C at the surface, which represents denatured proteins, while the metal is characterized by Co and Cr

material persists, separates body and counterbody, and allows for the extreme shear rates. Because of the very small size of its constituents far below 100 nm TEM with EDS or EELS is the only microscopical method that renders a combined local information about structure, defect state, and chemistry.

Büscher et al. [47] as well as Pourzal et al. [48] showed that even at body temperature and at relatively small Hertzian contact pressures of about 50 MPa a high strength Co-Base alloy generates such tribomaterial inside the primary articulating surface of an artificial hip joint with thicknesses up to 500 nm (Fig. 2.11a, b). It consists of CoCrMo nanocrystals, human pseudosynovia (or bovine calf serum if tested in laboratory), and denatured-proteins, while the latter can be analyzed by EDS or EELS for their high Carbon content (Bright-Field and C-Map, Fig. 2.11b). It was found that such metal-organic composite is the main reason for the longevity of such selfmating metal-metal sliding contacts of austenitic alloys.

The metal part is a CoCrMo solid solution with a grain size ranging from 10 to 70 nm[3] and supported by a strain gradient that is stabilized by twins, stacking faults, and ε-martensite lathes forming rhomboid domains that become the smaller the closer to the surface. Such tribomaterial is a chaotic structure and might consist

white layer, fragmented layer, transfer layer, glaze layer, mixed layer, 3rd bodies, highly deformed layer,....

[3]It is important to notice that even though the tribomaterial appears in different thicknesses and mixtures nearly at any analyzed position while its grain size range was always very similar and independent of source (cast, wrought, new or retrieved hip joint), origin (1960s–2000s), make (heat treatment, low- or high Carbon content, manufacturer) of the CoCrMo alloy, and loading (simulator, one or twenty-two years in vivo).

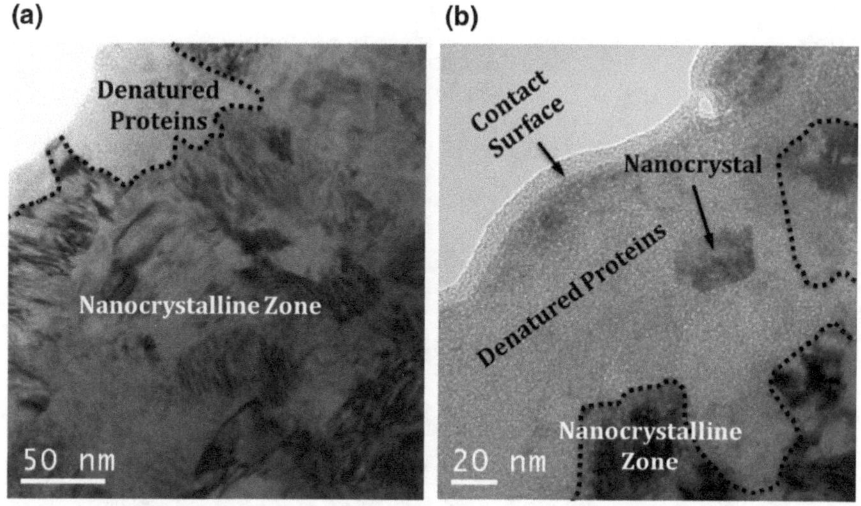

Fig. 2.12 BF Images of tribomaterial of **a** a simulator tested CoCrMo Head after 5×10^6 cycles depicting a nanocrystalline zone with denatured proteins on top while **b** shows a denatured protein film with embedded metal nanoparticles in a different position of the same sample from [48]. The dotted line marks the interface within the tribomaterial (compare to the scheme in Fig. 2.3) between nanocrystalline CoCrMo and the mixed zone of denatured proteins and CoCrMo nanosize particles

of a mixture of metal nanoparticles with a face-centered-cubic (austenite) and/or hexagonally-closed-packed (ε-martensite) lattice topped by and intermixed with denatured proteins (Fig. 2.12a) or denatured proteins with embedded metallic or metal-oxide nanoparticles (Fig. 2.12b).

The structure of the denatured protein constituent in such tribomaterial has been investigated by TEM as well. It is quite surprising to find graphitic structures which stem from denatured proteins like e.g. albumin but any artefacts during preparation and electron-beam as well as photon-based analyses has been experimentally ruled out [27]. Still the elementary mechanism of this transformation is not clear yet (Fig. 2.13).

2.1.2.2 The Tribomaterial of Ferritic and Martensitic Steels

Ferritic or perlitic steels are used in many tribosystems demonstrating their high wear resistance e.g. in wheel/rail contacts or in cylinders of combustion engines. The latter might be coated with a ferritic steel by plasma-based processes for the need of higher ignition pressures and/or because the crank case is made out of AlSi-based castings. Principally plasma-spray processes allow for a very high solidification rate and lead to quite chaotic microstructure of μm-size so-called splat grains, which internally consist of a very fine microstructure (Fig. 2.14) [49].

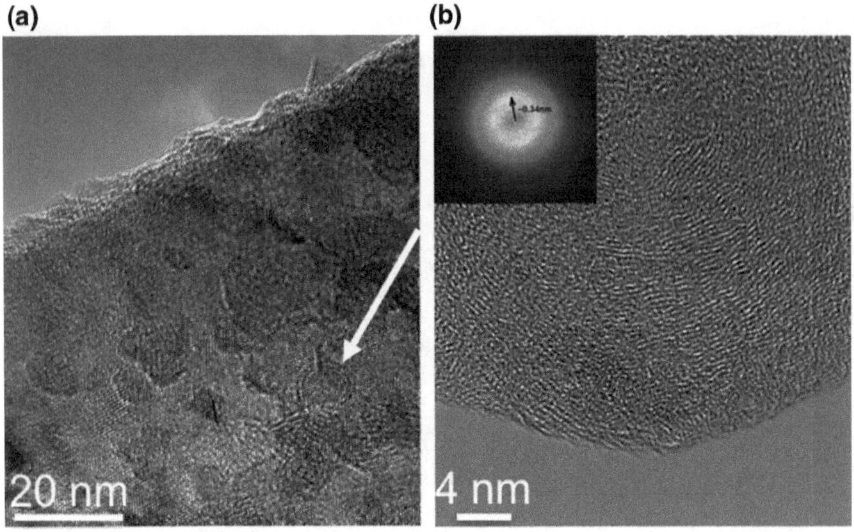

Fig. 2.13 HRTEM image of **a** Tribomaterial with nanocrystals and an embedded graphitic particle (white arrow); **b** graphitic structure at higher magnification from [27] and supporting online material

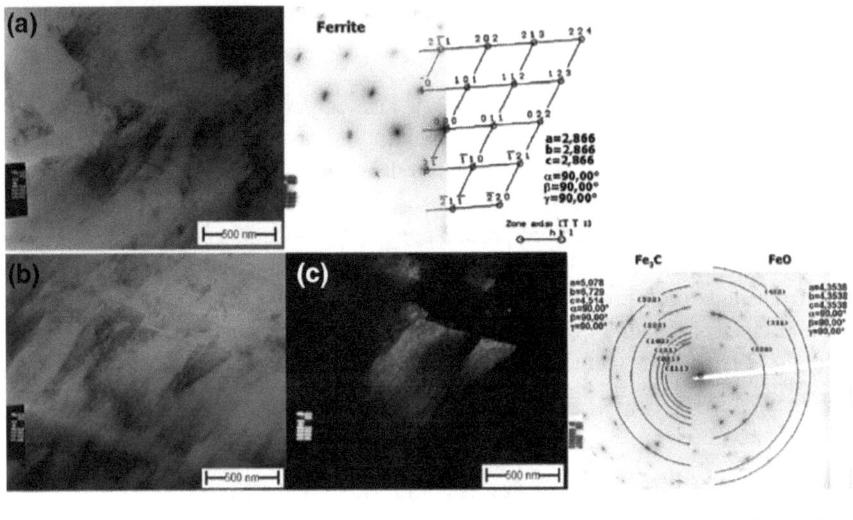

Fig. 2.14 TEM micrograph of the microstructure of a plasma-transferred-wire-arc coating (PTWA) of plain steel with 0, 1 weight-% carbon depicting **a** BF; a microcrystalline ferrite, and **b** BF and **c** DF nanocrystalline Fe_3C and FeO precipitates from [49]

Thus the incipient structure inside the splat grains appears like a precipitation hardened ferrite, while the alloying elements C and O stem either from the alloy by 0.1% C or from contamination by O during atmospheric spraying under just an inert-gas shield. Laboratory, engine, and field tests did reveal that such coating

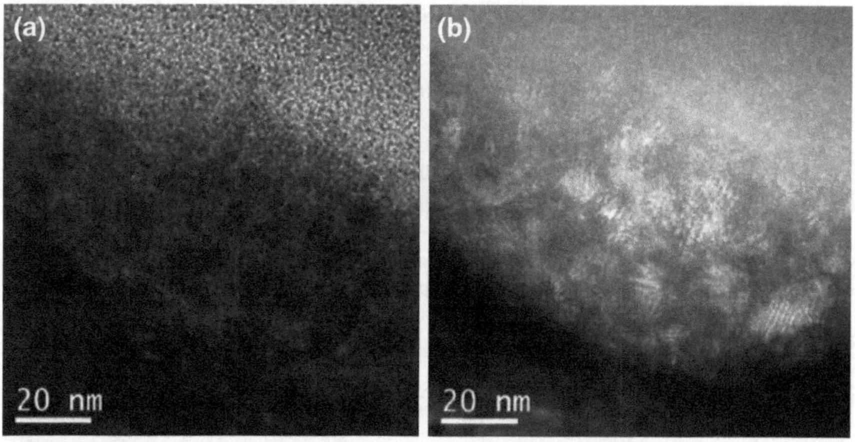

Fig. 2.15 **a** Bright field and **b** HRTEM micrographs of a PTWA coating surface in the combustion chamber after a field test of 60,000 km in a passenger car's diesel engine showing an amorphous layered structure with incorporated nanocrystalline particles from [49]

provides a sufficient resistance against wear and lower friction against Cr-coated piston rings. Beside the classical analyses of such tribosystems the TEM analyses revealed some new findings at that time. A cylinder-piston ring system consists of four different tribosystems as to the loading characteristics (top dead center, bottom dead center, combustion chamber above top dead center, and stroke) resulting in different microstructural alterations of the coating into tribomaterial and strain gradient. Thus any alterations must be related to the specific position inside the cylinder. Despite the fact that inside the combustion chamber above the top dead center no piston ring contact takes place the tribomaterial shows some amorphous layer with embedded nanocrystals of about 10 nm in size (Fig. 2.15a), while it ranges between 100 and 300 nm within the underlying strain gradient (Fig. 2.15b).

By checking the surface below the bottom dead center it was found that after honing neither a strain gradient nor tribomaterial could be observed. The incipient microcrystalline structure must have been refined about 1000 fold just by the cyclic impacts of the ignition at higher temperatures. An EFETM analysis revealed that the amorphous layer consists mainly of C, the nanosize particles contain higher amounts of O and Fe, while the base material contains Fe. From this on can derive that combustion products form an amorphous layer on top of the oxidized steel surface, which is mechanically and thermally loaded by the ignition leading to grain refinement. Whether the oxide particles stem from the precipitates or represent oxidized wear particles was not differentiated (Fig. 2.16).

At the top dead center the piston ring is sliding over the surface but generates cyclic impacts during ignition. Still the microstructural and chemical alterations are not very different from the combustion chamber, but now measurable wear of some μm took place at this position (Fig. 2.17). The tribomaterial appears like a nanostructured multilayer coating while the strain gradient shows ultra-file grains.

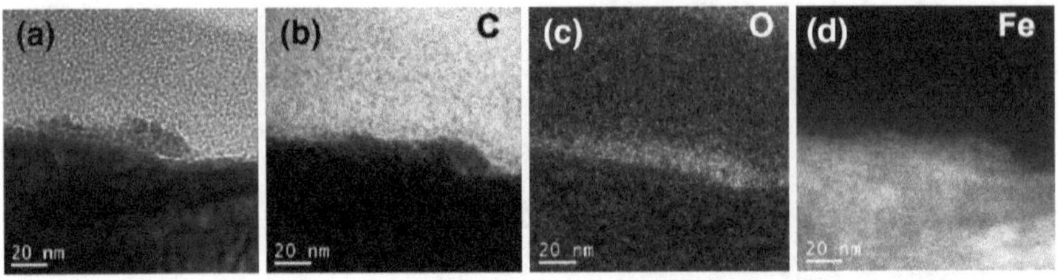

Fig. 2.16 The EFTEM analysis of the amorphous layer with embedded nanocrystals in the combustion chamber of Fig. 2.15; **a** bright field, **b** C-map, **c** O-map, **d** Fe-map

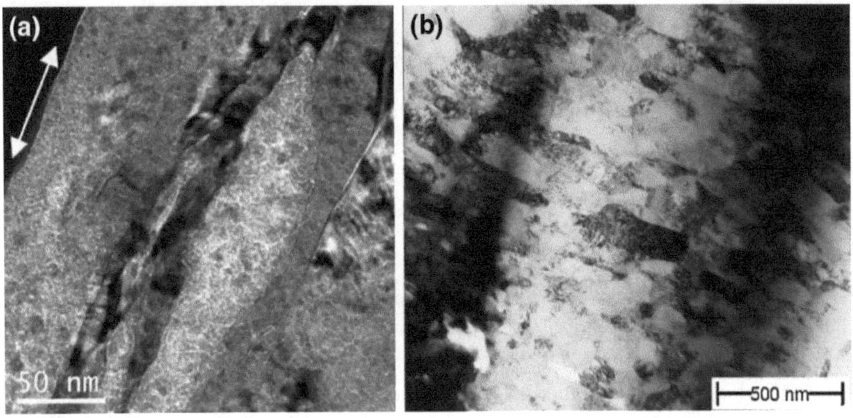

Fig. 2.17 HRTEM bright field micrographs of a PTWA coating surface at top dead center showing **a** a nanostructured tribomaterial supported by **b** an ultrafine crystalline strain gradient from [49]. The white arrow represents the relative movement of the Piston Ring

It is quite interesting that the latter appears like a former subgrain or dislocation cell structure generated by cyclic loading that is more or less immediately recrystallized by elevated subsurface temperatures during the ignition phase. This was supported by the fact that at the much colder bottom dead center the dislocation cells were still visible [49].

An EELS analysis reveals that this layered structure is mainly brought about by O and C, while interestingly enough O is not always combined with but also exchanged by Fe in this nanostructure. From this one can conclude that this layer is a mixture of combustion products and near-surface Fe-base coating. Still any constituents of the lubricant are missing (Fig. 2.18).

Thus, from such TEM micrographs one might get a very good impression about the chaotic nature of tribomaterial but in order to demystify it additional analyses e.g. by XPS, AES, or RS are necessary.

Martensitic steels have a much higher strength—mostly characterized by hardness—than ferritic ones but they might generate tribomaterial as well (Fig. 2.19a, b). Now it is important to notice that such tribomaterial must not necessarily be a

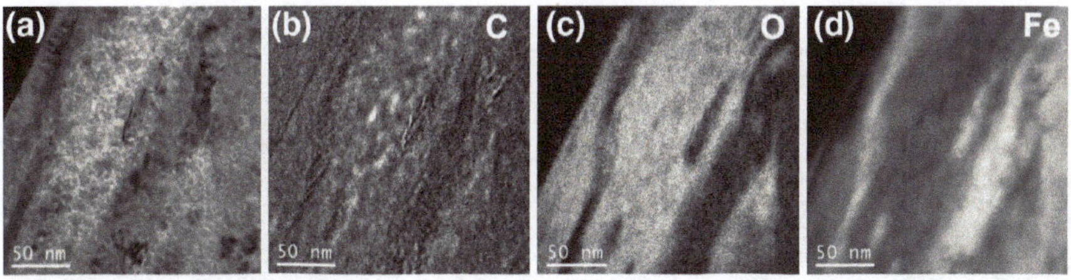

Fig. 2.18 The EELS analysis of the nanostructured tribomaterial at the top dead center of Fig. 2.17; **a** bright field, **b** C-map, **c** O-map, **d** Fe-map from [49]

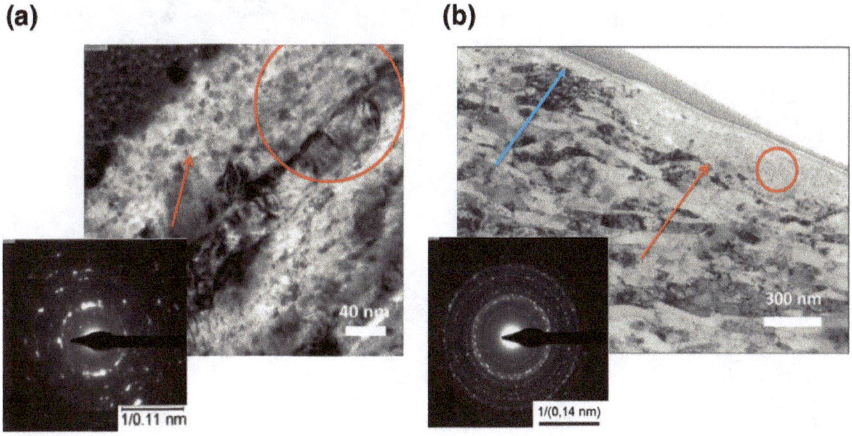

Fig. 2.19 Nanocrystalline tribomaterial after boundary lubricated sliding wear of the polished cold work tool steel 56NiCrMoV7 of Fig. 2.6 but now worn under a normal load of 400 N from [39]

nanocrystalline layer of a certain thickness (Fig. 2.19c, blue arrow), but it can also appear as a certain near-surface volume of different thicknesses depending on the loading history of that specific location (Fig. 2.19d, red arrow). Still the interface between the strain gradient and the tribomaterial is a sharp line.

In order to get a 1st glimpse—still not a full insight—on the loading history of a specific location for TEM analyses on can combine the depth of the wear scar in comparison to the incipient surface with EBSD analyses of cross sections as shown in Fig. 2.20. Due to the limited resolution of EBSD analyses even at 30 nm step size any nanocrystalline tribomaterial is characterized by no-counts (black areas in Figs. 2.20a, b). But from such a figure one can read that the nanocrystalline surface zone from milling (Fig. 2.20a, c), which represents the secondary-shear zone generated during chip formation, has completely been worn away and a totally new one has been generated by further sliding wear (Fig. 2.20b, d). By this procedure it is unequivocally shown the strain gradient in Fig. 2.7 as well as the tribomaterial in Fig. 2.20d are no remains of the last production step but generated by wear.

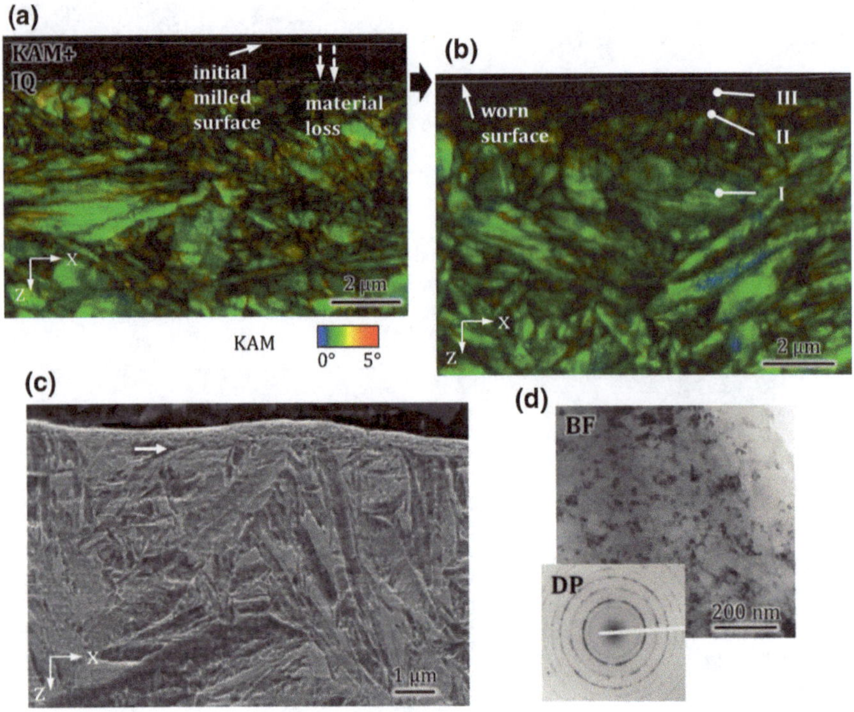

Fig. 2.20 Cross sections depicting the microcrystalline martensitic structure of a carburized 17CrNiMo7-6. **a** EBSD after milling, **c** SEM after milling, **b** EBSD after boundary lubricated sliding wear, **d** TEM BF and DP after Wear. Compare the BFs and DPs of (**d**) with those of 7a (bulk) and 7b (strain gradient)

At ultra-mild wear this becomes quite tricky, because at extremely small wear rates e.g. after milling and finishing the surface levels before wear (Fig. 2.21a) and after wear (Fig. 2.21b) are the same. Thus one cannot distinguish from such EBSD analyses alone whether there were any tribologically driven alterations of the near-surface nanocrystalline microstructure. Now after milling and finishing there is a nanocrystalline zone on the surface (Fig. 2.21c, d) which also looked absolutely the same after wear. It was quite interesting to see that the polished microcrystalline surface of the self-mating counterbody became nanocrystalline during wear and generated a distinct strain gradient similar to that in Fig. 2.20b, d. Thus, one can conclude that the frictional work was dissipated mostly—if not solely—by the counterbody which also corresponded to the measurable differences of the amounts of wear of both bodies.

Thus the near-surface structures and the wear behavior after machining in comparison to machining and finishing leads to the conclusion that not any nanocrystalline layer generated by manufacturing processes is a good precursor of tribomaterial and diminishes the wear rate. It strongly depends on the existence and structure of a

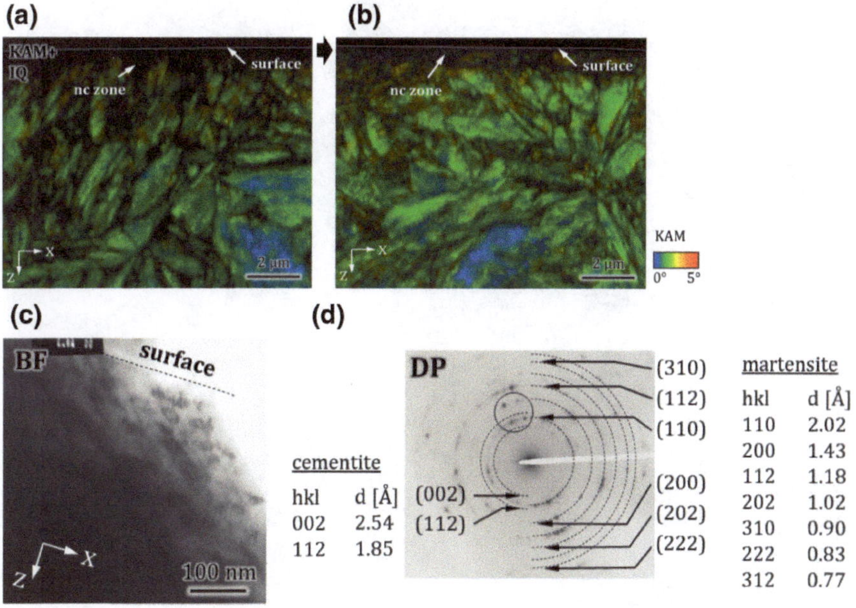

Fig. 2.21 Cross sections depicting the microcrystalline martensitic structure of a carburized 17CrNiMo7-6. **a** EBSD after milling and finishing, **b** EBSD after boundary lubricated sliding wear, **c** TEM BF image and **d** DP after milling and finishing

supporting sub-surface strain gradient as well as it is for that of the counterbody. It is absolutely necessary, therefore, to investigate both body and counterbody by means of similar methods in order to understand wear of a tribosystem.

2.1.3 Wear Particle Analyses by TEM

In the past wear particles were often believed being representatives for a characteristic segment of the contacting surfaces before they failed [33]. Thus wear particle analyses were always a crucial part of exploring and explaining sub-mechanisms like e.g. "triboxidation" [20, 34]. But by time it was understood that such particles might distinctly change their structure and chemistry after they detach from surfaces [50]. Today the preparation of wear particles from e.g. the lubricant play an extremely important role in order to avoid artefacts, that would mislead the analyses. In general it is known that depending on its size and nature such particles would oxidize immediately after detachment. It is not clear in any case, whether they might stem from an oxide film that detached or whether they detached as metallic particle and reacted with the interfacial medium because of their pyrophoricity. Either way wear

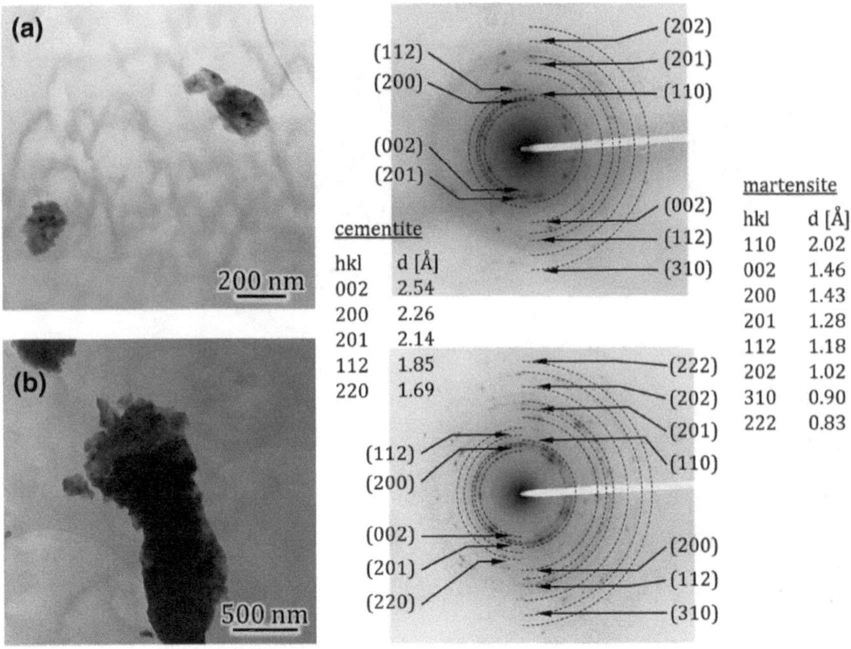

Fig. 2.22 TEM micrographs and corresponding diffraction patterns of wear particles extracted from the lubricant after a self-mating wear test of carburized 17CrNiMo7-6 under boundary lubrication

particles contain information about the tribological behavior of the specific system and should always be analyzed in combination with the surfaces they stem from [25].

E.g. nanometer small particles detaching from carburized 17CrNiMo7-6 under boundary lubrication exhibit a compact granular shape and a highly deformed internal nanostructure (Fig. 2.22).

This resembles that of the tribomaterial from which one can conclude that the wear particles stem from the near-surface region. By diffraction patterns any oxidation could be ruled out as it was found under dry sliding wear conditions of such steels. While such nanosize particles tend to agglomerate (Fig. 2.23) either inside the wear tracks or outside during preparation for TEM analyses any evaluation of agglomerates needs further analyses beyond TEM in order to be related to a certain mechanism or submechanism.

In biomedical engineering wear particles are of major concern, because most of them are in the nm-range, likely causing adverse tissue reactions, and finally leading to failure of artificial hip joints [51–53]. Doorn et al. analyzed such particles by TEM isolated from periprosthetic tissue which stem from both the CoCrMo-head of such hip joints as well as the TiAl6V4-metal backing [54]. Before his work particle sizes were reported somewhat between 10 nm and 50 μm, while his particles were sub-μm size from 10 nm to about 400 nm, while the agglomerates of them were in the μm-range. He could show that most of the wear particles had a metallic

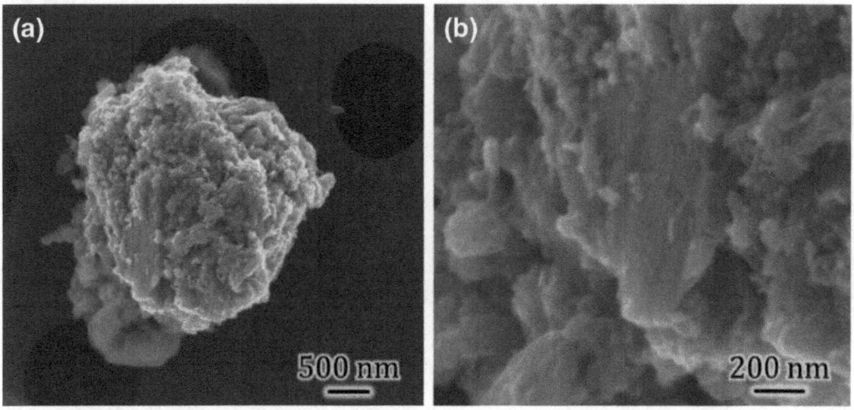

Fig. 2.23 SEM micrographs of agglomerated wear Debris **a** in full and **b** in detail extracted from the lubricant after a self-mating wear test of carburized 17CrNiMo7-6 under boundary lubrication

core while some were of Cr- or Ti-oxides. Catelas et al. compared wear particles taken from hip simulator tests and periprosthetic tissue by TEM and EDS [55]. They concluded that most particles from the hip simulator resemble those in vivo and—if not agglomerated—are smaller than 150 nm. According to the EDS analyses they were either Cr-oxides, Cr-carbides, or they stem from the CoCrMo metal matrix. The latter appeared needle-shaped and were later related to strain-induced ε-martensite lathes by Büscher et al. [56]. Billi et al. offered a different protocol for particle preparation and found additional ultrafine particles of about 200–500 nm in size that mostly contained Mo and Cr [57]. Acc. to Stemmer et al. such particles originate from intermetallic σ-phases that might precipitate in CoCrMo alloys together with Cr-carbides of the same size forming "so-called" mixed hard phases [58]. Pourzal et al. used the Catelas-protocol and compared wear particles by TEM, EDS and EFTEM [59]. Three different groups of wear particles—named I, II, and III—could be distinguished by size, chemical composition, and lattice structure.

All type I and II wear particles contain oxygen and chromium (Fig. 2.24). While type II are Cr-oxides the type I might also contain some cobalt. None showed any distinct amount of carbon, which would be characteristic for carbides. Surprisingly some type I particles depicted an amorphous structure with a crystalline Co-rich core (Fig. 2.25) while type III particles were 5–15 nm small fragments of type II fully oxidized into Cr_2O_3.

All these particles would originate from the tribomaterial and can be related to "mechanical mixing". The oxides could either be parts of the passive film generated by "tribocorrosion" that detaches under load or by nanosize particles that oxidize because of their pyrophoricity. A possible hypothesis would be that single grains within the nc-zone are transported due to the plastic flow of entire grains to the mechanically mixed zone where carbon-rich denatured organic material as well as phosphates settle at the grain boundaries. During ongoing plastic flow by "indenta-

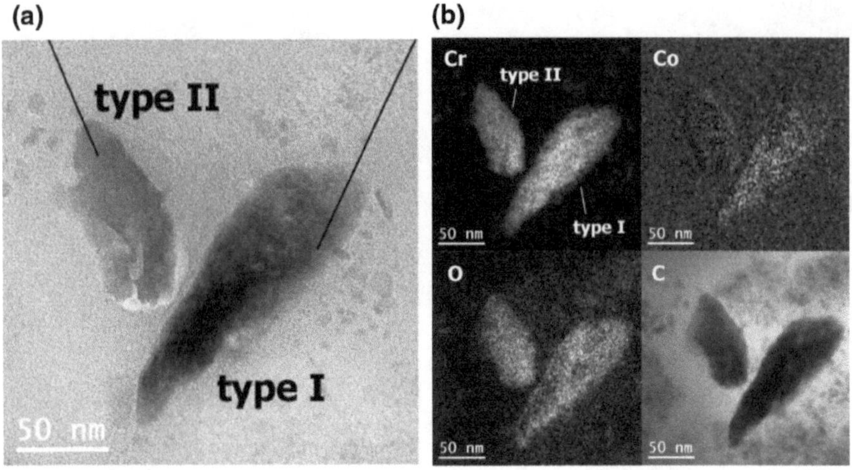

Fig. 2.24 TEM micrograph of **a** type I and type II wear particles and **b** corresponding EFTEM mapping from [59]

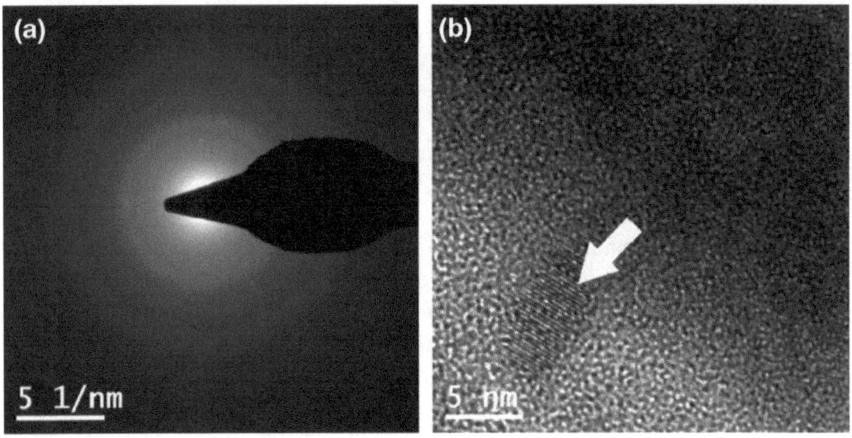

Fig. 2.25 **a** Diffraction pattern and **b** HRTEM micrograph of the core of a type I wear particle from [59]. The region marked by the white arrow is a Co-rich nanocrystal being embedded into an amorphous matrix

tion" and/or "microploughing", such grains could be transported to the uppermost surface where they would stick in a film of denatured proteins. Eventually, the grains would be pulled out of the tribomaterial and form single particles. Interestingly enough Pourzal showed by AFM-measurements that at this point, the particles would still exhibit a relatively high amount of cobalt as well as chromium, and in addition be protected by a shell of organic material [60]. Thus, it must be mentioned here that this organic shell is removed by the current preparation protocols of wear particles and was never found by TEM.

2.2 Conclusions and Outlook

TEM proved to be a suitable tool to understand the pathways of frictional energy dissipation into metallic materials. It can render information on the uppermost tribomaterial, its structure, and chemical composition from which the sequence of its generation can be modelled. Further TEM shows, whether a wear particle is generated inside the tribomaterial or inside the strain gradient. It allows for structural and chemical analyses of nm-size particles and brings about information of their composition and the interaction inside and outside of the contact interface.

Elementary mechanisms and processes can only be analyzed by TEM—even in situ [61]—rendering a combination of structural and chemical information at a particular position at high resolution. It is one of the essential tools to understand metallurgical features and to allow for a well-aimed development of more wear resistant metallic materials.

Still TEM is not a stand-alone method even though today it is combined with EDS, EFTM, EELS, and EBSD. In order to understand the relevance of any appearance and whether it contributes to the characteristics of a tribosystem further techniques can and should be involved like e.g. SEM, AFM, XPS, AES, or RS.

Future work should be based on what has been investigated in the past and continue in connecting mechanisms and submechanisms to the elementary processes. Due to the fact that most papers focus on steady-state the predominantly plastic interaction during run-in should be analyzed in more detail. It has been proven that a combination of TEM and MD might help to understand peculiar and unexpected tribological behavior [62]. Even though with TEM we can get information at high resolution as to location in a wear track we still lack of that high-resolution information as to the time-line.

Attachments

Preparation of Samples

This chapter exemplifies processes to prepare samples for TEM analyses of near-surface and subsurface areas of worn specimens and parts of metallic alloys.

Specimens from the Strain Gradient of Austenitic or Ferritic Metal Alloys

For the preparation of cross sections the method developed by Büscher et al. [63] or Hahn [64] can be used. Here, two corresponding segments of body and counterbody were taken and glued onto each other's articulating surface by means of a suitable adhesive (Epoxy G1, Gatan, Munich, Germany). The sample was fixed by a slotted pipe with a diameter of 2.5 mm and positioned in a brass tube of 3 mm diameter. The tube was cut to slices of 400 μm thickness. The cross section of one slice precisely exposed the segments of cups and head lying on each other separated by a small gap of hardened epoxide adhesive. Using grinding, dimple grinding (Model 656, Gatan, Munich, Germany) and ion milling (PIPS 691, Gatan, Munich, Germany) the sample was thinned to the desired thickness of about 40 nm. Afterward the specimens were investigated by means of transmission electron microscopy (TEM 400 Phillips, Eindhoven, Netherlands). In order to observe chemical changes in the uppermost surface layers, a high resolution TEM (Tecnai F20 Phillips, Eindhoven, Netherlands) with EDS and electron energy loss spectroscopy (EELS) was used.

Specimens from the Worn Surface of Martensitic Steels by FIB [39]

Sample preparation by means of focused ion beam (FIB), was performed using a dual beam FIB/SEM system (Helios NanoLab 600, FEI, Eindhoven, Netherlands). The cross-section procedure contained several steps. First, a protective Pt-layer was deposited at the area of interest on the sample surface. After this, bulk material was removed in a wedge-shaped trench by Ga+ ions, on one side of the area. Using decreasing energies of the ion beam, the sidewall was polished at a small glancing angle in subsequent polishing steps. The processing parameters are given in Table 2.3. Micrographs of the cross-sections were obtained in SE mode at 2 kV accelerating voltage. Microstructural features emerge due to electron channeling contrast (ECC).

Specimens from the Worn Surface of Martensitic Steels by Ion Milling

TEM cross-section samples of the wear tracks were prepared using an ion polishing system (EM-09100IS, Jeol, Akishima, Japan). Therefore, small samples parallel to the sliding direction were cut, as pictured in Fig. 2.4 (1–4) and a silicon wafer was

Table 2.3 FIB cross sectioning parameters

	Acceleration voltage (kV)	Ion current (nA)	Stage tilt angle (°)	Shape
Fast milling	30	6.5	52	Regular cross section
Milling	30	2.8	52	Cleaning cross section
Polishing	30	0.46	53	Cleaning cross section
Polishing	30	93×10^{-3}	54	Cleaning cross section
Cleaning	5	47×10^{-3}	55	Rectangle

Table 2.4 Ion slicer parameter

	Acceleration voltage (kV)	Gun tilt angle (°)	Interval	Time (h)
Milling	5	1	77	5
Cleaning	2	1.5	77	0.25

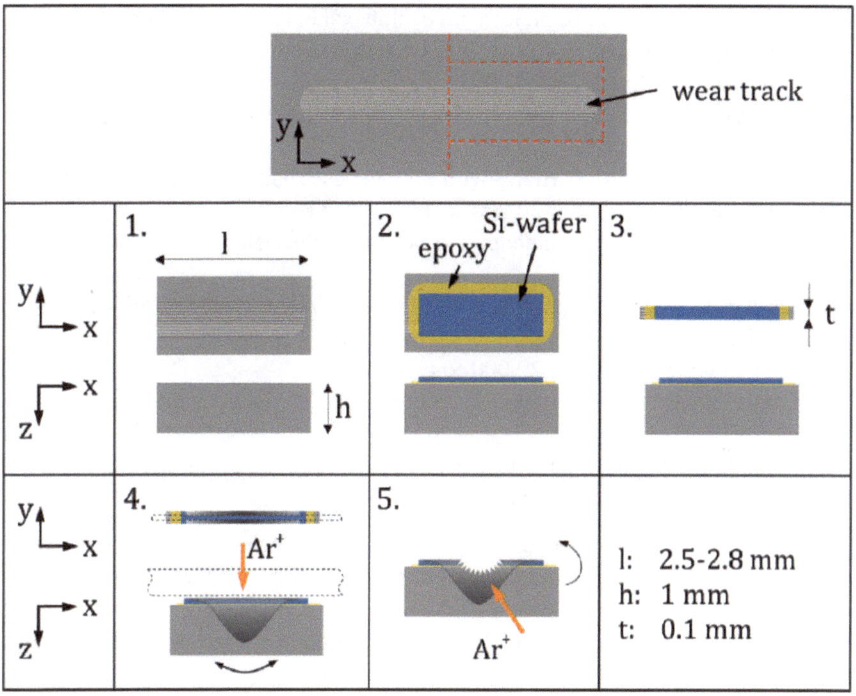

Fig. 2.26 Cross section preparation out of a wear track. 1 Cutting, 2 Glueing a Si-Wafer on top of the wear track 3 grinding down to a thickness of about 100 μm, 4 ion polishing by IS, 5 final thinning by PIPS from [65]

applied to the surface as described in 4.2. Afterwards, the samples were ground with 1200 grit SiC paper to a thickness of 100 μm. Using an ion slicer, the cross-sections were then polished on both sides with Ar+ ions, while a thin ridge of the sample was masked by a metal foil. The process parameters are shown in Table 2.4. In order to gain electron transparency, an additional ion-milling process was necessary (Fig. 2.26). Therefore, the specimens were successively thinned on both sides using an ion-mill (PIPSII Model No. 695, Gatan, USA). The ion-mill was operated at accelerating voltages between 5–0.5 kV and gun angles between 5° and 10°.

Wear Particle Preparation from Lubricant (Gear Oil)

Particle Isolation

(1) Centrifuge lubricant samples and carefully remove supernatants (centrifugation time depends on the viscosity of the medium). Do not touch the pellet at the bottom of the tubes.
(2) Resuspend and wash particles with cyclohexane, use shaker (Vortex-Genie-2, Scientific Industries, Bohemia, NY, USA) and ultrasonic cleaner.
(3) Centrifuge for 15 min at 20,000 g.
(4) Carefully discard supernatants. Do not touch the pellet at the bottom of the tubes!
(5) Resuspend and wash particles with acetone, use shaker and ultrasonic cleaner.
(6) Centrifuge for 15 min at 20,000 × g.
(7) Carefully discard supernatants. Do not touch the pellet at the bottom of the tubes!
(8) Resuspend and wash particles with isopropanol, use shaker and ultrasonic cleaner.
(9) Centrifuge for 15 min at 20,000 × g.
(10) Carefully discard supernatants. Do not touch the pellet at the bottom of the tubes!
if necessary repeat steps 2–8.
(11) Store particles in isopropanol or ethanol at 4 °C.

Cyclohexane and acetone are used in order to dissolve the lubricant. These solvents (particularly cyclohexane) should not be stored in the tubes for too long since they can damage the tubes! They would also damage the carbon film on the Cu-grids. Therefore, it is necessary to suspend the particles in a less aggressive solvent, such as isopropanol or ethanol, before applying them on Cu-grids.

Particle embedding

1st day:

(1) Centrifuge for 15 min at 20,000 × g.
(2) Prepare acetone/epoxy mixture
Mix epoxy and hardener at a ratio of 100:50 (Epoxy 3000 Quick, Cloeren Technology GmbH, Wegberg, Germany)
Add 100% acetone at a ratio of 1:1.
(3) Carefully discard supernatants. Do not touch the pellet at the bottom of the tubes! Add 0.5 ml of the acetone/epoxy mixture.
(4) Place tubes in a rotator (Thermomixer compact, Eppendorf, Hamburg, Germany) for 24 h.

2nd day:

(1) Centrifuge for 15 min at 20,000 × g.
(2) Place tubes, with open lids, under vacuum for 1 h in order to remove the acetone.
(3) Place tubes, with open lids, at 80 °C for 24 h for polymerization.

3rd day:
Remove tubes to obtain solid pieces of resin with particles embedded at the bottom.
Section with diamond blade to a thickness of approx. 90 nm.

Wear Particle Preparation from Biomedical Lubricant (Bovine Calf Serum)

The protein rich testing fluid (bovine serum) from both tests containing the wear particles was stored at −20 °C. Proteins disturb the image contrast, increase the degree of agglomeration and inhibit the identification of single particles in SEM and TEM. Compared to rather inert polyethylene particles, the isolation of metallic wear particles requires a more sophisticated protocol using enzymatic digestion to remove organic components, especially proteins, without damaging the particles. Particles in the present study were isolated following the protocol developed by Catelas et al. [55, 66] but with minor changes. This protocol was shown to minimize particle damage. It consists of several washing steps as well as incubation steps with two different enzymes, papain and proteinase K. Papain is a cysteine protease enzyme. Its main mechanism of action is destruction of peptide bond. Proteinase K is a broad-spectrum serine protease. This enzyme is known for its protein denaturing abilities, especially in presence of reagents like etylenediaminetetraacetic acid (EDTA). The amount of testing fluid retrieved from the laboratory tests ranged from 60 ml (hip simulator tests) to 300 ml (sliding wear test rig). Fluid samples were aliquoted into 40 ml tubes and centrifuged at 18,000 × g for 15 min using a Sorvall RC-5B superspeed centrifuge (particles from the different aliquots of the same testing fluid sample were recombined after the first overnight incubation). Supernatants were discarded except 1.5 ml in order to resuspend the pellets and transfer them in microcentrifuge tubes. All centrifugations were performed for 10 min at 18,000 × g. First, the pellets were resuspended in 2.5% sodium dodecyl sulfate (SDS) (v/v distilled water) and boiled for 10 min. This was followed by one wash in 80 % acetone and three washes in 1 ml of 250 mMol sodium phosphate buffer solution (PBS) containing 25 mMol EDTA at pH 7.4. After 30 s of sonification, papain (3.2 Units in 1 ml PBS/EDTA) was added. The samples were then incubated overnight on an Eppendorf thermomixer at 65 °C under slight motion. After the incubation, pellets from aliquots of the same initial fluid sample were combined in microcentrifuge tubes. Pellets were then resuspended in 2.5% SDS, boiled again for 10 min and then washed twice in 1 ml of 50 mM Tris-HCl pH 7.6. Before adding proteinase K (3 Units in 1 ml Tris-HCl), the pellets were sonicated for 30 s. A second overnight incubation was conducted at 55 °C under slight motion. After the incubation, the samples were boiled again in 2.5% SDS. The pellets were then washed once in 1 ml of 50 mMol Tris-HCl, once in 3 % SDS (v/v 80% acetone) and once in distilled, deionized water. At the end of the isolation protocol, the pelleted particles were stored in 100% ethanol at 4 °C. In a few cases, the pellets still contained denatured organic components. For those cases, the second overnight incubation was repeated. This may have been due to the use of 40 ml initial aliquots instead of 15 ml. Particles were then embedded in epoxy resin for TEM analysis. Prior to embedding, the particles were centrifuged for 20 min at 18,000 × g. The supernatant was removed and the particles were resuspended in a solution of 0.5 ml acetone and 0.5 ml liquid epoxy resin (Epoxy 3000, Cloeren Technology GmbH, Wegberg, Germany). Microcentrifuge tubes were then placed on a thermomixer overnight under slight motion at room temperature to allow pellet infiltration by the resin. The tubes were further centrifuged for 20 min at 18,000 × g and degassed for 1 h to evaporate the acetone. One ml of epoxy resin was added and the tubes were placed in a vacuum chamber for 1 h to remove potential air bubbles and trace of remaining acetone. The samples were then placed in an oven at 60 °C for one hour to harden the epoxy resin. Once hardened, the transparent epoxy cone at the tip of the tubes showed the embedded particles. Approximately 100 nm thick sections were cut using a diamond knife and placed on carbon film coated copper grid nets (formvar carbon-film S162-4, Plano GmbH, Wetzlar, Germany) for TEM analysis.

References

1. M. Knoll, E. Ruska, Das Elektronenmikroskop. Z. fuer Phys. **78**, 318–339 (1932)
2. R.F. Deacon, J.F. Goodman, Lubrication by lamellar solids. Proc. R. Soc. Lond. Math. Phys. Eng. Sci. **243**, 464–482 (1958)
3. F.P. Bowden, A.J.W. Moore, D. Tabor, The ploughing and adhesion of sliding metals. J. Appl. Phys. **14**, 80–91 (1943)
4. R.S. Plumb, W.A. Glaeser, Wear prevention in molds used to mold boron filled elastomers. Wear **46**, 219–229 (1966)
5. D.A. Rigney, W.A. Glaeser, The significance of near surface microstructure in the wear process. Wear **46**, 241–250 (1978)
6. P.A. Higham, F.H. Stott, B. Bethune, Mechanisms of wear of the metal surface during fretting corrosion of steel on polymers. Corros. Sci. **18**, 3–13 (1978)
7. P. Heilmann, J. Don, T.C. Sun, D.A. Rigney, W.A. Glaeser, Sliding wear and transfer. Wear **91**, 171–190 (1983)
8. S.B. Newcomb, W.M. Stobbs, A transmission electron microscopy study of the white-etching layer on a rail head. Mater. Sci. Eng. **66**, 195–204 (1984)
9. N. Ohmae, T. Nakai, T. Tsukizoe, Prevention of fretting by ion plated film. Wear **30**, 299–309 (1974)
10. J.M. Martin, J.L. Mansot, I. Berbezier, H. Dexpert, The nature and origin of wear particles from boundary lubrication with a zinc dialkyl dithiophosphate. Wear **93**, 117–126 (1984)
11. K.H.Z. Gahr, Formation of wear debris by the abrasion of ductile metals. Wear **74**, 353–373 (1981)
12. J.J. Wert, F. Srygley, C.D. Warren, R.D. McReynolds, Influence of long-range order on deformation induced by sliding wear. Wear **134**, 115–148 (1989)
13. W.M. Rainforth, R. Stevens, J. Nutting, Deformation structures induced by sliding contact. Philos. Mag. A Phys. Condens. Matter Struct. Defects Mech. Prop., **66**, 621–641 (1992)
14. C. Greiner, Z. Liu, L. Strassberger, P. Gumbsch, Sequence of stages in the microstructure evolution in copper under mild reciprocating tribological loading. ACS Appl. Mater. Interfaces **8**, 15809–15819 (2016)
15. H. Czichos, D. Dowson, Tribology: a systems approach to the science and technology of friction, lubrication and wear. Tribol. Int. **11**, 259–260 (1978)
16. H. Czichos, Wear mechanisms in tribological systems, in *Overview and Classification*, ed. by K.F. Ehmann (Publ by ASME, New Orleans, LA, USA, 1993), pp. 239–241

17. A. Fischer, Well-founded selection of materials for improved wear resistance. Wear **194**, 238–245 (1996)
18. K.H. Zum Gahr, Microstructure and Wear of Materials (Elsevier Science Publishers, Amsterdam, The Netherlands, 1987)
19. M.E. Sikorski, The adhesion of metals and factors that influence it. Wear **7**, 144–162 (1964)
20. T.F.J. Quinn, NASA Interdisciplinary Collaboration in Tribology. A Review of Oxidational Wear (Georgia Inst. Technol, 1983)
21. S. Jahanmir, N.P. Suh, E.P. Abrahamson II, The delamination theory of wear and the wear of a composite surface. Wear **32**, 33–49 (1975)
22. D. Landolt, S. Mischler, M. Stemp, Electrochemical methods in tribocorrosion: a critical appraisal. Electrochim. Acta **46**, 3913–3929 (2001)
23. M.A. Wimmer, J. Loos, M. Heitkemper, A. Fischer, The acting wear mechanisms on metal-on-metal hip joint bearings—in-vitro results. Wear **250**, 129–139 (2001)
24. D.A. Rigney, J.E. Hammerberg, Mechanical mixing and the development of nanocrystalline material during the sliding of metals., in *Advanced Materials in the 21st Century: The 1999 Julia R. Weertman Symposium, The Minerals, Metals & Materials Society*, ed. by Y.W. Chung, D.C. Dunand, P. Liaw, G.B. Olson (Warrendale, PA, USA, 1999), pp. 465–474
25. A. Fischer, S. Weiss, M.A. Wimmer, The tribological difference between biomedical steels and CoCrMo-alloys. J. Mech. Behav. Biomed. Mater. **1**, 50–62 (2012)
26. M.A. Wimmer, M.P. Laurent, M.T. Mathew, C. Nagelli, Y. Liao, L.D. Marks, J.J. Jacobs, A. Fischer, The effect of contact load on CoCrMo wear and the formation and retention of tribofilms. Wear **332–333**, 643–649 (2015)
27. Y. Liao, R. Pourzal, M.A. Wimmer, J.J. Jacobs, A. Fischer, L.D. Marks, Graphitic tribological layers in metal-on-metal hip replacements. Science **334**, 1687–1690 (2011)
28. M.A. Wimmer, A. Fischer, R. Buscher, R. Pourzal, C. Sprecher, R. Hauert, J.J. Jacobs, Wear mechanisms in metal-on-metal bearings: the importance of tribochemical reaction layers. J. Orthop. Res. **28**, 436–443 (2010)
29. R. Valiev, Nanostructuring of metals by severe plastic deformation for advanced properties. Nat. Mater. **3**, 511–516 (2004)
30. G. Schmaltz, Technische Oberflächenkunde; Feingestalt und Eigenschaften von Grenzflächen technischer Körper, insbesondere der Maschinenteile, J. (Springer, Berlin, Germany, 1936)
31. R. Büscher, Gefügeumwandlungen und Partikelbildung in künstlichen Metall/Metall-Hüftgelenken, Werkstofftechnik, PhD-Thesis, Universität Duisburg-Essen, Germany, s.a. VDI-Fortschr.Ber., Reihe 17, Nr.256, (VDI-Verlag, Düsseldorf, Germany, 2005)
32. R. Glardon, S. Chavez, I. Finnie, Simuation of sliding wear by cyclic plastic deformation under combined stresses. J. Eng. Mater. Technol. Trans. ASME **106**, 248–252 (1984)
33. W.A. Glaeser, Transmission electron microscopy on wear debris from bronze bearings. Wear **43**, 393–394 (1977)
34. W.J. Saleski, R.M. Fisher, R.O. Ritchie, G. Thomas, The nature and origin of sliding wear debris from steels, in *Wear of Materials '83* ed. by K.C. Ludema, (ASME, 345 East 47th Street, New York, N.Y. 10017, USA, Reston, VA, USA, 1983), pp. 434–445
35. M. Schymura, R. Stegemann, A. Fischer, Crack propagation behavior of solution annealed austenitic high interstitial steels. Int. J. Fatigue **79**, 25–35 (2015)
36. N. Jost, I. Schmidt, Friction-induced martensitic transformation in austenitic manganese steels. Wear **111**, 377–389 (1986)
37. Z.M. He, Q.C. Jiang, S.B. Fu, J.P. Xie, Improved work-hardening ability and wear resistance of austenitic manganese steel under non-severe impact-loading conditions. Wear **120**, 305–319 (1987)
38. C.W. Shao, P. Zhang, R. Liu, Z.J. Zhang, J.C. Pang, Q.Q. Duan, Z.F. Zhang, A remarkable improvement of low-cycle fatigue resistance of high-Mn austenitic TWIP alloys with similar tensile properties: importance of slip mode. Acta Mater. **118**, 196–212 (2016)
39. A. Brink, Einlaufverhalten von geschmierten Stahl-Stahl PAarungen unter Berücksichtigung der Mikrostruktur., PhD-Thesis, Institut für Angewandte Materialien - Computational Materials Science (Karlsruhe Institute of Technology, Karlsruhe, Germany, 2015)

40. D. Stickel, A. Fischer, The influence of topography on the specific dissipated friction power in ultra-mild sliding wear: experiment and simulation. Tribol. Int. **91**, 48–59 (2015)
41. J.F. Archard, W. Hirst, An examination of a mild wear process, Proc. R. Soc. Lond. Ser. A. Math. Phys. Sci. **238**, 515–530 (1957)
42. M. Godet, The third-body approach: a mechanical view of wear. Wear **100**, 437–452 (1984)
43. A. Fischer, D. Stickel, C. Schoss, R. Bosman, M. Wimmer, The growth rate of tribomaterial in bovine serum lubricated sliding contacts. Lubricants **4**, 21 (2016)
44. N. Beckmann, P.A. Romero, D. Linsler, M. Dienwiebel, U. Stolz, M. Moseler, P. Gumbsch, Origins of folding instabilities on polycrystalline metal surfaces. Phys. Rev. Appl. **2**, 064004 (2014)
45. W.M. Rainforth, R. Stevens, J. Nutting, Deformation structures induced by sliding contact. Philos. Mag. A **66**, 621–641 (1992)
46. D.A. Rigney, J.E. Hammerberg, Unlubricated sliding behavior of metals. MRS Bull. **23**, 32–36 (1998)
47. R. Büscher, A. Fischer, The pathways of dynamic recrystallization in all-metal hip joints. Wear **259**, 887–897 (2005)
48. R. Pourzal, R. Theissmann, M. Morlock, A. Fischer, Micro-structural alterations within different areas of articulating surfaces of a metal-on-metal hip resurfacing system. Wear **267**, 689–694 (2009)
49. M. Hahn, R. Theissmann, B. Gleising, W. Dudzinski, A. Fischer, Microstructural alterations within thermal spray coatings during highly loaded diesel engine tests. Wear **267**, 916–924 (2009)
50. D.A. Rigney, The role of characterization in understanding debris generation, in *Tribology Series; Wear Particles: Frorn the Cradle to the Grave, Proceedings of the 18th Leeds-Lyon Symposium on Tribology* ed. by D. Dowson, C.M. Taylor, T.H.C. Childs, M. Godet, G. Dalmaz (Lyon, France, 1992), pp. 405–412
51. I. Catelas, J.J. Jacobs, Biologic activity of wear particles. Instr. Course Lect. **59**, 3–16 (2010)
52. H.G. Willert, H. Bertram, G. Hans Buchhorn, Osteolysis in alloarthroplasty of the hip: the role of ultra-high molecular weight polyethylene wear particles. Clin. Orthop. Relat. Res. 95–107 (1990)
53. H.G. Willert, G.H. Buchhorn, C.H. Lohmann, *Hypersensitivity to CoCrMo-debris from metal/metal hip endoprostheses*, in Transactions—7th World Biomaterials Congress, Sydney, (2004), p. 486
54. P.R. Doorn, P.A. Campbell, J. Worrall, P.D. Benya, H.A. McKellop, H.C. Amstutz, Metal wear particle characterization from metal on metal V total hip replacements: transmission electron microscopy study of periprosthetic tissues and isolated particles. J. Biomed. Mater. Res. **42**, 103–111 (1998)
55. I. Catelas, J.B. Medley, P.A. Campbell, O.L. Huk, J.D. Bobyn, Comparison of in vitro with in vivo characteristics of wear particles from metal-metal hip implants. J. Biomed. Mater. Res. Part B Appl. Biomater. **70**, 167–178 (2004)
56. R. Büscher, G. Täger, W. Dudzinski, B. Gleising, M.A. Wimmer, A. Fischer, Subsurface microstructure of metal-on-metal hip joints and its relationship to wear particle generation. J. Biomed. Mater. Res. Part B Appl. Biomater. **72**, 206–214 (2005)
57. F. Billi, P. Campbell, Nanotoxicology of metal wear particles in total joint arthroplasty: a review of current concepts. J. Appl. Biomater. Biomech. **8**, 1–6 (2010)
58. P. Stemmer, R. Pourzal, Y. Liao, L. Marks, M. Morlock, J.J. Jacobs, M.A. Wimmer, A. Fischer, Microstructure of retrievals made from standard cast HC-CoCrMo alloys, in *ASTM STP 1560 Metal-on-Metal Total Hip Replacement Devices*, ed. by S.M. Kurtz, A.S. Greenwald, W.M. Mihalko, J.E. Clemson (ASTM, West Conshohocken, 2013), pp. 251–267
59. R. Pourzal, I. Catelas, R. Theissmann, C. Kaddick, A. Fischer, Characterization of wear particles generated from CoCrMo alloy under sliding wear conditions. Wear **271**, 1658–1666 (2011)
60. R. Pourzal, *Possible pathways of particle formation in CoCrMo sliding wear*. Ph.D.-thesis University Duisburg-Essen, Duisburg, Germany, VDI Verlag, s.a. Fortschr Ber VDI Z, 17(285), Düsseldorf, Germany (2011)

61. Y. Liao, L. Marks, Direct observation of layer-by-layer wear. Tribol. Lett. **59**, 1–11 (2015)
62. P. Stoyanov, P. Stemmer, T.T. Järvi, R. Merz, P.A. Romero, M. Scherge, M. Kopnarski, M. Moseler, A. Fischer, M. Dienwiebel, Friction and wear mechanisms of tungsten-carbon systems: a comparison of dry and lubricated conditions. ACS Appl. Mater. Interfaces **5**, 6123–6135 (2013)
63. R. Büscher, B. Gleising, W. Dudzinski, A. Fischer, Transmission electron microscopy examinations on explanted metal-on-metal hip joints. Prakt. Metallogr./Pract. Metallogr. **42**, 15–34 (2005)
64. M. Hahn, Mikrostrukturelle Veränderungen in der Zylinderlaufbahn von PKW Dieselmotoren aus Grauguss und mittels thermischer Spritzverfahren hergestellter Stahlschichten. Dissertation Universität Duisburg-Essen, 2013s.a. Fortschr.-Ber. VDI Reihe 5: Grund- und Werkstoffe/Kunststoffe, Nr. 750 (VDI-Verlag, Düsseldorf, Germany, 2013)
65. P. Stemmer, *The divergent pathways and mechanisms of energy dissipation at the interfaces of martensitic tribocouples*. Ph.D.-thesis, Materials Science and Engineering, University of Duisburg-Essen, Germany, DuEPublico ID: 42437 (2016)
66. I. Catelas, J. Dennis Bobyn, J.B. Medley, J.J. Krygier, D.J. Zukor, A. Petit, O.L. Huk, Effects of digestion protocols on the isolation and characterization of metal-metal wear particles. I. Analysis of particle size and shape. J. Biomed. Mater. Res. **55**, 320–329 (2001)

Part II
Tribochemical Characterization Techniques

Chapter 3
Near Edge X-Ray Absorption Fine Structure Spectroscopy: A Powerful Tool for Investigating the Surface Structure and Chemistry of Solid Lubricants

Filippo Mangolini and J. Brandon McClimon

Abstract Synchrotron-based spectroscopic techniques have been critical tools for developing a better understanding of the structure and properties of materials and material surfaces as well as their evolution in response to energetics inputs, such as mechanical strains present in tribological contacts. Among these techniques, near edge X-ray absorption fine structure (NEXAFS) spectroscopy is one of the most powerful tools thanks to its elemental specificity, surface sensitivity, and ability to provide important information about local bonding configurations, such as hybridization, chemical states, and bond orientations. In addition, when coupled with imaging methods like photoemission electron microscopy and magnetically-guided imaging, NEXAFS spectroscopy enables chemical imaging of materials with high spatial resolution. This capability can be critical when investigating materials after tribological experiments, where chemical changes and structural transformations occur in the first few atomic layers and spatial inhomogeneities can be present across small length scales. The present contribution first describes the principles of NEXAFS spectroscopy, followed by experimental methods for the acquisition and processing of NEXAFS data. Finally, the potential of this analytical method for fundamental and applied research in tribology is demonstrated by discussing case studies in the area of solid lubricating carbon-based thin films.

F. Mangolini (✉)
Department of Mechanical Engineering, Materials Science and Engineering Program, The University of Texas at Austin, Austin, TX 78712, USA
e-mail: Filippo.Mangolini@austin.utexas.edu

J. B. McClimon
Department of Materials Science and Engineering, University of Pennsylvania, Philadelphia, PA 19104, USA
e-mail: mcclimon@seas.upenn.edu

M. Dienwiebel and M.-I. De Barros Bouchet (eds.), *Advanced Analytical Methods in Tribology*, Microtechnology and MEMS,
https://doi.org/10.1007/978-3-319-99897-8_3

3.1 Introduction

Fundamental research in tribology hinges on the development of novel experimental methods capable of assessing, characterizing, and monitoring the phenomena occurring at buried sliding interfaces. It is exactly thanks to instrumental developments and progress in research that over the last four decades the state of knowledge in tribology has been shifting from empirically-based to scientifically-based. In particular, advancements in surface analysis have allowed researchers in tribology to systematically investigate the local structure, chemistry, and mechanics of materials after tribological interactions take place, and correlate the resulting information with the observed tribological behavior, thus providing new and exciting opportunities to investigate the fundamentals of friction, wear, and lubrication [1–3].

Among the tools and techniques that allow for the exploration of the chemical, structural, and electronic properties of materials, in the last four decades synchrotron-based methods have been of utmost importance [4, 5]. In particular, the development of new generations of synchrotron radiation facilities (e.g., third generation synchrotrons), which are brilliance-optimized (i.e., they provide a beam with small cross section and a high degree of collimation), have tremendously expanded the material characterization toolbox and provided flexibility in the development of synchrotron endstations [6]. This has effectively given researchers the possibility of simultaneously accessing several advanced analytical methods in the same facility.

Among the several synchrotron-based techniques developed for the analysis of the properties of materials, surfaces, and interfaces, X-ray absorption fine structure (XAFS) spectroscopy is a unique tool for investigating, at the atomic and molecular scale, the local structure around a certain element contained in a material [7–10]. Traditionally, the XAFS signal is distinguished in two regions [11]: (a) the *near edge region*, which is within 30–50 eV above the absorption edge and usually referred to as X-ray absorption near edge structure (XANES) or near edge X-ray absorption fine structure (NEXAFS). The term XANES is more commonly employed for solids and inorganic complexes, while the term NEXAFS is preferentially employed for surface studies [5]; and (b) the *fine structure region*, which extends from the near edge region up to typically one thousand eV above the absorption edge and is usually referred to as extended X-ray absorption fine structure (EXAFS).

While the EXAFS signals are dominated by single scattering contributions that can be used for gaining information about the local geometric structure surrounding a given atomic species [8, 10, 12, 13], the absorption features characterizing the NEXAFS region are sensitive to changes in valence state, coordination chemistry, local structure, and ligand symmetry around the absorbing atom. One of the most important applications of NEXAFS spectroscopy since its introduction in the 1980s is the study of low atomic number (low-*Z*, i.e., carbon, oxygen, nitrogen, and fluorine) molecules and materials [5]. The capability of NEXAFS spectroscopy to provide information about low-*Z* elements derives from the strong directionality and short length of the covalent bonds between low-*Z* atoms, the strong dependence of the bond length on its hybridization, and the large backscattering amplitude of low energy electrons from low-Z atoms [5]. Because of this, resolvable, structure-dependent resonances are usually detected in NEXAFS spectra of low-Z elements, whose intensity is strongly affected by the orientation of the final state orbital with respect to the electric field vector of the incident photon beam [5].

Thanks to this elemental specificity, ability to obtain important information about local bonding environment (such as hybridization, chemical state, and bond orientation [5]), as well as the possibility to obtain spectra with high spatial resolution using imaging techniques such as photoemission electron microscopy [14–17] and magnetically-guided imaging [18], NEXAFS spectroscopy has become an attractive analytical tool for several research fields, such as tribology [18–20], catalysis [21–25], self-assembly at surfaces [26–29], nanomaterials [30–38], and polymer science [39–44].

The present contribution describes the basics of NEXAFS spectroscopy and the experimental methods for the acquisition and processing of NEXAFS data. Selected examples of application of this analytical method for fundamental and applied research in tribology is demonstrated by discussing case studies in the area of solid lubricating thin films, with a particular focus on carbon-based coatings.

3.2 Principles of Near Edge X-Ray Absorption Fine Structure (NEXAFS) Spectroscopy

To gain an understanding of the spectral features characterizing NEXAFS data, a solid theoretical basis of the principles of this spectroscopic technique is a prerequisite. In the following, a brief and simplified description of the basics of NEXAFS spectroscopy is presented. For a more comprehensive overview, the reader can refer to selected, focused monographs [4, 5].

When a collimated X-ray beam with flux Φ_0 (number of photons per unit time and unit cross-section) passes through a sample of thickness t, the flux is reduced according to Beer-Lambert's Law:

$$\Phi = \Phi_0 \exp(-\mu(\omega)t) \tag{3.1}$$

where Φ is the flux of the transmitted X-ray beam, and $\mu(\omega)$ is the linear attenuation coefficient, which depends on the energy of the X-ray photons and the sample composition and density.

For X-ray photons with energy between 0.1 and 40 keV (normally employed in XAFS experiments), two different mechanisms lead to X-ray attenuation, i.e, photoelectric absorption and scattering (coherent and incoherent). However, since photoelectric absorption dominates over other interactions (Fig. 3.1a) in the energy range from 0.1 to 40 keV, the attenuation coefficient in (3.1) can be approximated with the photoelectric absorption coefficient.

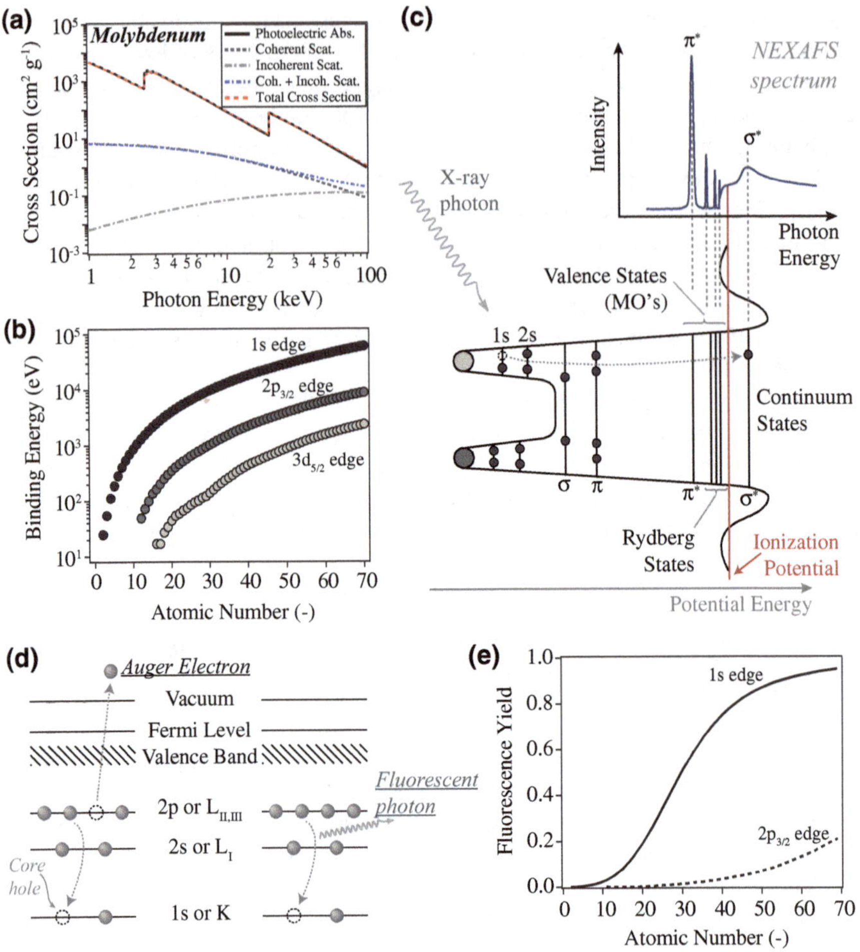

Fig. 3.1 **a** X-ray absorption cross-section (photoelectric absorption together with coherent and incoherent scattering and their sums) for molybdenum (data taken from the NIST XCOM database). The photoelectric absorption exhibits four edges: the $2p_{3/2}$ (L_{III}) at 2.52 keV, the $2p_{1/2}$ (L_{II}) 2.63 keV, the 2 s (L_I) at 2.87 keV, and the 1 s (K) at 19.99 keV; **b** binding energy of the levels 1s (K), $2p_{3/2}$ (L_{III}) and $3d_{5/2}$ (M_V) as a function of atomic number; **c** absorption of an X-ray photon with excitation of a core electron to an unoccupied molecular orbital; **d** de-excitation mechanisms following the emission of a core electron, i.e., emission of an Auger electron or X-ray photon (fluorescence); **e** average fluorescence yield as a function of the atomic number. Data taken from NIST X-ray Data Booklet

Even though the photoelectric absorption coefficient is overall a smooth function of the energy of the incident X-rays, sharp discontinuities (*absorption edges*) occur at well-defined photon energies and correspond to the extraction of an electron from a core level. For example, *K* edges, which are photoabsorption edges with the highest energy for a given element, result from the emission of electrons from 1s levels.

Table 3.1 Relationships between X-ray notation, core level notation, and quantum numbers for K, L, and M edges

X-ray notation	Core level notation	Quantum numbers			
		n	l	s	$j = \lvert l + s \rvert$
K	$1s_{1/2}$	1	0	+1/2; −1/2	1/2
L_1	$2s_{1/2}$	2	0	+1/2; −1/2	1/2
L_2	$2p_{1/2}$	2	1	−1/2	1/2
L_3	$2p_{3/2}$	2	1	+1/2	3/2
M_1	$3s_{1/2}$	3	0	+1/2; −1/2	1/2
M_2	$3p_{1/2}$	3	1	−1/2	1/2
M_3	$3p_{3/2}$	3	1	+1/2	3/2
M_4	$3d_{3/2}$	3	2	−1/2	3/2
M_5	$3d_{5/2}$	3	2	+1/2	5/2

Quantum numbers: n principal quantum number; l orbital angular momentum; s spin angular momentum; j total angular momentum (equal to $|l + s|$)

The relationship between the X-ray notation for high-energy edges, core electronic levels, and quantum numbers is provided in Table 3.1. The fact that the edge energy varies with the atomic number as shown in Fig. 3.1b makes NEXAFS spectroscopy element-specific.

The absorption edge is the result of the excitation of a core electron to a continuum or quasi-continuum of final states. Since the edge energy is related to the oxidation state of the absorbing atom (it raises by several eV per oxidation unit), the edge position can be used to determine the valence state of the photoabsorber [45]. Additionally, the shape of the edge can provide valuable information about the chemical environment and ligand geometry, thus making it useful for fingerprinting molecular species [46, 47]. Around the ionization potential (IP), which is usually superposed on transitions into Rydberg states, resonant transitions, which originate from the excitation of an electron from a core level to unoccupied molecular states, are detected. These transitions are superposed on a step-like absorption edge and only occur when the energy of the incoming X-ray photon matches the energy difference between the core level and the unoccupied molecular level. Empty molecular orbitals can be labeled according to their symmetry, i.e., π^* or σ^* orbitals. In the case of π-bonded diatomic subunits, the lowest unoccupied molecular orbital (LUMO) is usually a π^* orbital, while σ^* orbitals are normally detected at higher photon energies. While π^* resonances are observed below the ionization threshold because of electron-hole Coulomb interactions [5], σ^* spectral features are most often found above the vacuum level for the neutral molecule. A schematic showing the origin of NEXAFS features for a diatomic molecule is provided in Fig. 3.1c together with the effective electrostatic potential and the corresponding K-shell spectrum.

The width of the resonances detected in experimental NEXAFS data is a convolution of the natural width of the core hole (or lifetime of the excited state), the resolution of the instrument, as well as the vibrational motion of the molecule [48]. Compared to π^* resonances, transitions corresponding to orbitals with σ-symmetry are broader because of their large overlap with continuum states, which increases the decay probability of the electron into continuum states and decreases the lifetime of the core hole. As a rule of thumb, the higher is the energy of a resonance, the larger is its linewidth. σ^* resonances are often characterized by distinct high-energy tails, which derive from molecular vibrations along the bond direction.

The energy of σ^* resonances can be used to obtain information about bond lengths in organic molecules through the so-called *bond length with a ruler* method [5, 49–54], which relies on the determination of the core level IP together the position of the σ^* transition relative to the IP. Even though there is empirical evidence supported by theoretical predictions [54, 55] for the dependence of the position of σ^* resonances on the bond length, the validity of the use of these spectral features for bond length determination has been controversial [55–59].

Weak and sharp resonances are often detected between the characteristic π^* spectral features and the IP. These resonances are assigned to Rydberg orbitals. For the case of strongly chemisorbed molecules, pure Rydberg resonances are quenched due to their large spatial extent (they have most of their density at the periphery of the molecule). On the other hand, in the case of hydrocarbons or molecules with carbon-hydrogen bonds, Rydberg orbitals mix with hydrogen-derived antibonding orbitals with the same symmetry, thus leading to an increase of the intensity of the corresponding absorption feature [60, 61].

Finally, multi-electron features (also called shake-up structures, analogous to those observed in X-ray photoelectron spectroscopy, XPS) can also be detected in NEXAFS spectra. While these features are observed at lower kinetic energies (or higher binding energies) than the main photoelectron signal in XPS spectra, they are detected at higher photon energy than the primary resonance in X-ray absorption data.

The absorption process results in a photoelectron and a core hole. An atom with a core hole is unstable, and thus spontaneously tends to relax by filling the core hole with an electron from an upper level. The resulting relaxation energy can be released in two ways: (a) through the emission of X-ray photons (fluorescence); and (b) through the emission of an Auger electron (Fig. 3.1d). Both channels can be used for measuring the formation of a core hole following the absorption of X-rays and, thus, are a measure of the absorption cross section. The experimental methods used for detecting either photoemitted Auger electrons or fluorescent photons are described in Sect. 3.3.1.1. It is worth mentioning that these two de-excitation mechanisms are in competition. Their relative strength is usually referred to as fluorescence yield η_i:

$$\eta_i = \frac{X_i}{X_i + A_i} \tag{3.2}$$

where the subscript i represents a defined absorption edge (e.g., K or 1s), while X_i and A_i are the emission probabilities of a fluorescent photon and an Auger electron, respectively. The fluorescence yield depends on the atomic number (Fig. 3.1e). In particular, the Auger electron yield is much higher than the fluorescence yield for low-Z elements (e.g., carbon).

Even though X-ray photons penetrate deeply into the sample (several microns), electron detection provides much higher surface sensitivity compared to fluorescence yield due to the short mean free path of electrons in condensed matter (typically less than 2 nm for energies between 250 and 600 eV) [62, 63]. The information depth for electron yield detection, which is the specimen thickness measured normal to the surface from which a specified percentage of typically 95% of the detected signal originates, is thus usually less than 5 nm [5]. A detailed description of the influence of the detection mode on the resulting surface sensitivity is reported in Sect. 3.3.1.1.

Since synchrotron radiation is polarized, NEXAFS spectroscopy can be used to gain information about the orientation of molecules and molecular fragments [5, 26–28, 30, 31, 64, 65]. Through the investigation of the dependency of the intensity of a given resonance on the X-ray incidence angle (Θ) or, in other words, the direction of the electric field vector of the impinging X-rays, insights into the spatial orientation of the corresponding orbital (i.e, the direction of maximum amplitude of the excited atomic species) can be obtained.

Under the assumption of linearly polarized radiation, the equation that relates the initial (Ψ_i) and final (Ψ_f) states to the absorption cross section (σ_x) can be derived from a quantum mechanical description of the excitation process for a single electron in the dipole approximation (*Fermi's Golden Rule*):

$$\sigma_x = \left|\langle\Psi_f|\boldsymbol{e}\cdot\boldsymbol{p}|\Psi_i\rangle\right|^2\rho_f(E) \tag{3.3}$$

where $\boldsymbol{e}$ is the unit electric field, $\boldsymbol{p}$ the dipole transition operator, and $\rho_f(E)$ the density of final states [5].

In the case of linearly polarized radiation, the matrix element can be simplified $\left|\langle\Psi_f|\boldsymbol{e}\cdot\boldsymbol{p}|\Psi_i\rangle\right|^2 = \left|\boldsymbol{e}\langle\Psi_f|\boldsymbol{p}|\Psi_i\rangle\right|^2 = \left|\langle\Psi_f|\boldsymbol{p}|\Psi_i\rangle\right|^2$, where $\langle\Psi_f|\boldsymbol{p}|\Psi_i\rangle$ is known as transition dipole moment (TDM). In the case of a 1 s (K) initial state and a directional final state, the transition intensity becomes:

$$I \propto \left|\boldsymbol{e}\langle\Psi_f|\boldsymbol{p}|\Psi_{1s}\rangle\right|^2 \propto |\boldsymbol{e}\cdot\boldsymbol{O}|^2 \propto \cos^2\delta \tag{3.4}$$

where δ is the angle between the electric vector ($\boldsymbol{E}$) and the TDM direction. Therefore, the resonance intensity is maximum when the electric field vector is along the direction ($\boldsymbol{O}$) of the final state orbital (or the TDM direction). Note that σ^* orbitals have a maximum amplitude along the bond axis, while π^* orbitals have maximum amplitude normal to the bond direction.

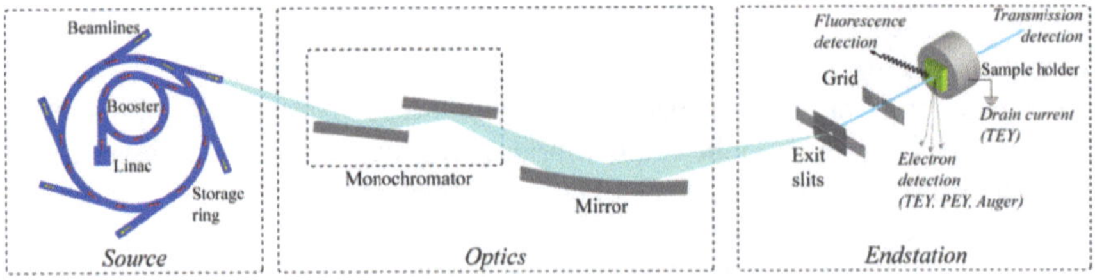

Fig. 3.2 Schematic of a synchrotron beamline for NEXAFS spectroscopy

3.3 Experimental Methods in NEXAFS Spectroscopy

3.3.1 Data Acquisition

3.3.1.1 Synchrotron Beamline for NEXAFS Spectroscopy

Since NEXAFS spectroscopy requires monochromatic X-rays with tunable energy, NEXAFS measurements are limited to synchrotron facilities. These facilities, which generate electromagnetic radiation by accelerating electric charges (electrons) and forcing them to move on a curved path, usually consist of a storage ring and a booster (Fig. 3.2). The latter accelerates charged particles (electrons) close to the speed of light before injecting them into the former, where they orbit. In the storage ring, synchrotron radiation is generated in bending magnets, which keep electrons on a closed path, and in wigglers and undulators placed in the linear sections [10, 66–68]. In the bending magnets and wigglers/undulators, the electrons are accelerated and thus emit radiation. The emitted radiation covers a wide range of the electromagnetic spectrum, with a broad peak in intensity at an energy that depends on the energy of the electrons. This synchrotron light with a high photon flux is conveyed to experimental chambers (endstations) through beamlines. The primary elements of a beamline are the monochromator and optical mirrors. The former, together with the exit slits, selects a single wavelength and thus provides a monochromatic X-ray beam with high-brilliance to experimentalists. The latter collimates and focuses the beam. Synchrotron X-ray sources have three main advantages compared to laboratory sources: (a) they provide monochromatic X-rays with tunable energy; and the emitted radiation has (b) high brilliance and flux, as well as (c) a high degree of polarization.

Before reaching the samples, the X-ray photon beam often passes through a gold grid (with typical transmission around 85%). The electron yield signal from this grid is used to monitor the beam intensity and becomes critical when processing NEXAFS data (see Sect. 3.3.2).

Even though the most straightforward way to acquire X-ray absorption spectra would be to measure the attenuation of a beam of X-rays transmitted through a sample, this method would require thin samples since the penetration depth of X-rays is only a few microns (depending on their energy). Additionally, this methodology would not allow for assessing the structural, chemical, and electronic properties of the near-surface regions of materials.

In light of this, the large majority of NEXAFS measurements is nowadays carried out by detecting the electrons or fluorescent photons emitted from a bulk sample as a consequence of the relaxation of the excited atom (see Sect. 3.3.2).

In the case of electron yield detection, three main modes are currently in use: Auger yield, partial electron yield, and total electron yield. Auger yield detection (AEY) uses an electron energy analyzer to count only those electrons with a defined kinetic energy E_a (corresponding to a specific Auger emission energy) chosen by the experimenter. This leads to a very high signal to background ratio at the cost of reduced signal intensity. However, during the emission from the sample surface, some of the electrons will suffer an energy loss and emerge with a kinetic energy lower than E_a. Since the intensity of inelastically-scattered electrons follows the one of elastically-scattered electrons, secondary electrons can be used to increase the signal-to-noise. This is used in partial electron yield (PEY) detection mode, where electrons of kinetic energy larger than a threshold energy E_p are detected. This results in boosting the signal intensity substantially, while also increasing the background. PEY measurements are normally carried out by selecting the potential (also called entrance grid bias, EGB) of the center grid of a standard three-grid high pass kinetic energy filter for the channeltron electron yield detector [26]. Upon increasing the values of the EGB voltage, only those Auger electrons that have suffered limited energy loss during the emission from the sample and have sufficiently high kinetic energy to reach the channeltron will be detected. Thus, the information depth from which Auger electrons are detected can be adjusted by changing the EGB (the higher the EGB, the smaller the information depth). Total electron yield (TEY) detection is the simplest detection scheme, which relies on measuring the drain current from an electrically isolated sample by means of a sensitive ammeter. Electrons of all energies and from all X-ray penetrations depths are captured, so this scheme has the lowest signal-to-background ratio and surface sensitivity. To summarize, the surface sensitivity of electron yield detection modes increases in the following order: AEY > PEY > TEY.

The fluorescence yield (FY) detection mode is not discussed in detail here because of its limited surface sensitivity. The FY detection mode is, however, the method of choice when investigating bulk samples or liquids, since the short mean free path of electrons in these environments hinders the acquisition of data in electron yield detection modes (either PEY or TEY).

3.3.1.2 Imaging Techniques

Photoemission Electron Microscopy (PEEM)

Photoemission electron microscopy (PEEM) is a full field imaging technique that is commonly combined with X-ray synchrotron radiation to give a surface sensitive spectroscopy with excellent spatial resolution (<50 nm [69], or even below 10 nm in aberration corrected instruments [70]) [14, 71–75]. Since the pioneering work of Tonner and Harp [76], several instrumental development have been reported, which have allowed for the use of PEEM in numerous fields, including tribology [14, 20, 77–82], thin film magnetism [83–85], catalysis [86], semiconductor [87], polymer science [88–90], geology [91–93], and microbiology [94].

A typical PEEM microscope consists of a sample held at a high negative voltage followed by a series of electron optics, and a detector (microchannel plane and phosphor screen). A schematic of this technique is shown in Fig. 3.3. Monochromatized X-rays impinge on the sample surface at a grazing angle. This geometry is required by the close proximity of the electron lens optics to the sample surface (a few millimeters). The sample is held at a large negative bias (between −15 and −20 kV). Because of this, specimens for PEEM measurements should be flat as any feature on the sample surface can locally concentrate the electric field, lead to distortions of the electron images, and, in the worst case, cause arcing between the sample and the electron lens optics. The photoemitted electrons (primary, secondary, and Auger) are accelerated by the applied field from the sample towards the electron optics column, which can be either electrostatic or magnetic and often includes a series of deflectors and stigmators. After the optics column, which accelerates, filters, and focuses the electrons, they are detected using conventional channeltrons and phosphor screens. Overall, the signal acquired in PEEM measurements is in most of the cases close to NEXAFS data acquired in TEY detection mode, even though detection schemes equivalent to PEY and AEY modes have been implemented by inserting retarding grids or energy analyzers upstream of the image detector [95, 96]. Upon varying the photon energy of the impinging X-rays in a similar manner to the approach used for the acquisition of NEXAFS spectra, a stack of images is captured, where each pixel represents a spatially-resolved NEXAFS spectrum.

Magnetically-Guided Imaging NEXAFS Spectroscopy

Magnetically-guided imaging NEXAFS spectroscopy combines NEXAFS spectroscopy, a full field parallel processing magnetic field electron yield optics detector, and a large incident soft X-ray beam (Fig. 3.4a). This technique, which was previously described in [18, 97], relies on a rapid parallel processing magnetic field electron yield optics detector, in which the emitted photoelectrons move along the magnetic field lines towards the partial electron yield grid in front of the channel plate electron detector. The detected electrons produce a series of two-dimensional NEXAFS lateral images as the incident soft X-ray energy is scanned across the selected absorption edge (Fig. 3.4b). The rejection of low energy electrons by the partial electron yield detector allows for the mitigation of surface charging. The image stack can be employed for the structural and chemical characterization of the near-surface region of materials with a lateral resolution of 50 μm and a field of view of $18 \times 13\ mm^2$. Unlike photoelectron emission microscopy (PEEM), magnetically-guided NEXAFS spectro-microscopy can handle insulators, conductors, and non-planar samples without charging effects [18, 97]. In particular, the possibility of analyzing curved surfaces provides significantly increased flexibility in the range of samples that can be analyzed, while making this technique particularly attractive for

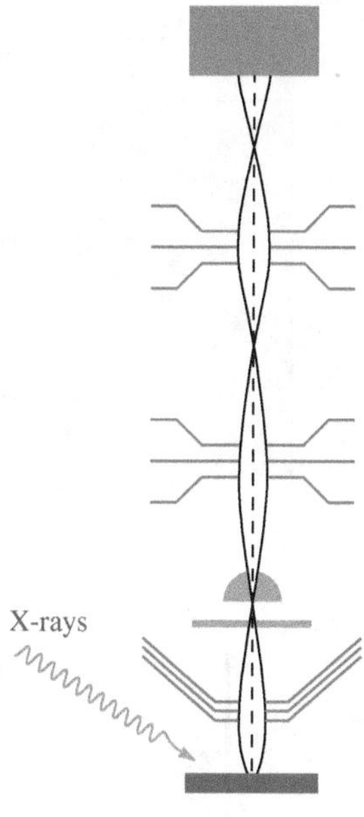

Fig. 3.3 Schematic of electron optical path through a photoemission electron microscope (PEEM). Adapted from [73]

the ex situ characterization of tribological samples, which often comprise a spherical counterface.

As in the case of the acquisition of NEXAFS spectra in PEY detection mode, changing the entrance grid bias (EGB) of the channel plate electron detector allows magnetically-guided imaging NEXAFS measurements with different surface sensitivity to be performed [26, 98, 99] (Fig. 3.4b).

3.3.1.3 Data Processing

The analysis of NEXAFS data is complicated by many experimental factors, including photon energy calibration, intensity normalization, and artifacts due to beam instabilities, signal offsets, and the beamline transmission function [5, 100–102]. In particular, the presence of an adventitious carbon contamination on X-ray optics in synchrotron beamlines makes the processing of carbon 1s NEXAFS spectra challenging, since the carbon contamination can result in artifacts in the as-collected data, which can even dominate true NEXAFS spectra. The common approach to remove major artifacts (including the beamline transmission function, flux monitor contamination, beam stability and signal offsets) with the aim to deconvolute the real NEXAFS spectrum for the sample under investigation relies upon normalization techniques. These techniques are usually based on the use of reference spectra acquired from metallic mesh materials (such as gold) placed in the beamline upstream from the analysis chamber (see Sect. 3.3.1.1). Even though several methods of normalization, energy calibration, and artifact removal have been published in

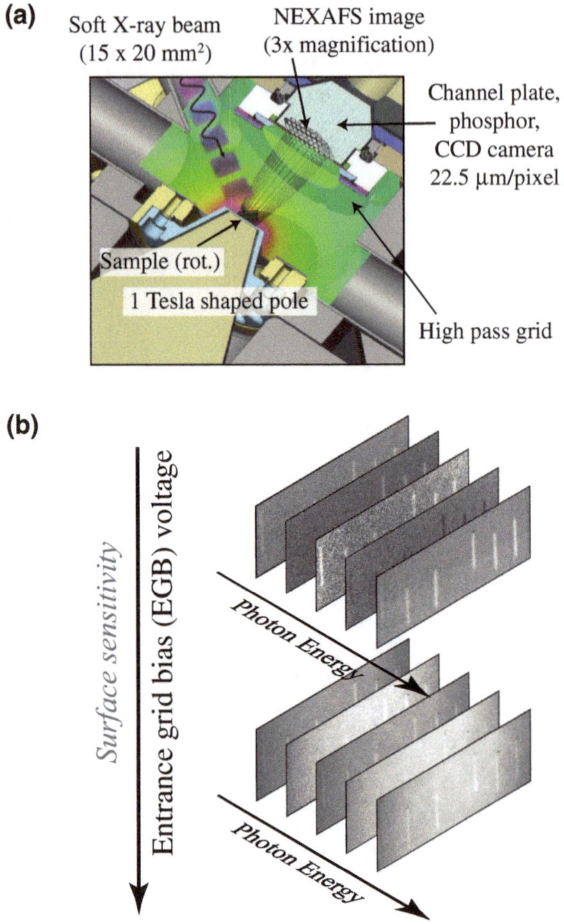

Fig. 3.4 **a** Schematic of the parallel processing imaging NEXAFS system at the NIST/Dow endstation of beamline U7A at the National Synchrotron Light Source (Brookhaven National Laboratory, Upton, NY, USA). A large incident soft X-ray beam ($15 \times 20\ mm^2$) illuminates the sample surface. The photoelectrons emitted from the specimen are magnetically guided onto a detector that forms a two-dimensional NEXAFS image over a $13 \times 18\ mm^2$ region, while providing a spatial resolution of 50 μm at all relevant X-ray incidence angles [18]; **b** principle of parallel processing imaging NEXAFS measurements. The rapid parallel processing magnetic field electron yield optics detector produces a series of two-dimensional NEXAFS lateral images as the incident soft X-ray energy is scanned across an absorption edge (e.g., the carbon 1s absorption edge). Changing the entrance grid bias (EGB) of the channeltron detector allows NEXAFS measurements with different surface sensitivity to be performed [26, 98, 99]

the literature, they have been reported in a fragmented manner, thus hindering their use and implementation.

Recently, Watts et al. reviewed and implemented the numerous methods for calibration, photon flux normalization, and artifact removal for NEXAFS spectra (with a particular focus upon carbon 1s spectra) reported in the literature [102]. Besides providing criteria for assessing the most appropriate method for the photon flux normalization, Watts et al. also presented the methodologies for the determination of appropriate background functions along with a way for their removal from experimental data. Additionally, the methods used for photon flux normalization were discussed together with the criteria for assessing the most appropriate one for a specific application.

While the approaches outlined by Watts et al. are effective for addressing the issues commonly encountered in the evaluation of NEXAFS data [102], the corrected spectra they yield represent the photo-absorption spectra of the specimens under the assumption of structural and compositional homogeneity within the nanometer-scale depth probed by NEXAFS spectroscopy (in the case of NEXAFS spectra of low-*Z* materials acquired in electron yield mode, the information depth, which is the specimen thickness measured normal to the surface from which a specified percentage of typically 95% of the detected signal originates, is usually less than 5 nm [5]). However, the assumption of chemical and structural homogeneity does not hold in the vast majority of solid surfaces, since complex surface-bound species and layers, e.g., natural oxide and contamination layers, are often present [103, 104]. Thus, this assumption can lead to significant errors when analyzing elements that are simultaneously present in multiple layers. In the case of carbon-based materials previously exposed to air, as a particular example, their carbon 1s NEXAFS spectra, even when corrected using any of the approaches outlined by Watts et al. [102], are a convolution of the spectrum of the sample of interest and the spectrum of the adventitious carbon contamination on its surface since the thickness of the latter (typically <2 nm [105]) is smaller than the information depth at the carbon 1s.

The authors of this chapter recently developed a methodology for accurately removing the contribution of thin overlayers (with thickness smaller than the information depth) from NEXAFS spectra of two-layered systems to reveal the photo-absorption NEXAFS spectrum of the substrate [99]. This method relies on the subtraction of the NEXAFS spectrum of the overlayer adsorbed on a reference surface (e.g., gold) from the spectrum of the two-layer system under investigation, where the thickness of the overlayer is independently determined by XPS. The approach was applied to NEXAFS data acquired for one of the most challenging cases: air-exposed hard carbon-based materials (namely ultrananocrystalline diamond (UNCD) and hydrogenated amorphous carbon (a-C:H)) with adventitious carbon contamination from ambient exposure. The possibility of removing the contribution of the adventitious carbon contamination from the as-acquired NEXAFS carbon 1s spectra allowed the computation of the contamination-corrected photo-absorption NEXAFS spectra for the materials under investigation (UNCD and a-C:H films) (Fig. 3.5), which provided qualitatively distinct interpretations and quantitatively distinct values regarding the sample's composition and bonding compared to the as-acquired data. In particular, the method, which also revealed that the adventitious contamination can be described as a layer containing carbon and oxygen ($[O]/[C] = 0.11 \pm 0.02$) with a thickness of 0.6 ± 0.2 nm and a fraction of sp^2-bonded carbon of 0.19 ± 0.03, significantly altered the calculated fraction of sp^2-hybridized carbon, thus demonstrating that the assumption of chemical and structural homogeneity could introduce large errors in the computed carbon hybridization state (between 5 and 20%) when analyzing carbon-based materials previously exposed to air. Additionally, upon removing the contribution of the carbonaceous contamination from the as-acquired data, the absorption features assigned to the C–O σ^* antibonding orbital (at 288.5–289.3 eV) [106] was greatly reduced (Fig. 3.5), which clearly indicated the importance of accounting for nanometer-thick adventitious contamination when investigating the surface chemistry of air-exposed carbon surfaces by NEXAFS spectroscopy.

3.4 Applications of NEXAFS Spectroscopy

3.4.1 NEXAFS Spectroscopic Investigation of the Surface Structure and Chemistry of Thin Films for Tribological Applications

As outlined above, NEXAFS spectroscopy can provide valuable information about the chemical, structural, and electronic properties of materials and materials surfaces. Additionally, it can be effectively used for the determination of molecular orientations. This wealth of detailed information NEXAFS yields has made it an attractive analytical tool for fundamental and applied studies in tribology [18–20], catalysis [21–25], self-assembly at surfaces [26–29], nanomaterials [30–38], and polymer science [39–44].

One of the most important applications of NEXAFS spectroscopy is the study of carbon-based materials. The strong directionality and short length of the covalent bonds between carbon atoms together with the strong dependence of the bond length on its hybridization and the large backscattering amplitude of low-energy electrons from carbon atoms results in the detection of resolvable absorption resonances in carbon 1s NEXAFS spectra. These absorption features are structure-dependent and their intensity depends on the orientation of the final state orbital with respect to the electric field vector of the impinging X-rays [5].

For carbon 1s NEXAFS spectroscopy, in particular, the resolvable energy difference between the resonant X-ray excitations of a core-level (1s) electron to unoccupied molecular orbitals (either π^* or σ^*) allows the identification of the bonding configuration and hybridization state of carbon atoms in the near-surface region for many materials, including diamond [18, 20, 33, 34, 36, 37, 77, 80, 82, 107–109], diamond-like carbon [78, 110–128], graphene [38, 129, 130], carbon nanotubes [131, 132], and polymers [31, 39–44], as well as the determination of the surface molecular orientation of nanomaterials [30, 31, 64, 65, 132] and adsorbates [26–28, 133].

As an example, the characteristic carbon 1s NEXAFS spectra (not corrected for the contribution of the contamination layer following the method outlined in [99]) of two tribologically-relevant, as-deposited thin films, i.e., hydrogenated amorphous carbon (a-C:H) and silicon- and oxygen-containing hydrogenated amorphous carbon

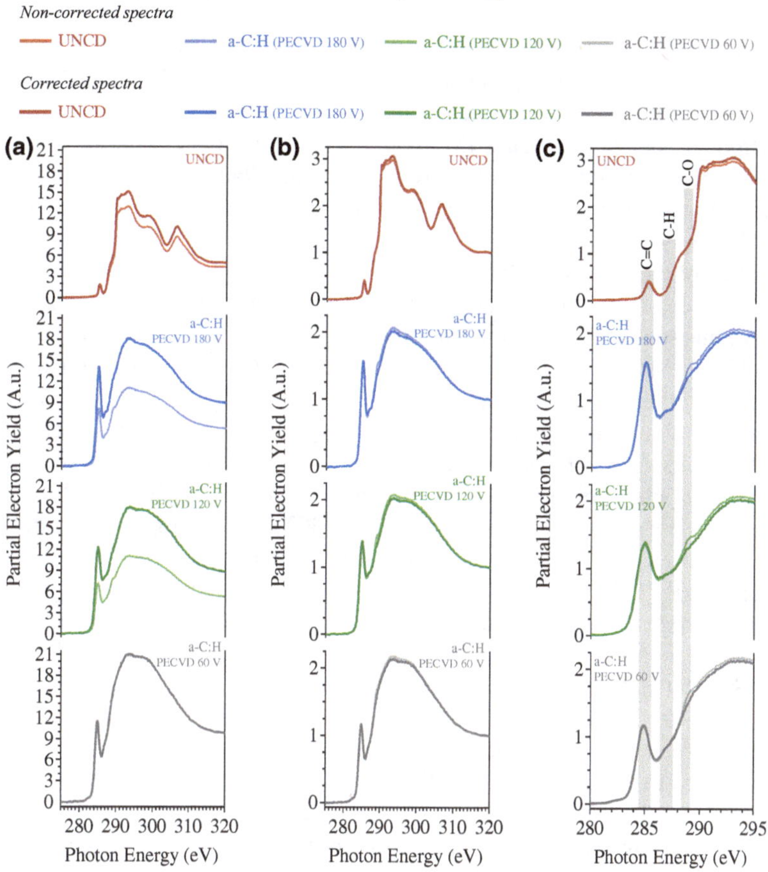

Fig. 3.5 Carbon 1s NEXAFS spectra (before and after the correction for the presence of a carbonaceous contamination layer) of UNCD and a-C:H films grown by PECVD with different acceleration bias voltages (60, 120, and 180 V). **a** Pre-edge normalized spectra (whole photon energy scale). The pre-edge normalization removes factors such as synchrotron ring current, beam dispersion character, and monochromator profile. The spectra were then normalized to the absorption current measured simultaneously from a gold mesh placed in the beamline upstream from the analysis chamber. The pre-edge region was then subtracted to zero; **b** pre- and post-edge normalized spectra (whole photon energy scale). The post-edge intensity depends on the number density of the absorbing atoms. By normalizing the post-edge to unity, variations in spectral intensity only arise from chemical changes and are independent of the total number density of absorbers; and **c** pre- and post-edge normalized spectra (zoomed absorption edge region). Upon removing the contribution from the carbonaceous contamination layer, the intensity of the C $1s \rightarrow \pi^*$ transition for disordered carbon−carbon bonds [5] at 285.0 ± 0.1 eV slightly decreased in the case of UNCD, whereas it increased for a-C:H films. Very significantly, for the a-C:H films, the broad absorption feature at 288.5–289.3 eV (assigned to the C–O σ^* antibonding orbital [106]) was greatly reduced. The spectral changes induced by the removal of the contribution of the carbonaceous contamination from the as-acquired data substantially affected the computation of the carbon hybridization state. From [99]

(a-C:H:Si:O), are shown in Fig. 3.6. More details about these thin films can be found in [134], respectively. For comparison, the NEXAFS spectra of ultrananocrystalline diamond (UNCD) and highly ordered pyrolytic graphite (HOPG) are also displayed. The NEXAFS spectrum of as-deposited a-C:H, a-C:H:Si:O, and UNCD exhibits an absorption feature at 285.0 ± 0.1 eV, which is due to the C 1s→π^* transition for disordered carbon–carbon bonds [5, 118]. The width of this C 1s→π^* absorption feature provides information about the bond length distribution of sp^2 carbon–carbon bonds: the broader the peak, the larger the bond length distribution. It should be noted that in the case of HOPG, the C 1s→π^* feature is detected at 285.5 ± 0.1 eV. Thus, the position of the C 1s→π^* transition can be used to gain insights into the ordering of sp^2-bonded carbon atoms in amorphous carbon materials: the higher the photon energy of this spectral feature, the more ordered the sp^2-bonded carbon phase [112]. A broad hump between 288 and 310 eV, which is due to the C 1s→σ^* transition for disordered carbon–carbon σ bonds [5, 118], characterizes the NEXAFS spectrum of as-deposited amorphous carbon materials (Fig. 3.6a), whereas the spectrum of UNCD exhibits sharper C 1s→σ^* transitions that are characteristic of ordered sp^3-hybridized carbon–carbon bonds, namely the edge jump at ~289 eV, the exciton peak at ~289.3 eV, and the second band gap at 303 eV [20] (Fig. 3.6b). Similarly, in the case of HOPG sharp C 1s→σ^* features are detected at 291.8 ± 0.2 eV and 292.8 ± 0.3 eV [135]. The presence of a significant amount of carbon–hydrogen bonds in hydrogenated amorphous carbon materials and the hydrogen-termination of UNCD result in the detection of a shoulder at ~287.0 eV (for a-C:H and a-C:H:Si:O) and ~287.5 eV (for UNCD), which can be assigned to the C 1s→σ^* transition for C–H bonds [5, 80, 99]. The observed shift of the characteristic C–H absorption feature between the spectra of a-C:H and the spectrum of UNCD may be due to the different bonding states in these materials: while hydrogen terminates primarily sp^3-bonded carbon atoms at the UNCD surface, it is present in a range of bonding environments within the bulk of hydrogenated amorphous carbon films. Due to the sample exposure to air before the NEXAFS analysis, a broad absorption feature is usually detected at 288.8 ± 0.2 eV and could be assigned to the C 1s→σ^* transition for carbon–oxygen bonds, as well as to the C 1s→π^* transition for carboxyl groups [5, 106, 136]. In the case of a-C:H:Si:O, the characteristic C 1s→σ^* transition for carbon–silicon bonds [136] contributes to the spectral feature at 288.8 ± 0.2 eV. The presence of carbonyl groups on the surface of a-C:H and a-C:H:Si:O due their exposure to air also results in a weak shoulder at 286.7 ± 0.2 eV (mainly due to the C 1s→π^* transition for C=O bonds and with a small contribution from the C–O Rydberg orbitals [5, 106]) (Fig. 3.6b).

Besides providing valuable information about the surface chemistry of carbon-based materials, NEXAFS spectroscopy can also be used for the quantitative evaluation of the hybridization state of carbon atoms [137]. The characterization of the carbon bonding configuration in carbon-based materials, which is of paramount importance in the case of carbon-based materials due to the strong dependency of the tribological, mechanical, electrical, and optical properties on the carbon hybridization state [138], is a particularly challenging materials science problem, which has resulted in the use of some of the most powerful weapons in the materials characterization arsenal for the analysis of these materials, including NEXAFS spectroscopy [99, 111–113, 119, 122, 126, 138–142], Raman spectroscopy [113, 138, 143–146],

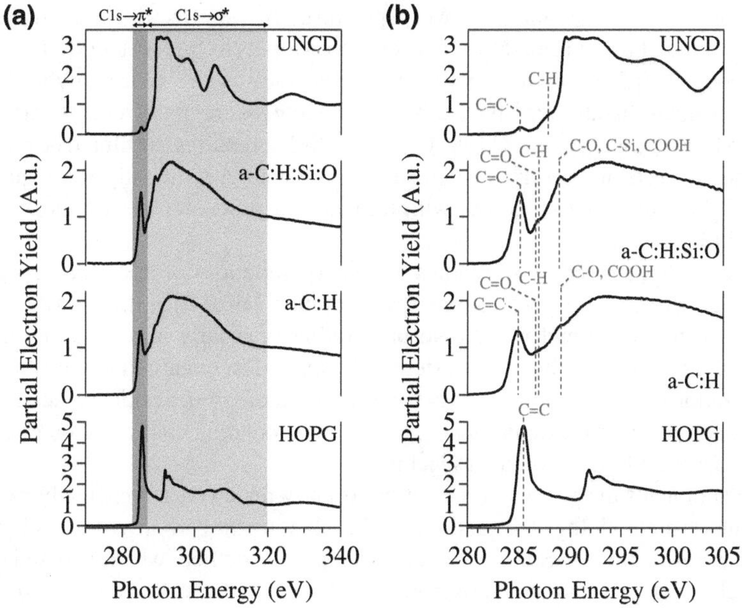

Fig. 3.6 **a** Carbon 1s NEXAFS spectra of HOPG acquired at an X-ray incidence angle (relative to the sample surface) of 40° together with carbon 1s NEXAFS spectra of a-C:H, a-C:H:Si:O, and UNCD acquired at an X-ray incidence angle of 55°; **b** zoomed view of the absorption edge region of the C 1s spectra. The spectra were not corrected for the contribution of the contamination layer following the method outlined in [99]

XPS [99, 121, 134, 146–153], Auger electron spectroscopy (AES, including X-ray induced AES, XAES) [134, 147, 149–151, 153–155], electron energy-loss spectroscopy (EELS) [145, 156–158], Fourier-transform infrared spectroscopy (FT-IR) [159], nuclear magnetic resonance (NMR) spectroscopy [160, 161], and X-ray reflectivity (XRR) [145]. Among the surface-sensitive techniques, electron spectroscopies (XPS, AES, XAES, NEXAFS, and EELS) are the mostly widely used for the characterization of carbon-based materials.

The characterization of the bonding configuration of carbon by XPS is usually carried out through the acquisition and fitting of the C 1s spectrum [121, 134, 146–149, 154, 155]. However, the validity of the methodology for the quantitative evaluation of the hybridization state of carbon on the basis of the C 1s signal, which relies on fitting it with two distinct features for sp^2- and sp^3-bonded carbon [121, 146, 154], has recently been questioned [134, 147], since the binding energy values of the C 1s transition for graphite (100% sp^2-bonded carbon), diamond (100% sp^3-bonded carbon), and UNCD ($94 \pm 1\%$ sp^3-bonded carbon [99]) are not significantly different [134, 147]. However, insights into the carbon hybridization state can be gained by XPS through the analysis of the plasmon band near the C 1s signal [134, 152], the π–π^* shake-up satellites [162, 163], or the X-ray induced C KVV Auger spectrum [134, 147, 150, 151, 153, 155].

Besides XPS, EELS and NEXAFS spectroscopy are effective methods for the determination of the hybridization state of carbon atoms in carbon-based films [138]. Even though EELS allows the fraction of carbon atoms in sp^2- and sp^3-hybridization state to be quantified through the analysis of the low loss region (from 0 to 40 eV) or the high loss region at the C 1s [138, 145, 156–158], it requires the film to be removed from the substrate or for a cross-section to be produced. Additionally, the acquisition of the C 1s is usually carried out with limited energy resolution of approximately 0.5 eV [138, 164].

NEXAFS spectroscopy overcomes these two limitations of EELS since spectroscopic data are acquired with an energy resolution lower than 0.1 eV (depending on the spectrometer) and no sample preparation is usually needed. Even though NEXAFS spectroscopy has been extensively applied to quantitatively evaluate the hybridization state of carbon atoms in the near-surface region, a critical assessment of the methodology for the quantification of the local bonding configuration of carbon on the basis of NEXAFS data was lacking.

The authors of the present contribution recently reviewed and critically assessed the common methodology for quantitatively evaluating the carbon hybridization state using carbon 1s NEXAFS measurements [137]. This method, which is based on the approach developed for the determination of the carbon hybridization state from the carbon core loss 1s edge in EELS [156, 165] and qualitatively supported by Car-Parrinello ab initio molecular dynamics simulations [166], considers the relative intensity of the C 1s→π^* and C 1s→σ^* absorption features in NEXAFS spectra of carbon-based materials. Since no theory exists to predict the π^*/σ^* ratio, a reference material (usually HOPG) with known sp^2 content is required for quantitative analysis. The equation used to compute the fraction of sp^2-bonded carbon (f_{sp^2}) is:

$$f_{sp^2} = \frac{I_{\text{sam}}^{\pi^*} / I_{\text{sam}}(\Delta E)}{I_{\text{ref}}^{\pi^*} / I_{\text{ref}}(\Delta E)} \tag{3.5}$$

where $I_{\text{sam}}^{\pi^*}$ and $I_{\text{ref}}^{\pi^*}$ are, respectively, the areas of the C 1s→π^* peaks for the sample and the reference (which arise exclusively from sp^2 bonds), whereas $I_{\text{sam}}(\Delta E)$ and $I_{\text{ref}}(\Delta E)$ are the areas under the NEXAFS spectrum between two integration limits (x_1 and x_2) for the sample and the reference, respectively. A graphical representation of the methodology is shown in Fig. 3.7.

A survey of the published literature revealed inconsistencies in applying this method. First of all, a wide range of integration limit values (x_1 and x_2) has been used for computing the area of the C 1s→σ^* transition (289–295 eV [123], 294–301 eV [111, 122], 293–302 eV [139], 288.6–325 eV [80], 289–325 eV [20], and 288.6–320 eV [99]). Secondly, HOPG reference spectra acquired at different X-ray beam incidence angles have been used in the literature (for example, HOPG spectra were acquired at 55° with respect to the sample surface [80], while others were acquired at 45° [99, 139]). This is particularly critical in the case of HOPG due to the strong dependence of the π^*/σ^* ratio on the orientation of the basal planes with respect to the incoming X-ray beam [5] (Fig. 3.8a).

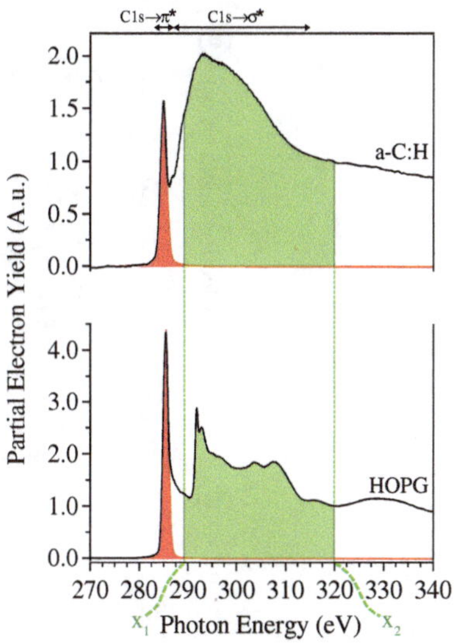

Fig. 3.7 Carbon 1s NEXAFS spectra of a-C:H and HOPG. Spectra are pre- and post-edge normalized. The NEXAFS spectrum of a-C:H was acquired at a photon incidence angle of 55° with respect to the sample surface (the so-called "magic angle") to suppress the effects related to the X-ray polarization [5], whereas the NEXAFS spectrum of HOPG was collected with the X-ray beam incident at an angle of 45° to the sample surface. The quantitative evaluation of the carbon hybridization state in carbon-based materials is based on the relative intensity of the C 1s→π^* and C 1s→σ^* absorption features (red- and green-shaded areas, respectively). A reference material with a known fraction of sp^2-bonded carbon (e.g., HOPG) is required for quantitative analysis. The area of the C 1s→π^* transition is determined by fitting this feature with a Gaussian synthetic peak [80, 99], whereas the area of the C 1s→σ^* absorption feature is computed by numerically integrating the spectrum between two limits x_1 and x_2. From [137]

The analysis of the dependence of total resonance intensity for the π^* and σ^* orbital on the angle of incidence of the impinging X-ray beam allowed the authors to derive the experimental conditions (i.e., the critical X-ray incidence angle, θ_c) to be used for the acquisition of HOPG reference spectra to which the π^* and σ^* states contribute equally (i.e., the conditions under which a molecular orbital oriented normal to the substrate surface contributes the same NEXAFS intensity as the orbital oriented within the substrate plane).

$$\theta_c = \cos^{-1}\sqrt{\frac{1}{2P}} \tag{3.6}$$

where P is the polarization factor in the plane of the electron beam orbit ($P = |E^{\parallel}|^2/\left(|E^{\parallel}|^2 + |E^{\perp}|^2\right)$) [5], which is equal to 0.85 for the beamlines used in the present work (NIST/Dow endstation of beamline U7A and at the Oak Ridge National Laboratory endstation of beamline U12A at the National Synchrotron Light Source (NSLS), Brookhaven National Laboratory, Upton, NY, USA) [26, 167]. In the case

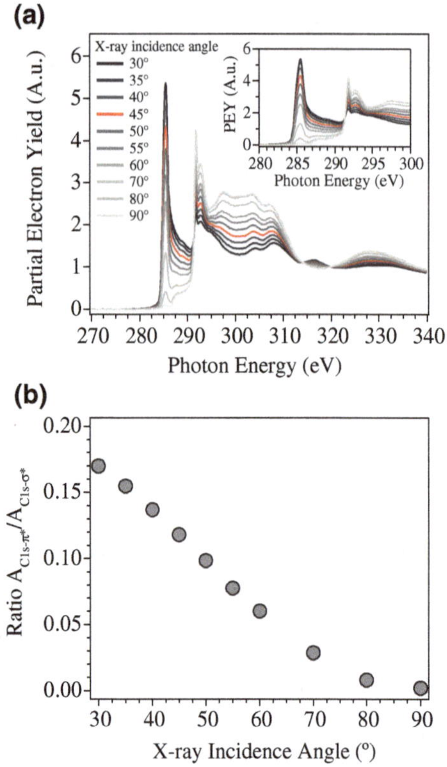

Fig. 3.8 **a** Carbon 1s NEXAFS spectra of HOPG acquired at different X-ray incidence angles. Spectra are pre- and post-edge normalized. Inset: zoomed view of the absorption edge region of the C 1s spectra. Spectra displayed without any offset to allow for comparisons; **b** ratio of the integrated intensity of the C $1s \rightarrow \pi^*$ (computed by fitting this signal with a Gaussian synthetic peak) and C $1s \rightarrow \sigma^*$ (computed by numerically integrating the spectrum between 288.6 and 320 eV) absorption features as a function of the X-ray incidence angle. Adapted from [137]

of linearly polarized X-rays ($P = 1$), θ_c is equal to 45°, while for elliptically polarized X-rays with polarization factor P equal to 0.85 (as in the case of U7A and U12A), θ_c is equal to 40°. It should be noted that there are claims in the literature that this angle should be the magic angle (i.e., 55° with respect to the sample surface). The analysis reported in [137] demonstrated that this is not correct and provided a formula for computing the X-ray incidence angles that should be used for acquiring HOPG spectra to be employed as reference in the determination of the carbon local bonding configuration from NEXAFS data (using (5)).

Since the π^*/σ^* ratio in HOPG spectra strongly depends on the X-ray incidence angle (θ) [5] (Fig. 3.8b), uncertainty in the angular position of the HOPG relative to the impinging X-ray photons can lead to errors in the quantitative analysis. The derivation of an analytical expression that describes the dependence of the π^*/σ^* ratio on the X-ray incidence angle allowed for the calculation of the uncertainty of in the fraction of sp^2-bonded carbon due to uncertainties in the X-ray incidence angle (Fig. 3.9a, b). The uncertainty increases with both the fraction of sp^2-bonded carbon (Fig. 3.9a) and the uncertainty in the angle of X-ray incidence (Fig. 3.9b). In the case of the beamlines used in this work, the uncertainty in the X-ray incidence angle was 1°, which translates into an uncertainty in the computed fraction of sp^2-

bonded carbon linearly increasing with the fraction of sp^2-bonded carbon up to 3.6% for 100% sp^2-bonded carbon, which is modest. However, a larger uncertainty in X-ray incidence angle of 5° produces a much more significant error of 8.9% for 50% sp^2-bonded carbon, and 17.9% for 100% sp^2-bonded carbon. These findings clearly demonstrated the importance of minimizing any uncertainty in the angular position of HOPG relative to the impinging X-ray beam.

As pointed out above, a wide range of integration limits (x_1 and x_2) has been used in the literature for computing the integrated intensity of the C 1s→σ^* absorption feature in NEXAFS spectra. The investigation of the evolution of the computed fraction of sp^2-bonded carbon for an a-C:H film as a function of the integration limits used to calculate the area of the C 1s→σ^* transition is shown in Fig. 3.9c, d. For highlighting the influence of the integration limits on the results of the quantification, the error in the calculated fraction of sp^2-bonded carbon relative to the fraction obtained using the integration limits employed in [99] ($x_1 = 288.6$ eV; $x_2 = 320$ eV) is also shown in Fig. 3.9c, d. The outcomes of this analysis indicated that for the quantification of the carbon hybridization state using carbon 1s NEXAFS spectroscopy, the low-photon-energy integration limit (x_1) can be chosen arbitrarily between 286.6 and 295 eV, whereas the high-photon-energy integration limit (x_2) should be taken at photon energies above 310 eV. In fact, large variations, as large as 10%, in the computed fraction of sp^2-bonded carbon occurred upon decreasing the high-photon-energy integration limit below 310 eV.

3.4.2 NEXAFS Spectroscopic Investigation of the Thermally-Induced Structural Evolution of Solid Lubricating Films

The thermally-induced structural evolution of amorphous carbon (a-C) thin films is a subject of fundamental and technological interest [112, 168–170]. As one particular example, the investigation of the effect of annealing temperature on the structure of tetrahedral amorphous carbon (ta-C) has important implications regarding the use of ta-C as an engineering material. In fact, thermal annealing at elevated temperatures (~650 °C) for a period of minutes is necessary in the case of ta-C to relieve the high compressive stress (2–8 GPa) typical for these films [112, 171, 172]. As another example, knowledge of the temperature effects on the structure of a-C coatings, which are used as overcoat material in hard-disk drives to protect the magnetic medium against corrosion and wear [173], is fundamental for developing one of the most promising solutions for increasing the magnetic storage density, namely heat-assisted magnetic recording (HAMR) [174]. The intensive localized heating employed in HAMR for data encoding raises concerns about the structural changes occurring in a-C overcoats.

Grierson et al. investigated the thermally-induced evolution of the structure of ta-C by NEXAFS spectroscopy [112]. The results provided evidence that sp^3-bonded

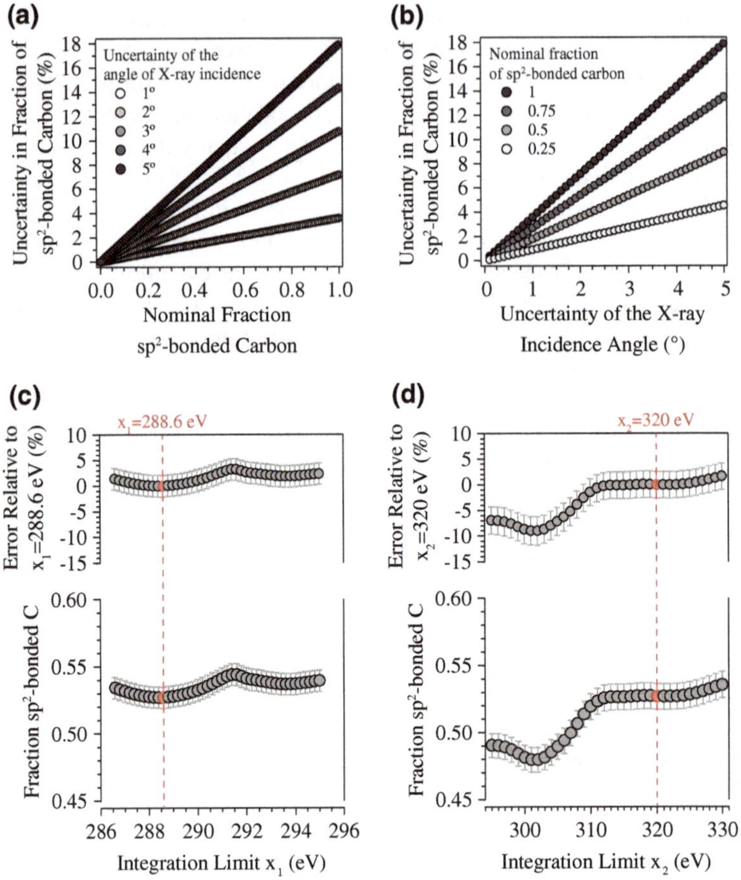

Fig. 3.9 Uncertainty in the computed fraction of sp^2-bonded carbon due to uncertainties in the X-ray incidence angle used for the acquisition of the HOPG reference spectrum as a function of: **a** the nominal fraction of sp^2-bonded carbon; and **b** the uncertainty in the X-ray incidence angle. **c** Influence of the lower-photon-energy integration limit (x_1) used to calculate the area of the C 1s→σ^* transition on the computed fraction of sp^2-bonded carbon for a hydrogenated amorphous carbon film (the high-photon-energy limit x_2 was kept fixed at 320 eV). Error bars represent the standard deviation calculated from multiple independent measurements; **d** influence of the high-photon-energy integration limit (x_2) used to calculate the area of the C 1s→σ^* transition on the computed fraction of sp^2-bonded carbon for a hydrogenated amorphous carbon film (the low-photon-energy limit x_1 was kept fixed at 288.6 eV). Error bars represent the standard deviation calculated from multiple independent measurements. For highlighting the influence of the integration limits on the results of the quantification, the error in the calculated fraction of sp^2-bonded carbon relative to the fraction obtained using the integration limits employed in [99] (x_1 = 288.6 eV; x_2 = 320 eV) is shown in the upper part of the graphs. Adapted from [137]

carbon–carbon bonds are directly converted to sp^2-hybridized carbon-carbon bonds upon annealing. Additionally, the broadening of the C 1s→π^* absorption feature clearly indicated an increase in bond length distribution with temperature, while the shift of this peak to higher photon energies suggested an increase in the degree of ordering of the sp^2 carbon atoms. The quantification of the fraction of carbon atoms in sp^2-hybridization state as a function of temperature allowed Grierson et al. to apply Sullivan et al.'s model [171] to determine an activation energy range for the sp^3-to-sp^2 conversion of carbon hybridization. Sullivan's model describes the transformation of sp^3- to sp^2-bonded carbon as a series of first order chemical reaction. Because of bond length and angle disorder in a-C materials, a distribution of activation energies for sp^3-to-sp^2 conversion of carbon hybridization is also included in the model. The activation energy range calculated by Grierson et al. (3.5 ± 0.9 eV) was in agreement with the activation energy for bulk sp^3-to-sp^2 conversion of carbon hybridization in ta-C derived from Raman spectroscopic measurements (3.3 eV) [172].

More recently, the authors of the present contribution investigated in situ the thermally-induced structural evolution of a class of doped hydrogenated amorphous carbon materials, namely silicon- and oxygen-containing hydrogenated amorphous carbon (a-C:H:Si:O—also referred to as silicon oxide-doped diamond-like carbon (SiO_x-DLC) or diamond-like nanocomposite (DLN)) [141, 142, 175]. a-C:H:Si:O is a promising class of multicomponent materials for several applications since the incorporation of silicon and oxygen reduces their residual stress [176, 177], while not significantly affecting the mechanical properties [177]. In addition, a-C:H:Si:O films may present other interesting properties, such as enhanced thermal stability [178, 179] and good tribological behavior (i.e., low friction and wear) across a broader range of conditions and environments compared to hydrogenated amorphous carbon (a-C:H) films [177, 180, 181]. The superior thermal stability of a-C:H:Si:O compared to a-C:H was suggested to arise from the fourfold coordination of silicon, which stabilizes the carbon atoms in the sp^3 hybridization state (thus inhibiting their conversion into sp^2-bonded carbon at elevated temperatures) [182–185]. Yet, a fundamental understanding of the origin of the superior thermal and thermo-oxidative stability of a-C:H:Si:O has not been achieved. This hampers the possibility to tailor the deposition process with the aim of depositing coatings with improved properties.

The thermally-induced structural evolution of a-C:H:Si:O was investigated in situ by NEXAFS and XPS spectroscopy under high vacuum conditions (pressure $<1.0 \times 10^{-8}$ Torr). Carbon 1s NEXAFS spectra (corrected for the contribution of the contamination layer following the method outlined in [99]) of a-C:H:Si:O before and after annealing are displayed in Fig. 3.10a, b. The NEXAFS spectrum of as-deposited a-C:H:Si:O exhibited an absorption feature at 285.0 ± 0.1 eV, which is due to the C $1s \rightarrow \pi^*$ transition for disordered carbon-carbon bonds [5, 118]. A broad hump between 288 and 310 eV, which is due to the C $1s \rightarrow \sigma^*$ transition for disordered carbon–carbon σ bonds [5, 118], characterized the NEXAFS spectrum of as-deposited a-C:H:Si:O. The presence of a significant amount of carbon–hydrogen and carbon–silicon bonds in a-C:H:Si:O also resulted in the detection of distinct absorption features at 287.0 ± 0.1 eV (assigned to the C $1s \rightarrow \sigma^*$ transition for C–H bonds [5, 80, 99]) and 288.9 ± 0.1 eV (C $1s \rightarrow \sigma^*$ transition for C–Si bonds [136]). It has to be emphasized that carbon–oxygen bonds are assumed to be only present in the contamination layer. Under this assumption the composition calculated by XPS more closely agrees with

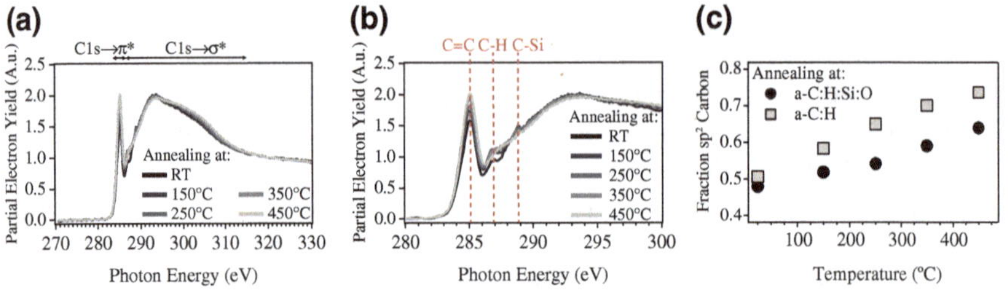

Fig. 3.10 **a** Pre- and post-edge normalized carbon 1s NEXAFS spectra of a-C:H:Si:O acquired before annealing, and after annealing at different temperatures in vacuum (pressure $<1.0 \times 10^{-8}$ Torr); **b** zoomed view of the absorption edge region of the carbon 1s spectra. Spectra displayed without any offset to allow for comparisons; and **c** fraction of sp^2-hybridized carbon versus annealing temperature calculated from NEXAFS spectra following the approach outlined in [137]. For comparison, the evolution of the fraction of sp^2-bonded carbon for a-C:H thin films annealed under similar experimental conditions [134] is also displayed (note: the as-deposited a-C:H thin film has a fraction of sp^2-bonded carbon comparable to the one of as-grown a-C:H:Si:O as well as a similar hydrogen concentration to a-C:H:Si:O, i.e., 26 ± 3 at.% for a-C:H and 34 ± 3 at.% for a-C:H:Si:O)

the results of secondary ion mass spectrometry (SIMS) and Rutherford backscattering spectrometry (RMS) (SIMS and RBS measurements performed by Evans Analytical Group, Sunnyvale, CA, USA). Thus, carbon–oxygen absorption features do not contribute to the contamination-corrected NEXAFS spectrum of a-C:H:Si:O shown in Fig. 3.10a, b, in contrast to as-acquired NEXAFS data (displayed in Fig. 3.6).

Upon annealing, the intensity of the peak at 285.0 ± 0.1 eV progressively increased, suggesting an increase in the fraction of sp^2-bonded carbon in the near-surface region. At the same time, the absorption feature at 288.9 ± 0.1 eV decreased in intensity with the annealing temperature, thus indicating the breakage of carbon–silicon bonds. The absorption feature assigned at 287.0 ± 0.1 eV also changed at elevated temperatures: its intensity first increased upon annealing at 150 °C, which might be due to hydrogen diffusion to the near-surface region upon low temperature annealing, and then progressively decreased, which suggests the scission of carbon–hydrogen bonds upon annealing. These spectral variations could be highlighted by computing the difference in NEXAFS spectra between subsequent annealing steps (not shown). The calculation of difference spectra also highlighted a progressive increase in spectral intensity on the high photon energy side (at ~285.5 eV) of the C $1s \rightarrow \pi^*$ transition, suggesting an increase in the degree of ordering of the sp^2 carbon phase [5, 112, 118]. This finding was corroborated by XPS measurements (not shown).

The fraction of sp^2-hybridized carbon, which was calculated following the methodology reported in [137], is displayed as a function of annealing temperature in Fig. 3.10c. For comparison, Fig. 3.10c also reports the evolution of the fraction of sp^2-bonded carbon for an a-C:H film obtained from in situ XPS annealing experiments performed under similar experimental conditions (the as-deposited a-C:H film has a fraction of sp^2-bonded carbon comparable to the one of as-grown a-C:H:Si:O as well as a similar hydrogen content to a-C:H:Si:O, i.e., 26 ± 3 at.% for a-C:H and

34 ± 3 at.% for a-C:H:Si:O) [134]. These results clearly indicate that doping a-C:H with silicon and oxygen at the modest level of, respectively, 6 ± 1 at.% and 3 ± 1 at.% slightly increases the thermal stability of a-C:H:Si:O compared to a-C:H in high vacuum [141, 142]. The analysis of the energetics of the thermally-induced structural evolution of a-C:H:Si:O is reported in [141], where atomistic insights (from molecular dynamics (MD) simulations) into the processes occurring in this material at elevated temperatures are also presented.

3.4.3 NEXAFS Spectroscopic Investigation of the Tribochemistry of Solid Lubricants

3.4.3.1 Photoemission Electron Microscopy (PEEM)

As outlined in Sect. 3.3.1.2.2, photoemission electron microscopy (PEEM) is a full field imaging technique commonly combined with X-ray synchrotron radiation. The high surface sensitive of this spectroscopic technique together with its excellent spatial resolution (<50 nm [69], or even below 10 nm in aberration corrected instruments [70]) [14, 71–75] has made it an increasingly attractive characterization technique in several fields, including tribology [14, 20, 77–82], thin film magnetism [83–85], catalysis [86], semiconductor [87], polymer science [88–90], geology [91–93], and microbiology [94].

Anders et al. employed PEEM to study lubricated and unlubricated hard disks after wear tests [78, 79]. The hard disk/slider interface involves mechanical and tribochemical processes between the hard carbon-based overcoat on the disk, the lubricant (a perfluoropolyether (ZDOL) fluid [173]), and the carbon coated or uncoated slider surface. By exploiting the excellent lateral resolution of PEEM, Anders et al. acquired NEXAFS spectra inside the wear track after the disks failed the tribological tests. The spectroscopic results indicated the formation of various new carbon–oxygen (mostly carboxylic) bonds in the wear tracks with a reduction of the amount of fluorine and carbon. These findings provided clear evidence that the chemical modifications inside the wear track were induced by the degradation of the lubricant, which was not solely a mechano-chemical process, but it was also accompanied by oxidation reactions.

Konicek et al. investigated by PEEM the surface-chemical phenomena governing the environmental dependence of the tribological properties of UNCD [20]. This material exhibits poor tribological performance in dry or vacuum conditions, which is hypothesized to be due to the high strength of the carbon–carbon bonds that rapidly form across the interface. This happens as a consequence of the removal of passivating surface species (like hydrogen) induced by the sliding process [186, 187]. The reduction of friction and wear as molecular species such as hydrogen, oxygen or water are present in the surrounding vapour is thought to be due to the dissociative adsorption of gaseous species, which saturate the dangling bonds formed on the surface by the applied stress [188–191]. These findings are consistent with molecular dynamics (MD) simulations [192–194] and density functional theory (DFT) calculations [195–197] as well as with the studies performed at the nanometer scale by atomic force microscopy (AFM) [198]. However, others proposed that low friction and wear are achieved in humid environments in the case of highly sp^3-bonded hydrogen-free carbon by the rehybridization of carbon atoms to ordered (graphitic) sp^2 bonding [199–201]. The debate regarding the origin of the impressively low friction and wear of highly sp^3-bonded hydrogen-free carbon-based films in humid environments may have been settled thanks to the work of Konicek et al., who, using PEEM combined with NEXAFS spectroscopy, provided a spectroscopic evidence of the passivation hypothesis [20, 82], as outlined in the following.

Konicek et al. performed tribological experiments using UNCD-coated silicon nitride (Si_3N_4) spheres and silicon wafers at low (0.1 N, 300 MPa mean Hertzian contact pressure) and high (1.0 N, 649 MPa) loads, in low (1%) and high (50%) relative humidity (RH). The tests were referred to as HD, LD, HW, and LW for high (H) or low (L) load and dry (D) or humid (W for wet) conditions. In all cases, friction coefficients below 0.05 were achieved at steady state, even though in the case of the experiments performed in dry environment at high load (HD), more sliding cycles were required to achieve this low friction coefficient.

Figure 3.11a displays an example of PEEM image captured at a photon energy of 289 eV (corresponding to the edge jump in the NEXAFS spectrum of UNCD, see Fig. 6) [20]. Carbon 1s NEXAFS spectra could be extracted from regions of interest (ROI) and compared to the spectrum of unworn UNCD (Fig. 3.11b). The authors observed increases in spectral intensity from 286 to 289 eV, an energy window in which the characteristic signals of C=O (~286.4 eV [5, 139]) and C–H (~287.5 eV [5]) are found. These changes in the intensity of NEXAFS spectra, which strongly depended on the environment and applied load (Fig. 3.11b), were accompanied by some conversion of sp^3-bonded carbon to sp^2-bonding, as evidenced by the increased intensity at 285 eV (due to the C 1s$\rightarrow\pi^*$ transition for disordered carbon–carbon bonds [5, 118]. In the case of as-deposited UNCD, this absorption peak is due to sp^2-bonded carbon present at grain boundaries, as surface contamination, and due to the reconstruction of diamond surfaces). However, upon sliding none of the C 1s$\rightarrow\pi^*$ absorption features exhibited the shift from 285 to 285.5 eV (characteristic position for the C 1s$\rightarrow\pi^*$ absorption peak for HOPG), thus indicating that graphitic (ordered) carbon is not formed upon sliding. Based on these findings, the authors concluded that the lubrication mechanism for diamond involving the formation of substantial graphitic carbon layers does not occur for UNCD under the broad range of conditions explored. Rather, the mechanism leading to low friction is passivation of dangling bonds.

This conclusion was substantiated by a subsequent study by the same authors [82], in which tribological experiments were performed using UNCD- and hydrogen-free tetrahedral amorphous carbon (ta-C)-coated silicon nitride spheres and silicon wafers. The experiments were carried out under different environmental conditions (relative humidity, RH, from 1 to 50%) and applied normal pressure (initial mean Hertzian contact pressure ranging from 240 to 649 MPa). For the ta-C, tracks created

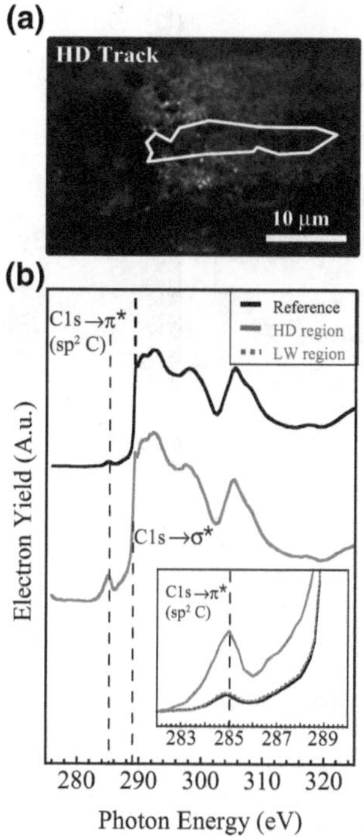

Fig. 3.11 **a** Example of PEEM image (acquired at 289.0 eV photon energies) of the wear track generated upon sliding a UNCD-coated silicon nitride (Si_3N_4) pin on UNCD with an applied load of 1.0 N and a relative humidity of 1.0% (HD); **b** carbon 1s NEXAFS spectra extracted from PEEM images. Top spectrum (black line): reference taken far from the wear track; bottom spectrum (gray line): spectrum extracted from the region drawn in (**a**). Inset: magnified plot (C1 s→π^* region) of the spectra shown in (**b**) together with a spectrum extracted from the track generated at 0.1 N and 50% RH (*LW.* PEEM image not shown). From [20]

at lower humidity and higher load experienced increased friction and wear, while for the UNCD friction trends were unclear but lower humidity did lead to greater wear rates.

Ex situ PEEM measurements were carried out to gain information about the chemical changes and structural transformations induced by the sliding process. Figure 3.12a displays a typical PEEM image (with a defined region of interest, ROI) obtained at 289.0 eV for the case of the track formed on ta-C at 0.5 N and 1.0% RH. This track was heavily worn. The extracted carbon 1s and oxygen 1s spectra from the selected ROI are also displayed in Fig. 3.12a. Reference spectra taken from the unworn region are reported together with the spectra obtained from the PEEM analyses of the track created upon sliding at 0.5 N and 50% RH (this track was only lightly worn). Two main differences could be noticed when comparing the carbon 1s NEXAFS spectra of the heavily-worn (0.5 N, 1.0% RH) and unworn regions: first, the C 1s→π^* transition for disordered carbon–carbon bonds at 285.0 eV [5, 118]

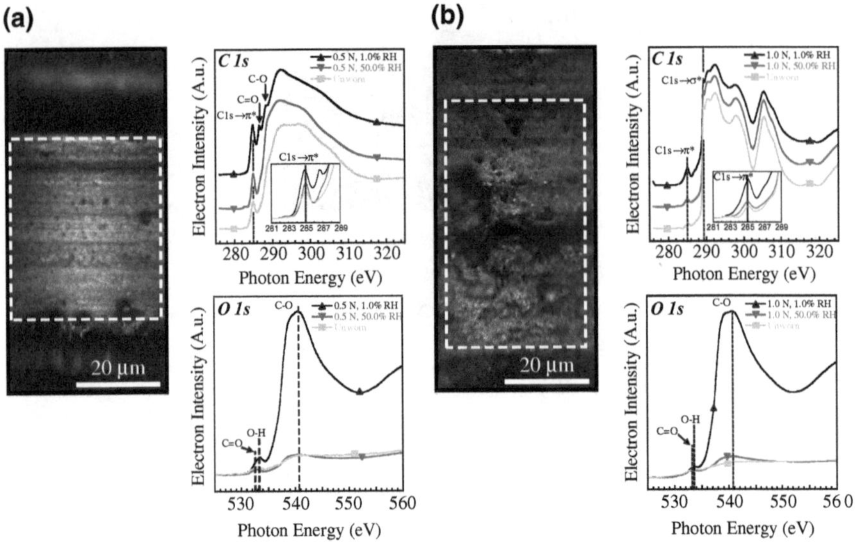

Fig. 3.12 **a** Example of PEEM image (acquired at 289.0 eV photon energy) of the wear track generated upon sliding a ta-C-coated silicon nitride (Si_3N_4) pin on ta-C with an applied load of 0.5 N and a relative humidity of 1.0%. The carbon 1s and oxygen 1s spectra extracted from the region of interest defined in PEEM images are shown. The spectra extracted from PEEM images collected from a sample after tribological testing at a relative humidity of 50% are also displayed; **b** example of PEEM image (acquired at 289.0 eV photon energies) of the wear track generated upon sliding a UNCD-coated silicon nitride (Si_3N_4) pin on UNCD with an applied load of 1.0 N and a relative humidity of 1.0%. The carbon 1s and oxygen 1s spectra extracted from the region of interest defined in PEEM images are shown. The spectra extracted from PEEM images collected from a sample tribologically-stressed at a relative humidity of 50% are also displayed. From [82]

increased in intensity in the former, which indicates the conversion of carbon atoms from sp^3- to sp^2-hybridization state upon sliding; secondly, a significant amount of oxidation is present in the track, as suggested by the peaks at ~286.7 and 288.6 eV, which was assigned to C–O Rydberg orbitals and C–O σ^* orbitals, respectively [106]. As for the case of the track generated at 0.5 N and 50.0% RH, which was lightly worn, these spectral differences occurred to a much lesser extent, indicating a small amount of rehybridization and some traces of oxidation. The oxygen 1 s spectra provided additional information. While the spectrum of the track formed at 0.5 N and 50.0% RH was very similar to the one extracted from the unworn area, in the case of the experiments performed at 0.5 N and 1.0% RH there was a substantially higher oxygen concentration in the wear track. Additionally, the oxygen was more σ bonded as opposed to π bonded, as demonstrated by the much larger edge jump at 538.0 eV and the C–O feature at ~541 eV. In the case of the tracks for at 0.5 N and 1.0% RH, the pre-edge peak shifted from 532.7 eV (characteristic to a π^* transition for double-bonded oxygen) to 533.4 eV, indicating the presence of O–H bonds [106].

The PEEM results obtained from the tribological tests carried out with UNCD were similar to those from the experiments performed with ta-C. Figure 3.12b shows a typical PEEM image (acquired at 289 eV) from a region entirely inside the UNCD track created at 1.0 N and 1.0% RH. The carbon 1s and oxygen 1s spectra extracted from the worn and unworn regions are reported, together with the spectra obtained from the experiments ran at 1.0 N and 50.0% RH. For the experiments carried out at 1.0 N and 1.0% RH, the increase in intensity of the peak at 285.0 eV, which is assigned to the C 1s→π^* transition for disordered carbon-carbon bonds [5, 118], provided evidence of sp^3-to-sp^2 rehybridization. Additionally, the spectral features appearing at ~286.4 and ~288.6 eV clearly indicated the oxidation of the near-surface region. The oxygen 1s data acquired on UNCD are also similar to those collected on ta-C: there is a weak oxygen signal in the unworn region, while this signal significantly increased in intensity in the case of the tracks generated at 1.0 N and 1.0% RH. While the spectrum extracted from the wear track generated at 1.0 N and 50.0% RH has a similar intensity to the one of the unworn region, the pre-edge peak shifted from 533.2 to 533.5 eV, indicating, as in the case of the ta-C experiments, more O–H bonding.

These results supported the conclusion that the lubrication mechanism of ta-C and UCND is related to the passivation of surface dangling bonds by dissociated water vapor. Upon sliding, dangling bonds are generated, but the overall tribological behavior is dominated by the competition between their passivation by the dissociative adsorption of water and the formation of covalent bonds across the sliding interface. Additionally, the experimental results provided clear evidence that no ordered graphitic carbon is formed upon sliding, even though some rehybridization of sp^3-bonded carbon to sp^2 bonding occurred. All together, the outcomes of this work indicated that the primary solid-lubrication mechanism for highly sp^3-bonded, nearly hydrogen-free carbon materials (either amorphous or polycrystalline) is the passivation of dangling bonds by OH and H from the dissociation of vapor-phase water. This was supported by ab initio density functional theory (DFT) calculations [195–197].

3.4.3.2 Magnetically-Guided Imaging NEXAFS Spectroscopy

As outlined in Sect. 3.3.1.2.2, magnetically-guided imaging NEXAFS spectroscopy is a spatially-resolved techniques that allows for the acquisition of spectroscopic data over a field of view of $18 \times 13\ mm^2$ with a lateral resolution of 50 μm.

Konicek et al. employed this surface-analytical method to perform chemical imaging of both countersurfaces of a self-mated sphere-on-flat UNCD contact [18]. Two wear tracks were created using a bidirectional linear reciprocating microtribometer placed in a sealed environment (5% RH): the first track was generated using a new, unconditioned sphere, whose surface was then conditioned during the first experiment and used for the second test. Conditioning of the sphere led to faster run-in, lower friction, and lower wear when producing the second wear track. By taking advantage of the capabilities of the rapid parallel processing magnetic field electron yield optics detector of NIST/Dow endstation of beamline U7A at the National Synchrotron Light Source (NSLS, Brookhaven National Laboratory, Upton, NY, USA), spectromicroscopic images could be collected on both the spherical and flat surfaces. The spectroscopic results acquired on the flat indicated that the carbon chemical state undergoes less modification when friction and wear are low. Additionally, the NEXAFS spectra extracted from the imaging data acquired on the film deposited on the sphere exhibited a C 1s$\rightarrow\pi^*$ transition at a photon energy of 285.2 eV both before and after sliding, thus suggesting the presence of both disordered and ordered carbon–carbon bonds [5, 118, 122, 139]. These results led the authors conclude that the higher ordering of the sp^2-bonded carbon for the material present in the near-surface region of the sphere compared to the material deposited on the flat was not responsible for low friction and short run-in. Rather, conditioning of the sphere, which could likely remove asperities and passivate surface dangling bonds, was proposed to be the reason leading to immediate run in when sliding on a new portion of the flat and resulting in lower friction with less chemical modification of the substrate.

The authors of the present contribution recently employed magnetically-guided imaging NEXAFS spectroscopy to investigate the tribological properties of silicon- and oxygen-containing hydrogenated amorphous carbon (a-C:H:Si:O). a-C:H:Si:O thin films are promising materials for several applications (e.g., microelectromechanical systems, aerospace, overcoats for hard-disks), since the incorporation of silicon and oxygen in the hydrogenated amorphous carbon matrix renders its tribological behavior less dependent on the environmental conditions compared to undoped carbon-based materials [177]. While the tribological properties of these films have been studied quite extensively in different environments (from dry to humid atmospheres) [177], only a few studies have focused on the mechanisms by which their excellent tribological performance is achieved, particularly under low pressure conditions.

Tribological tests were performed using a pin-on-flat configuration, where a steel pin was slid on an a-C:H:Si:O surface. The results revealed a strong environmental dependence of the tribological performance of a-C:H:Si:O: under high vacuum conditions, the friction coefficient quickly reached values above 1, but a significant reduction in friction was observed when hydrogen or oxygen was leaked in the experimental chamber at a pressure of at least 50 or 10 mbar, respectively [98]. Additionally, upon increasing the oxygen pressure in the chamber from 10 to 1000 mbar, the coefficient of friction increased from 0.02 ± 0.01 to 0.06 ± 0.01, whereas upon increasing the hydrogen pressure from 50 to 2000 mbar, the coefficient of friction decreased from 0.08 ± 0.01 to 0.02 ± 0.01.

To investigate the structural transformations and chemical reactions occurring in the near-surface region of a-C:H:Si:O upon sliding under different environmental conditions, imaging NEXAFS measurements were performed. Figure 3.13a shows a schematic of an a-C:H:Si:O sample on which multiple tribological experiments were performed at different oxygen and hydrogen partial pressures. Using the same experimental apparatus employed by Konicek et al. [18], a series of two-dimensional NEXAFS images were acquired as the energy of the incident soft X-rays was scanned across the carbon 1s edge. Figure 3.13b-e displays NEXAFS images collected at photon energies corresponding to the characteristic absorption features of the NEXAFS spectrum of a-C:H:Si:O [98], i.e., 285.1 $\pm$ 0.2 eV, 286.7 $\pm$ 0.2 eV, 287.6 $\pm$ 0.2 eV, and 288.8 $\pm$ 0.2 eV. These images were generated after pre- and post-edge normalizing the NEXAFS data to ensure that variations in spectral intensity only arise from chemical changes and are independent of the total carbon content. Independently of the sliding environment, the intensity of the signals at 285.1 $\pm$ 0.2 eV (assigned to the C 1s$\rightarrow\pi^*$ transition for disordered carbon–carbon sp^2 bonds [5, 118]) and 287.6 $\pm$ 0.2 eV (assigned to the C1 s$\rightarrow\sigma^*$ transition for C–H bonds [5, 80, 99]) increased inside the wear tracks, whereas the intensity of the spectral feature at 288.8 $\pm$ 0.2 eV (due to the C 1s$\rightarrow\sigma^*$ transition for carbon–silicon and carbon–oxygen bonds, as well as to the C 1s$\rightarrow\pi^*$ transition for carboxyl groups [5, 106, 136]) decreased in the wear tracks. As for the signal at 286.7 $\pm$ 0.2 eV (mainly due to the C 1s$\rightarrow\pi^*$ transition for C=O bonds and with a small contribution from the C–O Rydberg orbitals [5, 106]), its intensity significantly increased in the tracks generated upon sliding under oxygen environment.

NEXAFS spectra were extracted from the wear tracks generated at different hydrogen and oxygen partial pressure (Fig. 3.13f). Subtle changes in the intensity of the characteristic absorption features of a-C:H:Si:O could be discerned by comparing the spectra. To more easily investigate the structural transformations and chemical changes occurring in the near-surface region of a-C:H:Si:O upon sliding at different hydrogen and oxygen partial pressure, the difference between spectra obtained from the worn and unworn a-C:H:Si:O regions was calculated (Fig. 3.13g). Computing these difference spectra also allows the contribution of the carbonaceous contamination layer present on the sample surface due to its exposure to air to be eliminated. This is valid under the assumption that the adventitious contamination layer is laterally homogeneous in composition, density, and thickness across both worn and unworn region, i.e., is due to ambient exposure after the tribotesting. This approach avoids the need for correcting the as-acquired spectra for the presence of an adventitious carbonaceous contamination using a reference sample [99]. As for the tests performed in hydrogen, the intensity of the C 1s$\rightarrow\pi^*$ transition for disordered carbon–carbon sp^2 bonds at 285.1 $\pm$ 0.2 eV significantly changed upon sliding at different gas pressure: it increased for low gas pressure (50 and 200 mbar), which demonstrates a significant conversion from sp^3- to sp^2-bonded carbon, whereas it decreased for the highest hydrogen pressure (2000 mbar). Furthermore, a decrease in intensity of the signal at 288.8 $\pm$ 0.2 eV (assigned to the C 1s$\rightarrow\sigma^*$ transition for carbon–silicon and carbon–oxygen bonds, as well as to the C 1s$\rightarrow\pi^*$ transition for carboxyl groups [5, 106, 136]) was observed. Since the computation of difference spectra allows the contribution of the carbonaceous contamination layer present on the sample surface due to its exposure to air to be eliminated, the main contribution to the signal at 288.8 $\pm$ 0.2 eV derives from carbon–silicon bonds (carbon–oxygen bonds are assumed to be only present in the contamination layer. Under this assumption the composition calculated by XPS more closely agrees with the results of secondary ion mass spectrometry (SIMS) and Rutherford backscattering spectrometry (RMS). SIMS and RBS measurements performed by Evans Analytical Group, Sunnyvale, CA, USA). Finally, intense peaks were also observed in difference spec-

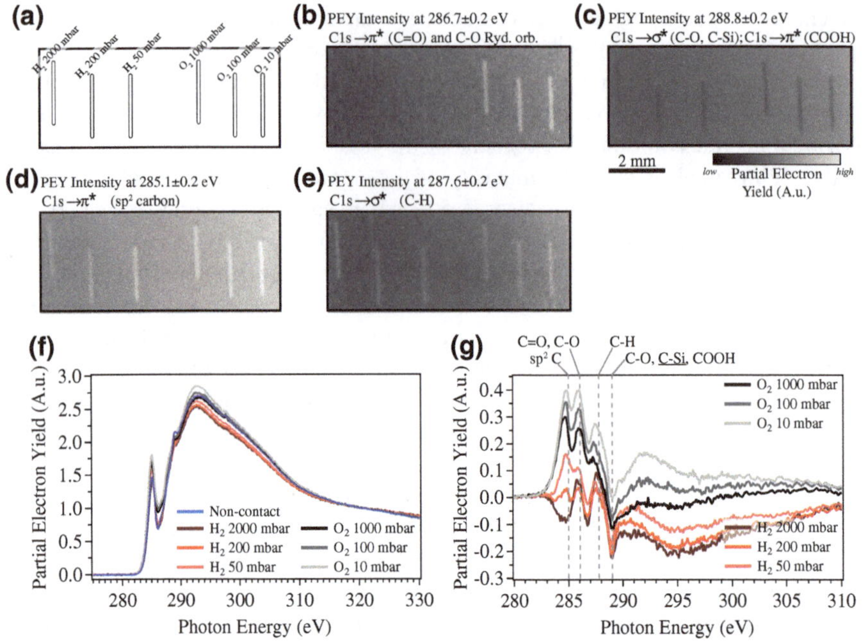

Fig. 3.13 **a** Schematic of an a-C:H:Si:O sample on which multiple tribological experiments were performed at different oxygen and hydrogen partial pressure using a steel counterbody; **b–e** NEXAFS maps acquired at different photon energies, i.e., 285.1 ± 0.2 eV (assigned to the C 1s→π^* transition for disordered carbon-carbon sp^2 bonds [5, 118]), 286.7 ± 0.2 eV (mainly due to the C 1s→π^* transition for C=O bonds and with a small contribution from the C–O Rydberg orbitals [5, 106]), 287.6 ± 0.2 eV (assigned to the C 1s→σ^* transition for C–H bonds [5, 80, 99]), and 288.8 ± 0.2 eV (due to the C 1s→σ^* transition for carbon–silicon and carbon–oxygen bonds, as well as to the C 1s→π^* transition for carboxyl groups [5, 106, 136]) are also displayed; **f** carbon 1s NEXAFS spectra of a-C:H:Si:O extracted from the contact and non-contact regions of the NEXAFS images reported in (**b–e**); **g** difference NEXAFS spectra (contact minus non-contact spectra) computed to highlight chemical differences between contact and non-contact regions

tra at 286.0 ± 0.2 eV and 287.8 ± 0.2 eV (respectively due to the C 1s→π^* transition for carbonyl groups [5, 106] and the C 1s→σ^* transition for C–H bonds [5, 80, 99]). The detection of a peak that can be assigned to carbon–oxygen bonds in the spectra extracted from the tracks produced in hydrogen can be due to some residual oxygen in the vacuum chamber. It is worth noticing that the relative intensity of these two peaks decreased with the hydrogen pressure, indicating a progressive increase in the concentration of carbon–hydrogen bonds in the near-surface region of a-C:H:Si:O upon increasing the hydrogen partial pressure during tribological tests against steel.

As for the tribological tests performed in oxygen, the difference carbon 1s NEXAFS spectra revealed a positive peak at 285.1 ± 0.2 eV, thus indicating an increase in the fraction of disordered carbon–carbon sp^2 bonds. Additionally, the peak at 288.8 ± 0.2 eV (assigned to the C 1s→σ^* transition for carbon–silicon and carbon–oxygen bonds, as well as to the C 1s→π^* transition for carboxyl groups [5,

106, 136]) decreased in intensity with gas pressure, while a positive peak appeared at 286.0 ± 0.2 eV (assigned to the C 1s→π^* transition for carbonyl groups [5, 106]), thus indicating an increase in the fraction of carbonyl groups in the near-surface region upon sliding in oxygen. The detection of a positive peak at 287.8 ± 0.2 eV, which can be assigned to carbon–hydrogen bonds [5, 80, 99], in the spectra extracted from the tracks produced in oxygen can be due to some residual hydrogen in the vacuum chamber. The relative intensity of the peaks at 286.0 ± 0.2 eV and 287.8 ± 0.2 eV did not significantly change with the oxygen partial pressure, but their absolute intensity decreased.

To gain information about the silicon oxidation state and quantitatively evaluate the composition of the near-surface region, in situ (*post mortem*) XPS analyses were performed (not shown). When sliding in hydrogen, a slight increase in the fraction of silicon atoms in low oxidation states was observed with gas pressure. On the contrary, in the case of the experiments performed in oxygen, the silicon oxidation state first decreased at the lowest gas pressure (10 mbar), and then increased upon increasing the gas pressure. The composition of the near-surface region of the wear tracks also changed with gas pressure: when sliding in hydrogen, no significant variations were observed up to 2000 mbar. At this pressure, the oxygen concentration in the near-surface region of the wear track slightly decreased. In the case of the experiments performed in oxygen, a progressive increase in oxygen and silicon concentration was observed in the wear track upon increasing the gas pressure during tribological experiments.

On the basis of the NEXAFS and XPS results, the authors developed a model for explaining the friction response of a-C:H:Si:O sliding against steel under different environmental conditions. Independent of the gas environment, a stress-induced conversion from sp^3- to sp^2-bonded (disordered) carbon–carbon bonds occurs in the near-surface region. When sliding in hydrogen, the newly-generated, strained sp^2 carbon layer reacts with hydrogen molecules to form a hydrogenated amorphous carbon material. Upon increasing the hydrogen pressure, the fraction of carbon–hydrogen bonds increases. This is proposed to progressively lower the shear strength of the material at the sliding interface, thus resulting in a decrease of friction with hydrogen pressure. When sliding in oxygen, the dissociative reaction of oxygen molecules with strained sp^2 carbon–carbon bonds leads to the formation of carbonyl groups, as indicated by imaging NEXAFS measurements. Additionally, increasing the oxygen pressure during tribological testing leads to an increase in oxygen concentration in the near-surface region of a-C:H:Si:O together with an increase in the fraction of silicon atoms in high oxidation states, which are proposed to increase friction with oxygen gas pressure by progressively increasing the shear strength of the material generated at the sliding interface.

3.5 Conclusions

Fundamental research in tribology hinges on the development of new characterization techniques to better analyze contacting, sliding interfaces. Synchrotron-based near edge X-ray absorption fine structure (NEXAFS) is a powerful surface-analytical tool thanks to its elemental specificity, surface sensitivity, and ability to provide important information about local bonding configurations, such as hybridization, chemical states, and bond orientations. Additionally, spatially-resolved NEXAFS spectroscopic techniques, such as photoemission electron microscopy (PEEM) and magnetically-guided imaging, enables chemical imaging of material surfaces with high spatial resolution. In light of this, NEXAFS spectroscopy has effectively been used for investigating tribological surfaces as well as for accessing spatial inhomogeneities in their surface chemistry and structure.

In the present contribution, the principles of NEXAFS spectroscopy were described together with the methods for acquiring and processing NEXAFS. Examples of the application of NEXAFS spectroscopy for fundamental and applied research in tribology were provided. In particular, case studies concerning the thermally- and tribologically-induced structural evolution of amorphous carbon-based materials were presented.

Acknowledgements This material is based upon work supported by the National Science Foundation under Grant No. DMR-1107642 and by the Agence Nationale de la Recherche under grant No. ANR-11-NS09-01 through the Materials World Network program. F.M. acknowledges support from The University of Texas at Austin Startup Funding, the Marie Curie International Outgoing Fellowship for Career Development within the 7th European Community Framework Program under contract no. PIOF-GA-2012-328776 and the Marie Skłodowska-Curie Individual Fellowship within the European Union's Horizon 2020 Program under contract no. 706289. The authors acknowledge support from the Advanced Storage Technology Consortium ASTC (grant 2011-012). The authors would like to thank Dr. C. Jaye and Dr. D. A. Fischer for the kind assistance with the NEXAFS measurements at the National Synchrotron Light Source. Use of the National Synchrotron Light Source, Brookhaven National Laboratory, was supported by the US Department of Energy, Office of Science, and Office of Basic Energy Sciences, under Contract No. DE-AC02-98CH10886. The authors would like to acknowledge Prof. R. W. Carpick (University of Pennsylvania, Philadelphia, USA) for fruitful discussions, valuable suggestions, and guidance. Finally, the authors would also like to thank Dr. K. D. Koshigan (Ecole Centrale de Lyon, Ecully-Cedex, France) and Dr. J. Fontaine (Ecole Centrale de Lyon, Ecully-Cedex, France) for performing tribological experiments on a-C:H:Si:O.

References

1. W.G. Sawyer, K.J. Wahl, Accessing inaccessible interfaces. In Situ Approaches Mater. Tribol. MRS Bull. **33**, 1145–1150 (2008)
2. W.G. Sawyer, N. Argibay, D.L. Burris, B.A. Krick, Mechanistic studies in friction and wear of bulk materials. Annu. Rev. Mater. Res. **44**(1), 395–427 (2014)
3. C. Donnet, in *Problem-Solving Methods in Tribology with Surface-Specific Techniques*, ed. by J.C. Rivière, S. Myhra. Handbook of Surface and Interface Analysis: Methods and Problem-

Solving, 2nd edn. (CRC Press, Taylor & Francis Group: Boca Raton, FL, 2009), pp. 351–388
4. S. Mobilio, F. Boscherini, C. Meneghini (eds.), *Synchrotron Radiation: Basics, Methods and Applications*. Springer (2015)
5. J. Stöhr, *NEXAFS Spectroscopy*. Springer (1992)
6. A. Balerna, S. Mobilio, in *Introduction to Synchrotron Radiation*, ed. by S. Mobilio, F. Boscherini, C. Meneghini. Synchrotron Radiation: Basics, Methods and Applications (Springer, Berlin Heidelberg, 2015), pp. 3–28
7. B.K. Agarwal, *X-ray Spectroscopy*. Springer (1991)
8. D.C. Koningsberger, R. Prins, *X-ray Absorption: Principles, Applications, Techniques of EXAFS, SEXAFS, and XANES* (Wiley, New York, 1988)
9. P. Fornasini, in *Introduction to X-Ray Absorption Spectroscopy*, ed. by S. Mobilio, F. Boscherini, C. Meneghini. Synchrotron Radiation: Basics, Methods and Applications (Springer, Berlin, Heidelberg, 2015), pp. 181–211
10. G. Bunker, *Introduction to XAFS: A Practical Guide to X-ray Absorption Fine Structure Spectroscopy* (Cambridge University Press, Cambridge, UK; New York, 2010)
11. P.A. Lee, G. Beni, New method for the calculation of atomic phase shifts: application to extended X-ray absorption fine structure (EXAFS) in molecules and crystals. Phys. Rev. B **15**(6), 2862–2883 (1977)
12. S. Gurman, Interpretation of EXAFS data. J. Synchrotron Radiat. **2**(1), 56–63 (1995)
13. B.K. Teo, *EXAFS: Basic Principles and Data Analysis*. Springer (1986)
14. S. Anders, H.A. Padmore, R.M. Duarte, T. Renner, T. Stammler, A. Scholl et al., Photoemission electron microscope for the study of magnetic materials. Rev. Sci. Instrum. **70**(10), 3973–3981 (1999)
15. E. Bauer, M. Mundschau, W. Swiech, W. Telieps, Surface studies by low-energy electron microscopy (LEEM) and conventional UV photoemission electron microscopy (PEEM). Ultramicroscopy **31**(1), 49–57 (1989)
16. W. Engel, M.E. Kordesch, H.H. Rotermund, S. Kubala, A. von Oertzen, A UHV-compatible photoelectron emission microscope for applications in surface science. Ultramicroscopy **36**(1–3), 148–153 (1991)
17. O. Renault, N. Barrett, A. Bailly, L.F. Zagonel, D. Mariolle, J.C. Cezar et al., Energy-filtered XPEEM with NanoESCA using synchrotron and laboratory X-ray sources: principles and first demonstrated results. Surf. Sci. **601**(20), 4727–4732 (2007)
18. A. Konicek, C. Jaye, M. Hamilton, W. Sawyer, D. Fischer, R. Carpick, Near-edge X-ray absorption fine structure imaging of spherical and flat counterfaces of ultrananocrystalline diamond tribological contacts: a correlation of surface chemistry and friction. Tribol. Lett. **44**(1), 99–106 (2011)
19. M. Nicholls, M.N. Najman, Z. Zhang, M. Kasrai, P.R. Norton, P.U.P.A. Gilbert, The contribution of XANES spectroscopy to tribology. Can. J. Chem. **85**(10), 816–830 (2007)
20. A.R. Konicek, D.S. Grierson, P.U.P.A. Gilbert, W.G. Sawyer, A.V. Sumant, R.W. Carpick, Origin of ultralow friction and wear in ultrananocrystalline diamond. Phys. Rev. Lett. **100**(23), 235502 (2008)
21. R. Lindsay, G. Thornton, Structure of atomic and molecular adsorbates on Low-Miller-Index ZnO surfaces using X-ray absorption spectroscopy. Top. Catal. **18**(1–2), 15–19 (2002)
22. M. Bauer, C. Gastl, X-Ray absorption in homogeneous catalysis research: the iron-catalyzed Michael addition reaction by XAS, RIXS and multi-dimensional spectroscopy. Phys. Chem. Chem. Phys. **12**(21), 5575–5584 (2010)
23. D.E. Ramaker, D.C. Koningsberger, The atomic AXAFS and $\Delta\mu$ XANES techniques as applied to heterogeneous catalysis and electrocatalysis. Phys. Chem. Chem. Phys. **12**(21), 5514–5534 (2010)
24. J.B. MacNaughton, L.-A. Naslund, T. Anniyev, H. Ogasawara, A. Nilsson, Peroxide-like intermediate observed at hydrogen rich condition on Pt(111) after interaction with oxygen. Phys. Chem. Chem. Phys. **12**(21), 5712–5716 (2010)
25. T. Anniyev, H. Ogasawara, M.P. Ljungberg, K.T. Wikfeldt, J.B. MacNaughton, L.-A. Naslund et al., Complementarity between high-energy photoelectron and L-edge spectroscopy for

probing the electronic structure of 5d transition metal catalysts. Phys. Chem. Chem. Phys. **12**(21), 5694–5700 (2010)
26. J. Genzer, E.J. Kramer, D.A. Fischer, Accounting for Auger yield energy loss for improved determination of molecular orientation using soft x-ray absorption spectroscopy. J. Appl. Phys. **92**(12), 7070–7079 (2002)
27. M. Gliboff, L. Sang, K.M. Knesting, M.C. Schalnat, A. Mudalige, E.L. Ratcliff et al., Orientation of phenylphosphonic acid self-assembled monolayers on a transparent conductive oxide: a combined NEXAFS, PM-IRRAS, and DFT study. Langmuir **29**(7), 2166–2174 (2013)
28. F. Cheng, L.J. Gamble, D.G. Castner, XPS, TOF-SIMS, NEXAFS, and SPR characterization of nitrilotriacetic acid-terminated self-assembled monolayers for controllable immobilization of proteins. Anal. Chem. **80**(7), 2564–2573 (2008)
29. S. Turgman-Cohen, D.A. Fischer, P.K. Kilpatrick, J. Genzer, Asphaltene adsorption onto self-assembled monolayers of alkyltrichlorosilanes of varying chain length. ACS Appl. Mater. Interfaces **1**(6), 1347–1357 (2009)
30. T. Hemraj-Benny, S. Banerjee, S. Sambasivan, M. Balasubramanian, D.A. Fischer, G. Eres et al., Near-edge X-ray absorption fine structure spectroscopy as a tool for investigating nanomaterials. Small **2**(1), 26–35 (2006)
31. A.D. Winter, E. Larios, F.M. Alamgir, C. Jaye, D. Fischer, E.M. Campo, Near-edge X-ray absorption fine structure studies of electrospun poly(dimethylsiloxane)/poly(methyl methacrylate)/multiwall carbon nanotube composites. Langmuir **29**(51), 15822–15830 (2013)
32. T. Breuer, G. Witte, Diffusion-controlled growth of molecular heterostructures: fabrication of two-, one-, and zero-dimensional C_{60} nanostructures on pentacene substrates. ACS Appl. Mater. Interfaces **5**(19), 9740–9745 (2013)
33. H.-J. Lee, K.-S. Lee, J.-M. Cho, T.-S. Lee, I. Kim, D.S. Jeong et al., Novel aspect in grain size control of nanocrystalline diamond film for thin film waveguide mode resonance sensor application. ACS Appl. Mater. Interfaces **5**(22), 11631–11640 (2013)
34. Y.S. Li, Y. Tang, Q. Yang, J. Maley, R. Sammynaiken, T. Regier et al., Ultrathin W–Al dual interlayer approach to depositing smooth and adherent nanocrystalline diamond films on stainless steel. ACS Appl. Mater. Interfaces **2**(2), 335–338 (2010)
35. K.J. Sankaran, Y.-F. Lin, W.-B. Jian, H.-C. Chen, K. Panda, B. Sundaravel et al., Structural and electrical properties of conducting diamond nanowires. ACS Appl. Mater. Interfaces **5**(4), 1294–1301 (2013)
36. A. Saravanan, B.-R. Huang, K.J. Sankaran, S. Kunuku, C.-L. Dong, K.-C. Leou et al., Bias-enhanced nucleation and growth processes for ultrananocrystalline diamond films in Ar/CH_4 plasma and their enhanced plasma illumination properties. ACS Appl. Mater. Interfaces **6**(13), 10566–10575 (2014)
37. W.S. Yeap, X. Liu, D. Bevk, A. Pasquarelli, L. Lutsen, M. Fahlman et al., Functionalization of boron-doped nanocrystalline diamond with N3 dye molecules. ACS Appl. Mater. Interfaces **6**(13), 10322–10329 (2014)
38. S. Zhong, J.Q. Zhong, H.Y. Mao, R. Wang, Y. Wang, D.C. Qi et al., CVD graphene as interfacial layer to engineer the organic donor-acceptor heterojunction interface properties. ACS Appl. Mater. Interfaces **4**(6), 3134–3140 (2012)
39. J. Kikuma, B.P. Tonner, XANES spectra of a variety of widely used organic polymers at the C K-edge. J. Electron Spectrosc. Relat. Phenom. **82**(1–2), 53–60 (1996)
40. B. Watts, S. Swaraj, D. Nordlund, J. Luning, H. Ade, Calibrated NEXAFS spectra of common conjugated polymers. J. Chem. Phys. **134**(2), 024702 (2011)
41. H. Ade, A.P. Hitchcock, NEXAFS microscopy and resonant scattering: composition and orientation probed in real and reciprocal space. Polymer **49**(3), 643–675 (2008)
42. D. Park, J.A. Finlay, R.J. Ward, C.J. Weinman, S. Krishnan, M. Paik et al., Antimicrobial behavior of semifluorinated-quaternized triblock copolymers against airborne and marine microorganisms. ACS Appl. Mater. Interfaces **2**(3), 703–711 (2010)
43. H.S. Sundaram, Y. Cho, M.D. Dimitriou, J.A. Finlay, G. Cone, S. Williams et al., Fluorinated amphiphilic polymers and their blends for fouling-release applications: the benefits of a triblock copolymer surface. ACS Appl. Mater. Interfaces **3**(9), 3366–3374 (2011)

44. A.F. Tillack, K.M. Noone, B.A. MacLeod, D. Nordlund, K.P. Nagle, J.A. Bradley et al., Surface characterization of polythiophene: fullerene blends on different electrodes using near edge X-ray absorption fine structure. ACS Appl. Mater. Interfaces **3**(3), 726–732 (2011)
45. S.P. Cramer, T.K. Eccles, F.W. Kutzler, K.O. Hodgson, L.E. Mortenson, Molybdenum x-ray absorption edge spectra. The chemical state of molybdenum in nitrogenase. J. Am. Chem. Soc. **98**(5), 1287–1288 (1976)
46. G. Meitzner, G.H. Via, F.W. Lytle, J.H. Sinfelt, Analysis of x-ray absorption edge data on metal catalysts. J. Phys. Chem. **96**(12), 4960–4964 (1992)
47. D.H. Pearson, C.C. Ahn, B. Fultz, White lines and d-electron occupancies for the $3d$ and $4d$ transition metals. Phys. Rev. B **47**(14), 8471–8478 (1993)
48. D. Hübner, F. Holch, M.L.M. Rocco, K.C. Prince, S. Stranges, A. Schöll et al., Isotope effects in high-resolution NEXAFS spectra of naphthalene. Chem. Phys. Lett. **415**(1–3), 188–192 (2005)
49. A.P. Hitchcock, C.E. Brion, K-shell excitation of HF and F_2 studied by electron energy-loss spectroscopy. J. Phys. B: At. Mol. Phys. **14**(22), 4399–4413 (1981)
50. F. Sette, J. Stöhr, A.P. Hitchcock, Determination of intramolecular bond lengths in gas phase molecules from K shell shape resonances. J. Chem. Phys. **81**(11), 4906–4914 (1984)
51. J.S. Stevens, A. Gainar, E. Suljoti, J. Xiao, R. Golnak, E.F. Aziz et al., Chemical speciation and bond lengths of organic solutes by core-level spectroscopy: ph and solvent influence on p-aminobenzoic acid. Chemistry **21**(19), 7256–7263 (2015)
52. J. Stöhr, F. Sette, A.L. Johnson, Near-edge X-ray-absorption fine-structure studies of chemisorbed hydrocarbons: bond lengths with a ruler. Phys. Rev. Lett. **53**(17), 1684–1687 (1984)
53. A. Gainar, J.S. Stevens, C. Jaye, D.A. Fischer, S.L. Schroeder, NEXAFS sensitivity to bond lengths in complex molecular materials: a study of crystalline saccharides. J. Phys. Chem. B **119**(45), 14373–14381 (2015)
54. V.L. Shneerson, D.K. Saldin, W.T. Tysoe, On the dependence with bond lengths of the observed energies of NEXAFS resonances of diatomic molecules. Surf. Sci. **375**(2–3), 340–352 (1997)
55. N. Haack, G. Ceballos, H. Wende, K. Baberschke, D. Arvanitis, A.L. Ankudinov et al., Shape resonances of oriented molecules: ***ab initio*** theory and experiment on hydrocarbon molecules. Phys. Rev. Lett. **84**(4), 614–617 (2000)
56. D. Arvanitis, N. Haack, G. Ceballos, H. Wende, K. Baberschke, A.L. Ankudinov et al., Shape resonances of oriented molecules. J. Electron Spectrosc. Relat. Phenom. **113**(1), 57–65 (2000)
57. B. Kempgens, H.M. Köppe, A. Kivimäki, M. Neeb, K. Maier, U. Hergenhahn et al., On the correct identification of shape resonances in NEXAFS. Surf. Sci. **425**(1), L376–L380 (1999)
58. M.N. Piancastelli, D.W. Lindle, T.A. Ferrett, D.A. Shirley, Reply to the "Comment on 'The relationship between shape resonances and bond lengths"'. J. Chem. Phys. **87**(5), 3255 (1987)
59. M.N. Piancastelli, D.W. Lindle, T.A. Ferrett, D.A. Shirley, The relationship between shape resonances and bond lengths. J. Chem. Phys. **86**(5), 2765–2771 (1987)
60. K. Weiss, P.S. Bagus, C. Wöll, Rydberg transitions in X-ray absorption spectroscopy of alkanes: the importance of matrix effects. J. Chem. Phys. **111**(15), 6834–6845 (1999)
61. S.G. Urquhart, R. Gillies, Rydberg-valence mixing in the carbon 1 s near-edge X-ray absorption fine structure spectra of gaseous alkanes. J. Phys. Chem. A **109**(10), 2151–2159 (2005)
62. D. Briggs, J.T. Grant (eds.), *Surface Analysis by Auger and X-Ray Photoelectron Spectroscopy* (IM Publications, Chichester (UK), 2003)
63. D. Briggs, M.P. Seah (eds.), *Practical Surface Analysis* (Wiley, New York, 1990)
64. T. Maruyama, Y. Ishiguro, S. Nartitsuka, W. Norimatsu, M. Kusunoki, K. Amemiya et al., Near-edge X-ray absorption fine structure study of vertically aligned carbon nanotubes grown by the surface decomposition of SiC. Jpn. J. Appl. Phys. **51**(Copyright (c) 2012 The Japan Society of Applied Physics), 055102
65. T. Maruyama, S. Sakakibara, S. Naritsuka, K. Amemiya, Initial stage of carbon nanotube formation process by surface decomposition of SiC: STM and NEXAFS study. Diam. Relat. Mater. **20**(10), 1325–1328 (2011)

66. G. Margaritondo, in *Characteristics and Properties of Synchrotron Radiation*, ed. by S. Mobilio, F. Boscherini, C. Meneghini. Synchrotron Radiation: Basics, Methods and Applications (Springer, Berlin, Heidelberg, 2015), pp. 29–63
67. G. Aquilanti, L. Vaccari, J.R. Plaisier, A. Goldoni, in *Instrumentation at Synchrotron Radiation Beamlines*, S. Mobilio, F. Boscherini, C. Meneghini. Synchrotron Radiation: Basics, Methods and Applications (Springer, Berlin, Heidelberg, 2015), pp. 65–104
68. D. Attwood, *Soft X-Rays and Extreme Ultraviolet Radiation: Principles and Applications* (2007)
69. A. Scholl, H. Ohldag, F. Nolting, J. Stöhr, H.A. Padmore, X-ray photoemission electron microscopy, a tool for the investigation of complex magnetic structures (invited). Rev. Sci. Instrum. **73**(3), 1362 (2002)
70. T. Schmidt, A. Sala, H. Marchetto, E. Umbach, H.J. Freund, First experimental proof for aberration correction in XPEEM: resolution, transmission enhancement, and limitation by space charge effects. Ultramicroscopy **126**, 23–32 (2013)
71. E. Bauer, *Surface Microscopy with Low Energy Electrons* (Springer, New York, 2014)
72. B.H. Frazer, M. Girasole, L.M. Wiese, T. Franz, G. De Stasio, Spectromicroscope for the Photoelectron imaging of nanostructures with X-rays (SPHINX): performance in biology, medicine and geology. Ultramicroscopy **99**(2–3), 87–94 (2004)
73. J. Feng, A. Scholl, in *Photoemission Electron Microscopy (PEEM)*, ed. by P.W. Hawkes, J.C.H. Spence. Science of Microscopy (Springer, New York, NY, 2007), pp. 657–695
74. G. De Stasio, M. Capozi, G.F. Lorusso, P.A. Baudat, T.C. Droubay, P. Perfetti et al., MEPHISTO: performance tests of a novel synchrotron imaging photoelectron spectromicroscope. Rev. Sci. Instrum. **69**(5), 2062–2066 (1998)
75. G. De Stasio, L. Perfetti, B. Gilbert, O. Fauchoux, M. Capozi, P. Perfetti et al., MEPHISTO spectromicroscope reaches 20 nm lateral resolution. Rev. Sci. Instrum. **70**(3), 1740–1742 (1999)
76. B.P. Tonner, G.R. Harp, Photoelectron microscopy with synchrotron radiation. Rev. Sci. Instrum. **59**(6), 853–858 (1988)
77. A.V. Sumant, D.S. Grierson, J.E. Gerbi, J. Birrell, U.D. Lanke, O. Auciello et al., Toward the ultimate tribological interface: surface chemistry and nanotribology of ultrananocrystalline diamond. Adv. Mater. **17**(8), 1039–1045 (2005)
78. S. Anders, T. Stammler, W. Fong, D.B. Bogy, C.S. Bhatia, J. Stöhr, Investigation of slider surfaces after wear using photoemission electron microscopy. J. Vac. Sci. Technol. A: Vac. Surf. Films **17**(5), 2731–2736 (1999)
79. S. Anders, T. Stammler, W. Fong, C.-Y. Chen, D.B. Bogy, C.S. Bhatia et al., Study of tribochemical processes on hard disks using photoemission electron microscopy. J. Tribol. **121**(4), 961–967 (1999)
80. A.V. Sumant, P.U.P.A. Gilbert, D.S. Grierson, A.R. Konicek, M. Abrecht, J.E. Butler et al., Surface composition, bonding, and morphology in the nucleation and growth of ultra-thin, high quality nanocrystalline diamond films. Diam. Relat. Mater. **16**(4–7), 718–724 (2007)
81. D.S. Grierson, A.V. Sumant, A.R. Konicek, M. Abrecht, J. Birrell, O. Auciello et al., Tribochemistry and material transfer for the ultrananocrystalline diamond-silicon nitride interface revealed by x-ray photoelectron emission spectromicroscopy. J. Vac. Sci. Technol. B: Microelectron. Nanometer Struct. **25**(5), 1700–1705 (2007)
82. A.R. Konicek, D.S. Grierson, A.V. Sumant, T.A. Friedmann, J.P. Sullivan, P.U.P.A. Gilbert et al., Influence of surface passivation on the friction and wear behavior of ultrananocrystalline diamond and tetrahedral amorphous carbon thin films. Phys. Rev. B **85**(15), 155448 (2012)
83. C.M. Schneider, G. Schönhense, Investigating surface magnetism by means of photoexcitation electron emission microscopy. Rep. Prog. Phys. **65**(12), 1785 (2002)
84. M.R. Freeman, B.C. Choi, Advances in magnetic microscopy. Science **294**(5546), 1484 (2001)
85. F. Nolting, A. Scholl, J. Stöhr, J.W. Seo, J. Fompeyrine, H. Siegwart et al., Direct observation of the alignment of ferromagnetic spins by antiferromagnetic spins. Nature **405**(6788), 767–769 (2000)

86. M. Kim, M. Bertram, M. Pollmann, Oertzen Av, A.S. Mikhailov, H.H. Rotermund et al., Controlling chemical turbulence by global delayed feedback: pattern formation in catalytic CO oxidation on Pt(110). Science **292**(5520), 1357 (2001)
87. S. Aggarwal, A.P. Monga, S.R. Perusse, R. Ramesh, V. Ballarotto, E.D. Williams et al., Spontaneous ordering of oxide nanostructures. Science **287**(5461), 2235 (2000)
88. F.-J. Meyer zu Heringdorf, M.C. Reuter, R.M. Tromp, Growth dynamics of pentacene thin films. Nature **412**(6846), 517–520 (2001)
89. C. Morin, H. Ikeura-Sekiguchi, T. Tyliszczak, R. Cornelius, J.L. Brash, A.P. Hitchcock et al., X-ray spectromicroscopy of immiscible polymer blends: polystyrene–poly(methyl methacrylate). J. Electron Spectrosc. Relat. Phenom. **121**(1–3), 203–224 (2001)
90. H. Ade, D.A. Winesett, A.P. Smith, S. Anders, T. Stammler, C. Heske et al., Bulk and surface characterization of a dewetting thin film polymer bilayer. Appl. Phys. Lett. **73**(25), 3775–3777 (1998)
91. G. De Stasio, P. Casalbore, R. Pallini, B. Gilbert, F. Sanità, M.T. Ciotti et al., Gadolinium in human glioblastoma cells for gadolinium neutron capture therapy. Can. Res. **61**(10), 4272 (2001)
92. G. De Stasio, B.H. Frazer, B. Gilbert, K.L. Richter, J.W. Valley, Compensation of charging in X-PEEM: a successful test on mineral inclusions in 4.4 Ga old zircon. Ultramicroscopy **98**(1), 57–62 (2003)
93. M. Labrenz, G.K. Druschel, T. Thomsen-Ebert, B. Gilbert, S.A. Welch, K.M. Kemner et al., Formation of sphalerite (ZnS) deposits in natural biofilms of sulfate-reducing bacteria. Science **290**(5497), 1744 (2000)
94. B. Gilbert, R. Andres, P. Perfetti, G. Margaritondo, G. Rempfer, G. De Stasio, Charging phenomena in PEEM imaging and spectroscopy. Ultramicroscopy **83**(1–2), 129–139 (2000)
95. A. Locatelli, E. Bauer, Recent advances in chemical and magnetic imaging of surfaces and interfaces by XPEEM. J. Phys.: Condens. Matter **20**(9), 093002 (2008)
96. C. Wiemann, M. Patt, I.P. Krug, N.B. Weber, M. Escher, M. Merkel et al., A new nanospectroscopy tool with synchrotron radiation: NanoESCA@Elettra. e-J. Surf. Sci. Nanotechnol. **9**, 395–399 (2011)
97. J.E. Baio, C. Jaye, D.A. Fischer, T. Weidner, Multiplexed orientation and structure analysis by imaging near-edge X-ray absorption fine structure (MOSAIX) for combinatorial surface science. Anal. Chem. **85**(9), 4307–4310 (2013)
98. K.D. Koshigan, F. Mangolini, J.B. McClimon, B. Vacher, S. Bec, R.W. Carpick et al., Understanding the hydrogen and oxygen gas pressure dependence of the tribological properties of silicon oxide–doped hydrogenated amorphous carbon coatings. Carbon **93**, 851–860 (2015)
99. F. Mangolini, J.B. McClimon, F. Rose, R.W. Carpick, Accounting for nanometer-thick adventitious carbon contamination in X-ray absorption spectra of carbon-based materials. Anal. Chem. **86**(24), 12258–12265 (2014)
100. A. Schöll, Y. Zou, T. Schmidt, R. Fink, E. Umbach, Energy calibration and intensity normalization in high-resolution NEXAFS spectroscopy. J. Electron Spectrosc. Relat. Phenom. **129**(1), 1–8 (2003)
101. B. Watts, H. Ade, A simple method for determining linear polarization and energy calibration of focused soft X-ray beams. J. Electron Spectrosc. Relat. Phenom. **162**(2), 49–55 (2008)
102. B. Watts, L. Thomsen, P.C. Dastoor, Methods in carbon K-edge NEXAFS: experiment and analysis. J. Electron Spectrosc. Relat. Phenom. **151**(2), 105–120 (2006)
103. M. Olla, G. Navarra, B. Elsener, A. Rossi, Nondestructive in-depth composition profile of oxy-hydroxide nanolayers on iron surfaces from ARXPS measurement. Surf. Interface Anal. **38**(5), 964–974 (2006)
104. M.A. Scorciapino, G. Navarra, B. Elsener, A. Rossi, Nondestructive surface depth profiles from angle-resolved X-ray photoelectron spectroscopy data using the maximum entropy method. I. A New Protocol. J. Phys. Chem. C **113**(51), 21328–21337 (2009)
105. M. Seah, Ultrathin SiO_2 on Si I. quantifying and removing carbonaceous contamination. J. Vac. Sci. Technol. A **21**(2), 34 (2003)

106. I. Ishii, A.P. Hitchcook, The oscillator strengths for C 1s and O 1s excitation of some saturated and unsaturated organic alcohols, acids and esters. J. Electron Spectrosc. Relat. Phenom. **46**(1), 55–84 (1988)
107. J.F. Morar, F.J. Himpsel, G. Hollinger, G. Hughes, J.L. Jordan, Observation of a C-1s core exciton in diamond. Phys. Rev. Lett. **54**(17), 1960–1963 (1985)
108. S.C. Ray, R.M. Erasmus, H. Tsai, M. nbsp, C. Pao et al., Hydrogenation effects of ultra-nanocrystalline diamond detected by X-ray absorption near edge structure and raman spectroscopy. Jpn. J. Appl. Phys. **51**(Copyright (c) 2012 The Japan Society of Applied Physics), 095201
109. A.V. Sumant, D.S. Grierson, J.E. Gerbi, J.A. Carlisle, O. Auciello, R.W. Carpick, Surface chemistry and bonding configuration of ultrananocrystalline diamond surfaces and their effects on nanotribological properties. Phys. Rev. B **76**(23), 235429 (2007)
110. J. Diaz, S. Anders, X. Zhou, E.J. Moler, S.A. Kellar, Z. Hussain, Combined near edge X-ray absorption fine structure and X-ray photoemission spectroscopies for the study of amorphous carbon thin films. J. Electron Spectrosc. Relat. Phenom. **101–103**, 545–550 (1999)
111. R. Gago, I. Jiménez, J.M. Albella, A. Climent-Font, D. Cáceres, I. Vergara et al., Bonding and hardness in nonhydrogenated carbon films with moderate sp^3 content. J. Appl. Phys. **87**(11), 8174–8180 (2000)
112. D.S. Grierson, A.V. Sumant, A.R. Konicek, T.A. Friedmann, J.P. Sullivan, R.W. Carpick, Thermal stability and rehybridization of carbon bonding in tetrahedral amorphous carbon. J. Appl. Phys. **107**(3), 033523–033525 (2010)
113. C. Lenardi, P. Piseri, V. Briois, C.E. Bottani, A.L. Bassi, P. Milani, Near-edge x-ray absorption fine structure and Raman characterization of amorphous and nanostructured carbon films. J. Appl. Phys. **85**(10), 7159–7167 (1999)
114. S.C. Ray, H.M. Tsai, J.W. Chiou, B. Bose, J.C. Jan, K. Krishna et al., X-ray absorption spectroscopy (XAS) study of dip deposited a-C:H(OH) thin films. J. Phys.: Condens. Matter **16**(32), 5713 (2004)
115. A. Saikubo, N. Yamada, K. Kanda, S. Matsui, T. Suzuki, K. Niihara et al., Comprehensive classification of DLC films formed by various methods using NEXAFS measurement. Diam. Relat. Mater. **17**(7–10), 1743–1745 (2008)
116. D. Wesner, S. Krummacher, R. Carr, T.K. Sham, M. Strongin, W. Eberhardt et al., Synchrotron-radiation studies of the transition of hydrogenated amorphous carbon to graphitic carbon. Phys. Rev. B **28**(4), 2152–2156 (1983)
117. J.G. Buijnsters, R. Gago, A. Redondo-Cubero, I. Jimenez, Hydrogen stability in hydrogenated amorphous carbon films with polymer-like and diamond-like structure. J. Appl. Phys. **112**(9), 093502–093507 (2012)
118. G. Comelli, J. Stöhr, C.J. Robinson, W. Jark, Structural studies of argon-sputtered amorphous carbon films by means of extended x-ray-absorption fine structure. Phys. Rev. B **38**(11), 7511–7519 (1988)
119. J. Díaz, S. Anders, X. Zhou, E.J. Moler, S.A. Kellar, Z. Hussain, Analysis of the π^* and σ^* bands of the x-ray absorption spectrum of amorphous carbon. Phys. Rev. B **64**(12), 125204 (2001)
120. J. Diaz, O.R. Monteiro, Z. Hussain, Structure of amorphous carbon from near-edge and extended x-ray absorption spectroscopy. Phys. Rev. B **76**(9), 094201 (2007)
121. J. Diaz, G. Paolicelli, S. Ferrer, F. Comin, Separation of the sp^3 and sp^2 components in the C 1s photoemission spectra of amorphous carbon films. Phys. Rev. B **54**(11), 8064–8069 (1996)
122. Gago, I. Jiménez, J.M. Albella, Detecting with X-ray absorption spectroscopy the modifications of the bonding structure of graphitic carbon by amorphisation, hydrogenation and nitrogenation. Surf. Sci. 482–485, Part 1: 530–536 (2001)
123. R. Gago, M. Vinnichenko, H.U. Jäger, A.Y. Belov, I. Jiménez, N. Huang et al., Evolution of sp^2 networks with substrate temperature in amorphous carbon films: experiment and theory. Phys. Rev. B **72**(1), 014120 (2005)
124. H.-S. Jung, H.-H. Park, I.R. Mendieta, D.A. Smith, Determination of bonding structure of Si, Ge, and N incorporated amorphous carbon films by near-edge x-ray absorption fine structure and ultraviolet Raman spectroscopy. J. Appl. Phys. **96**(2), 1013–1018 (2004)

125. C. Lenardi, E. Barborini, V. Briois, L. Lucarelli, P. Piseri, P. Milani, NEXAFS characterization of nanostructured carbon thin-films exposed to hydrogen. Diam. Relat. Mater. **10**(3–7), 1195–1200 (2001)
126. C. Lenardi, M. Marino, E. Barborini, P. Piseri, P. Milani, Evaluation of hydrogen chemisorption in nanostructured carbon films by near edge X-ray absorption spectroscopy. Eur. Phys. J. B – Condens. Matter Complex Syst. **46**(3), 441–447 (2005)
127. O.R. Monteiro, M.-P. Delplancke-Ogletree, Investigation of non-hydrogenated DLC: Si prepared by cathodic arc. Surf. Coat. Technol. **163–164**, 144–148 (2003)
128. V. Palshin, R. Tittsworth, C. Fountzoulas, E. Meletis, X-ray absorption spectroscopy, simulation and modeling of Si-DLC films. J. Mater. Sci. **37**(8), 1535–1539 (2002)
129. B.J. Schultz, C.J. Patridge, V. Lee, C. Jaye, P.S. Lysaght, C. Smith et al., Imaging local electronic corrugations and doped regions in graphene. Nat. Commun. **2**, 372 (2011)
130. D. Pacilé, M. Papagno, A.F. Rodríguez, M. Grioni, L. Papagno, Ç.Ö. Girit et al., Near-edge X-ray absorption fine-structure investigation of graphene. Phys. Rev. Lett. **101**(6), 066806 (2008)
131. S. Banerjee, T. Hemraj-Benny, M. Balasubramanian, D.A. Fischer, J.A. Misewich, S.S. Wong, Ozonized single-walled carbon nanotubes investigated using NEXAFS spectroscopy. Chem. Commun. **7**, 772–773 (2004)
132. S. Banerjee, T. Hemraj-Benny, S. Sambasivan, D.A. Fischer, J.A. Misewich, S.S. Wong, Near-edge X-ray absorption fine structure investigations of order in carbon nanotube-based systems†. J. Phys. Chem. B **109**(17), 8489–8495 (2005)
133. D.A. Fischer, K. Efimenko, R.R. Bhat, S. Sambasivan, J. Genzer, Mapping surface chemistry and molecular orientation with combinatorial near-edge X-ray absorption fine structure spectroscopy. Macromol. Rapid Commun. **25**(1), 141–149 (2004)
134. F. Mangolini, F. Rose, J. Hilbert, R.W. Carpick, Thermally induced evolution of hydrogenated amorphous carbon. Appl. Phys. Lett. **103**(16), 161605 (2013)
135. P.E. Batson, Carbon 1s near-edge-absorption fine structure in graphite. Phys. Rev. B **48**(4), 2608–2610 (1993)
136. A. Wada, T. Ogaki, M. Niibe, M. Tagawa, H. Saitoh, K. Kanda et al., Local structural analysis of a-SiCx: H films formed by decomposition of tetramethylsilane in microwave discharge flow of Ar. Diam. Relat. Mater. **20**(3), 364–367 (2011)
137. F. Mangolini, J.B. McClimon, R.W. Carpick, Quantitative evaluation of the carbon hybridization state by near edge X-ray absorption fine structure spectroscopy. Anal. Chem. **88**(5), 2817–2824 (2016)
138. J. Robertson, Diamond-like amorphous carbon. Mater. Sci. Eng.: R: Rep. **37**(4–6), 129–281 (2002)
139. S. Osswald, G. Yushin, V. Mochalin, S.O. Kucheyev, Y. Gogotsi, Control of sp^2/sp^3 carbon ratio and surface chemistry of nanodiamond powders by selective oxidation in air. J. Am. Chem. Soc. **128**(35), 11635–11642 (2006)
140. F.L. Coffman, R. Cao, P.A. Pianetta, S. Kapoor, M. Kelly, L.J. Terminello, Near-edge X-ray absorption of carbon materials for determining bond hybridization in mixed sp^2/sp^3 bonded materials. Appl. Phys. Lett. **69**(4), 568–570 (1996)
141. F. Mangolini, J. Hilbert, J.B. McClimon, J.R. Lukes, R.W. Carpick, Thermally induced structural evolution of silicon- and oxygen-containing hydrogenated amorphous carbon: a combined spectroscopic and molecular dynamics simulation investigation. Langmuir **34**(9), 2989–2995 (2018)
142. F. Mangolini, B.A. Krick, T.D.B. Jacobs, S.R. Khanal, F. Streller, J.B. McClimon et al., Effect of silicon and oxygen dopants on the stability of hydrogenated amorphous carbon under harsh environmental conditions. Carbon **130**, 127–136 (2018)
143. A.C. Ferrari, J. Robertson, Raman spectroscopy of amorphous, nanostructured, diamond-like carbon, and nanodiamond. Philos. Trans. Royal Soc. Lond. Ser. A: Math. Phys. Eng. Sci. **2004**(362), 2477–2512 (1824)
144. F. Rose, N. Wang, R. Smith, Q.-F. Xiao, H. Inaba, T. Matsumura et al., Complete characterization by Raman spectroscopy of the structural properties of thin hydrogenated diamond-like carbon films exposed to rapid thermal annealing. J. Appl. Phys. **116**(12), 123516 (2014)

145. A.C. Ferrari, B. Kleinsorge, G. Adamopoulos, J. Robertson, W.I. Milne, V. Stolojan et al., Determination of bonding in amorphous carbons by electron energy loss spectroscopy, Raman scattering and X-ray reflectivity. J. Non-Cryst. Solids **266–269 Part 2**, 765–768 (2000)
146. J. Filik, P.W. May, S.R.J. Pearce, R.K. Wild, K.R. Hallam, XPS and laser Raman analysis of hydrogenated amorphous carbon films. Diam. Relat. Mater. **12**(3–7), 974–978 (2003)
147. S. Kaciulis, Spectroscopy of carbon: from diamond to nitride films. Surf. Interface Anal. **44**(8), 1155–1161 (2012)
148. A. Mezzi, S. Kaciulis, Surface investigation of carbon films: from diamond to graphite. Surf. Interface Anal. **42**(6–7), 1082–1084 (2010)
149. J.C. Lascovich, V. Rosato, Analysis of the electronic structure of hydrogenated amorphous carbon via Auger spectroscopy. Appl. Surf. Sci. **152**(1–2), 10–18 (1999)
150. S. Kaciulis, A. Mezzi, P. Calvani, D.M. Trucchi, Electron spectroscopy of the main allotropes of carbon. Surf. Interface Anal. **46**(10–11), 966–969 (2014)
151. B. Lesiak, J. Zemek, P. Jiricek, L. Stobinski, A. Jóźwik, The line shape analysis of electron spectroscopy spectra by the artificial intelligence methods for identification of C sp^2/sp^3 bonds. Phys. Status Solidi (b) **247**(11–12), 2838–2842 (2010)
152. G. Speranza, N. Laidani, Measurement of the relative abundance of sp^2 and sp^3 hybridised atoms in carbon based materials by XPS: a critical approach. Part I. Diam. Relat. Mater. **13**(3), 445–450 (2004)
153. J. Zemek, J. Zalman, A. Luches, XAES and XPS study of amorphous carbon nitride layers. Appl. Surf. Sci. **133**(1–2), 27–32 (1998)
154. Y. Mizokawa, T. Miyasato, S. Nakamura, K.M. Geib, C.W. Wilmsen, Comparison of the CKLL first-derivative auger spectra from XPS and AES using diamond, graphite SiC and diamond-like-carbon films. Surf. Sci. **182**(3), 431–438 (1987)
155. Y. Mizokawa, T. Miyasato, S. Nakamura, K.M. Geib, C.W. Wilmsen, The C KLL first-derivative x-ray photoelectron spectroscopy spectra as a fingerprint of the carbon state and the characterization of diamond like carbon films. J. Vac. Sci. Technol. A: Vac. Surf. Films **5**(5), 2809–2813 (1987)
156. S.D. Berger, D.R. McKenzie, P.J. Martin, EELS analysis of vacuum arc-deposited diamond-like films. Philos. Mag. Lett. **57**(6), 285–290 (1988)
157. J. Kulik, G.D. Lempert, E. Grossman, D. Marton, J.W. Rabalais, Y. Lifshitz, sp^3 content of mass-selected ion-beam-deposited carbon films determined by inelastic and elastic electron scattering. Phys. Rev. B **52**(22), 15812–15822 (1995)
158. Y. Wang, H. Chen, R.W. Hoffman, J.C. Angus, Structural analysis of hydrogenated diamond-like carbon films from electron energy loss spectroscopy. J. Mater. Res. **5**(11), 2378–2386 (1990)
159. M.J. Paterson, An investigation of the role of hydrogen in ion beam deposited a-C:H. Diam. Relat. Mater. **7**(6), 908–915 (1998)
160. C. Donnet, J. Fontaine, F. Lefebvre, A. Grill, V. Patel, C. Jahnes, Solid state ^{13}C and ^{1}H nuclear magnetic resonance investigations of hydrogenated amorphous carbon. J. Appl. Phys. **85**(6), 3264–3270 (1999)
161. J. Peng, A. Sergiienko, F. Mangolini, P.E. Stallworth, S. Greenbaum, R.W. Carpick, Solid state magnetic resonance investigation of the thermally-induced structural evolution of silicon oxide-doped hydrogenated amorphous carbon. Carbon **105**, 163–175 (2016)
162. G. Kovach, A. Karacs, G. Radnoczi, H. Csorbai, L. Guczi, M. Veres et al., Modified π-states in ion-irradiated carbon. Appl. Surf. Sci. **254**(9), 2790–2796 (2008)
163. J.A. Leiro, M.H. Heinonen, T. Laiho, I.G. Batirev, Core-level XPS spectra of fullerene, highly oriented pyrolitic graphite, and glassy carbon. J. Electron Spectrosc. Relat. Phenom. **128**(2–3), 205–213 (2003)
164. R.F. Egerton, An Introduction to EELS. *Electron Energy-Loss Spectroscopy in the Electron Microscope*. Springer, US, pp. 1–28 (2011)
165. P.J. Fallon, V.S. Veerasamy, C.A. Davis, J. Robertson, G.A.J. Amaratunga, W.I. Milne et al., Properties of filtered-ion-beam-deposited diamondlike carbon as a function of ion energy. Phys. Rev. B **48**(7), 4777–4782 (1993)

166. D.G. McCulloch, D.R. McKenzie, C.M. Goringe, Ab initio simulations of the structure of amorphous carbon. Phys. Rev. B **61**(3), 2349–2355 (2000)
167. K.E. Sohn, M.D. Dimitriou, J. Genzer, D.A. Fischer, C.J. Hawker, E.J. Kramer, Determination of the electron escape depth for NEXAFS spectroscopy. Langmuir **25**(11), 6341–6348 (2009)
168. S. Anders, J. Diaz, J.W. Ager Iii, R.Y. Lo, D.B. Bogy, Thermal stability of amorphous hard carbon films produced by cathodic arc deposition. Appl. Phys. Lett. **71**(23), 3367–3369 (1997)
169. S. Takabayashi, K. Okamoto, H. Sakaue, T. Takahagi, K. Shimada, T. Nakatani, Annealing effect on the chemical structure of diamondlike carbon. J. Appl. Phys. **104**(4), 043512–043516 (2008)
170. N. Wang, K. Komvopoulos, F. Rose, B. Marchon, Structural stability of hydrogenated amorphous carbon overcoats used in heat-assisted magnetic recording investigated by rapid thermal annealing. J. Appl. Phys. **113**(8), 083517–083517 (2013)
171. J.P. Sullivan, T. Friedmann, A. Baca, Stress relaxation and thermal evolution of film properties in amorphous carbon. J. Electron. Mater. **26**(9), 1021–1029 (1997)
172. A.C. Ferrari, S.E. Rodil, J. Robertson, W.I. Milne, Is stress necessary to stabilise sp^3 bonding in diamond-like carbon? Diam. Relat. Mater. **11**(3–6), 994–999 (2002)
173. C.M. Mate, *Tribology on the Small Scale—A Bottom Up Approach to Friction, Lubrication, and Wear* (Oxford University Press, Oxford, 2007)
174. M.H. Kryder, E.C. Gage, T.W. McDaniel, W.A. Challener, R.E. Rottmayer, J. Ganping et al., Heat assisted magnetic recording. Proc. IEEE **96**(11), 1810–1835 (2008)
175. J. Hilbert, F. Mangolini, J.B. McClimon, J.R. Lukes, R.W. Carpick, Si doping enhances the thermal stability of diamond-like carbon through reductions in carbon-carbon bond length disorder. Carbon **131**, 72–78 (2018)
176. C. Venkatraman, D. Kester, A. Goel, D. Bray, in *Diamond-Like Nanocomposite Coatings—A New Class of Materials*, ed. by T.S. Sudarshan, W. Reitz, J.J. Stiglich. Surface Modification Technologies IX (The Minerals, Metals & Materials Society, 1996)
177. T.W. Scharf, J.A. Ohlhausen, D.R. Tallant, S.V. Prasad, Mechanisms of friction in diamondlike nanocomposite coatings. J. Appl. Phys. **101**(6), 063521–063511 (2007)
178. W.J. Yang, Y.-H. Choa, T. Sekino, K.B. Shim, K. Niihara, K.H. Auh, Thermal stability evaluation of diamond-like nanocomposite coatings. Thin Solid Films **434**(1–2), 49–54 (2003)
179. C. Jongwannasiri, X. Li, S. Watanabe, Improvement of thermal stability and tribological performance of diamond-like carbon composite thin films. Mater. Sci. Appl. **4**, 630–636 (2013)
180. V.F. Dorfman, Diamond-like nanocomposites (DLN). Thin Solid Films **212**(1–2), 267–273 (1992)
181. D. Neerinck, P. Persoone, M. Sercu, A. Goel, D. Kester, D. Bray, Diamond-like nanocomposite coatings (a-C:H/a-Si:O) for tribological applications. Diam. Relat. Mater. **7**(2–5), 468–471 (1998)
182. F. Demichelis, C.F. Pirri, A. Tagliaferro, Influence of silicon on the physical properties of diamond-like films. Mater. Sci. Eng. B **11**(1–4), 313–316 (1992)
183. R. Hatada, S. Flege, K. Baba, W. Ensinger, H.J. Kleebe, I. Sethmann et al., Temperature dependent properties of silicon containing diamondlike carbon films prepared by plasma source ion implantation. J. Appl. Phys. **107**(8), 083307–083306 (2010)
184. G.J. Wan, P. Yang, R.K.Y. Fu, Y.F. Mei, T. Qiu, S.C.H. Kwok et al., Characteristics and surface energy of silicon-doped diamond-like carbon films fabricated by plasma immersion ion implantation and deposition. Diam. Relat. Mater. **15**(9), 1276–1281 (2006)
185. W.-J. Wu, M.-H. Hon, Thermal stability of diamond-like carbon films with added silicon. Surf. Coat. Technol. **111**(2–3), 134–140 (1999)
186. C. Donnet, A. Erdemir (eds.), *Tribology of Diamond-Like Carbon Films* (Springer, New York, 2008)
187. A. Erdemir, C. Donnet, Tribology of Diamond, Diamond-Like Carbon, and Related Films, in *Modern Tribology Handbook*, ed. by B. Bhushan (CRC Press, Boca Raton, 2001)
188. J. Andersson, R.A. Erck, A. Erdemir, Frictional behavior of diamondlike carbon films in vacuum and under varying water vapor pressure. Surf. Coat. Technol. **163–164**, 535–540 (2003)

189. Y. Tzeng, Very low friction for diamond sliding on diamond in water. Appl. Phys. Lett. **63**(26), 3586–3588 (1993)
190. M.N. Gardos, Surface chemistry-controlled tribological behavior of silicon and diamond. Tribol. Lett. **2**(2), 173–187 (1996)
191. M.N. Gardos, Tribological fundamentals of polycrystalline diamond films. Surf. Coat. Technol. **113**(3), 183–200 (1999)
192. J.A. Harrison, D.W. Brenner, Simulated tribochemistry: an atomic-scale view of the wear of diamond. J. Am. Chem. Soc. **116**(23), 10399–10402 (1994)
193. J.A. Harrison, C.T. White, R.J. Colton, D.W. Brenner, Investigation of the atomic-scale friction and energy dissipation in diamond using molecular dynamics. Thin Solid Films **260**(2), 205–211 (1995)
194. M.D. Perry, J.A. Harrison, Universal aspects of the atomic-scale friction of diamond surfaces. J. Phys. Chem. **99**(24), 9960–9965 (1995)
195. G. Zilibotti, M.C. Righi, M. Ferrario, Ab initio study on the surface chemistry and nanotribological properties of passivated diamond surfaces. Phys. Rev. B **79**(7), 075420 (2009)
196. O. Manelli, S. Corni, M.C. Righi, Water adsorption on native and hydrogenated diamond (001) surfaces. J. Phys. Chem. C **114**(15), 7045–7053 (2010)
197. Y. Qi, E. Konca, A.T. Alpas, Atmospheric effects on the adhesion and friction between non-hydrogenated diamond-like carbon (DLC) coating and aluminum—a first principles investigation. Surf. Sci. **600**(15), 2955–2965 (2006)
198. R.J.A. van den Oetelaar, C.F.J. Flipse, Atomic-scale friction on diamond (111) studied by ultra-high vacuum atomic force microscopy. Surf. Sci. **384**(1–3), L828–L835 (1997)
199. M.N. Gardos, B.L. Soriano, The effect of environment on the tribological properties of polycrystalline diamond films. J. Mater. Res. **5**, 2599–2609 (1990)
200. A. Erdemir, G.R. Fenske, A.R. Krauss, D.M. Gruen, T. McCauley, R.T. Csencsits, Tribological properties of nanocrystalline diamond films. Surf. Coat. Technol. **120–121**, 565–572 (1999)
201. S.E. Grillo, J.E. Field, The friction of CVD diamond at high Hertzian stresses: the effect of load, environment and sliding velocity. J. Phys. D Appl. Phys. **33**(6), 595 (2000)

Chapter 4
Tribochemistry of n-Alkane Thiols Examined by Gas-Phase Lubrication (GPL)

Maria-Isabel De Barros Bouchet and Jean-Michel Martin

Abstract Gas Phase Lubrication (GPL) is an approach allowing to simulate experimentally the reaction mechanism of lubrication additives on a surface thanks to a specific device, an environmentally controlled analytical tribometer (ECAT). Indeed it is shown that GPL is able to simulate the boundary lubrication regime by bringing directly the molecules on the surface in contact. The presence of in situ chemical analysis clearly gives some insights on the chemical dissociation pathway and reaction of additive molecules on the surface under shearing leading to the tribofilm formation. This is a huge advantage compared to standard liquid phase lubrication coupled with ex situ post mortem surface analyses. GPL approach also permits to investigate the critical role of the freshly nascent surfaces obtained by friction in UHV in the decomposition mechanism of the additive molecules. This brings new knowledge of the complex phenomena that can occur in lubricated friction experiments. Moreover, such experimental approach appears particularly suited for combination with numerical simulations, offering unique capabilities for investigating boundary lubricated interfaces. In this chapter, GPL is applied to investigate the tribochemical reactions of sulfur-based compounds, N-alkanethiols. Undoubtedly, the primordial role of nascent tribo-stressed metallic surfaces obtained after shearing in the dissociation of lubricant additives under boundary lubrication conditions is highlighted. The investigation into the reactivity of thiol molecules showed that they suffer dissociative adsorption on the asperities of the tribo-stressed nascent metallic surface, rapidly forming a thin protective film composed of iron sulfide reducing the friction.

M.-I. De Barros Bouchet (✉) · J.-M. Martin
Laboratoire de Tribologie et Dynamique des Systèmes, Ecole Centrale de Lyon, CNRS, UMR5513, Université de Lyon I, Lyon 69134, Ecully, France
e-mail: maria-isabel.de-barros@ec-lyon.fr

J.-M. Martin
e-mail: jean-michel.martin@ec-lyon.fr

M. Dienwiebel and M.-I. De Barros Bouchet (eds.), *Advanced Analytical Methods in Tribology*, Microtechnology and MEMS,
https://doi.org/10.1007/978-3-319-99897-8_4

4.1 Introduction

Currently, the development of new efficient and sustainable lubrication technologies, such as low SAPS (low sulfated ash, phosphorus and sulfur) lubricants or lubricants containing biodegradable additives, is crucial to meet both environmental and economic concerns of the 21st century [1]. As formulated lubricants are highly complex fluids due to the numerous additives mixed with the base oils, the reaction mechanisms of each, as well as possible synergistic or antagonist interactions between them, need to be well understood. Given the complexity of the reaction mechanisms at the origin of the properties of the lubricating additives and the physico-chemical phenomena that may impact these reactions, it is necessary for researchers to control and simplify tribological systems. Most studies in the literature were conducted with traditional tribometers to understand the action mode of the additives by controlling as many parameters as possible and using surface analysis techniques. However, these studies at atmospheric pressure still have some restrictions, such as the lack of control of the surface chemistry and post-mortem chemical analyses. To avoid these problems and since chemical analyses are crucial for understanding physico-chemical phenomena, studies combining tribometry and in situ surface analyses were carried out using ultra-high vacuum (UHV) chambers. Initially, such tribometers were used to investigate the friction of clean materials without contaminants [2], with the advantage that the surface of these materials can be modified in a controlled way by introducing molecules in the gaseous phase [3]. The coupling of tribological experiments with in situ analytical techniques (XPS, AES, SIMS, etc.) allowed the formation of well-defined surfaces with controlled molecular top layers and the determination of their absorption and decomposition properties [4, 5]. These tribometers have also been used to develop and test solid lubricants for aerospace applications [6, 7]. These new in situ approaches, aimed at further simplifying tribological systems by controlling the atmosphere, surfaces and products formed, appear to be perfectly suited for investigating the action mechanisms of lubricating additives. As in all simplified systems, the connection between these systems and reality remains to be established. In particular, it is necessary to study the link between liquid-phase lubrication and gas-phase lubrication. Many experiments, including those of Lauer [8], illustrate the possibility of lubricating a surface with gaseous molecules and suggest that the two above means of lubrication are not so different [9]. This point is particularly justified under boundary lubrication conditions, where friction mainly results from the additives adsorbed on the asperities of the sliding surfaces where there is no longer a liquid lubricant film and thus no effect related to the rheology of the fluid. Therefore, how the additives are brought to the sliding surfaces does not matter. The crucial point for boundary lubrication to be effective is their presence on the surface!

Consequently, we examined a new approach based on gas-phase lubrication (GPL) combined with in situ surface analyses to experimentally simulate reactions of the additives under boundary lubrication conditions. In the following sections of this chapter, the GPL approach is detailed, and its application in the study of the adsorption and tribochemistry of thiol compounds on steel-based substrates is presented. A specific experimental methodology allowed by GPL to separate the different sources of activation of the tribochemical reaction described in the literature (i.e., thermal effects, pressure effects, nascent surfaces, molecular shearing, exo-electron emission, etc.) is also presented.

4.2 Experimental

4.2.1 Gas-Phase Lubrication (GPL) Coupled with In Situ XPS/AES Surface Analyses

Under boundary lubrication conditions, surfaces asperities bear all the load of the contact. In these specific conditions, the contact can be considered 'dry' because there is practically no base oil fluid film at the contact. The friction is concentrated at the lubricant additives adsorbed on the surface of each body in contact. In GPL experiments, low-molecular-weight molecules, which are easily vaporized, are needed. First, polished surfaces are heated to approximately 100 °C under UHV to remove physisorbed molecules. Second, gaseous molecules are introduced and chemically or physically adsorbed on the surfaces. These model molecules must display the same chemical functionalities, i.e., long hydrocarbon chains containing additives, in order to simulate the tribochemical reactions that occur under boundary lubrication conditions. Third, friction experiments can be carried out under a controlled partial pressure of gas. In this way, the conditions are very similar to those existing under boundary lubrication conditions because the friction occurs on the molecules adsorbed on the contacting surfaces.

A device dedicated to GPL, coupling an environmentally controlled tribometer and XPS/AES surface analysis system, named ECAT (Environmentally Controlled Analytical Tribometer) was built in the LTDS laboratory (Fig. 4.1) [10]. This device is composed of three independent parts: a UHV chamber for tribometry, a surface analysis system and a UHV chamber for the preparation and cleaning of samples. ECAT allows friction experiments to be carried out under UHV or controlled gas pressures. Pure or mixed gases can be introduced, and the inlet gas pressure can be monitored from 10^{-9} to 2000 hPa with a membrane vacuum gauge. The purity and proportion of inlet gases can be controlled by a residual gas analyser. This analyser also permits the nature of molecules generated during the friction experiments to be analysed, providing complementary, interesting information about the tribochemical mechanism [11]. Friction experiments under a controlled environment were conducted with a reciprocating pin-on-flat tribometer over a wide range of contact pressures and sliding speeds. The flat temperature can be varied from liquid nitrogen temperature (approximately −100 °C) to 800 °C. For the experiments discussed in the following paragraphs, a normal load of 3.5 N was applied, corresponding to a maximum Hertzian contact pressure of 0.52 GPa and a track width of 114 μm. The track length was adjusted to 2 mm, and the sliding speed of the reciprocating motion

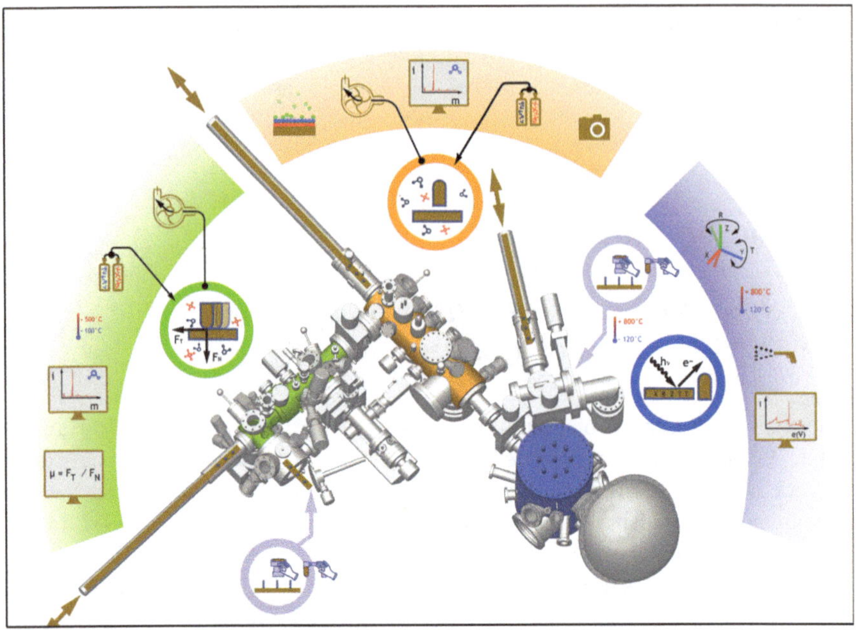

Fig. 4.1 ECAT device equipped with a tribometry chamber (green), XPS/AES surface analysis system (blue) and UHV preparation chamber (orange) [10]

was 0.5 mm/s. All experiments were performed at ambient temperature (flat temperature of approximately 25 °C). This tribometer can measure friction coefficients with both very low (milli range) and very high values (above 1). The applied normal load was adjusted automatically as the partial pressure changed.

In addition, this device makes it possible to carry out in situ XPS and Auger analyses. For these analyses, the tribometry chamber is connected to a surface analysis system by an intermediate UHV chamber dedicated to the cleaning (heat treatment and surface abrasion) and preparation (thermal evaporation) of samples (Fig. 4.1). This intermediate chamber prevents the worn surfaces from being contaminated by the gases used in the GPL experiments and allows transfer of the sample without exposure to air. XPS analyses were carried out with a focused (250 μm) and monochromatic Al X-ray source, and AES analyses were performed with an electron gun FEG1000 (spot size of approximately 0.1 μm). The photoelectrons emitted by the surface during XPS/AES analyses were detected by a VG 220i spectrometer. First, an XPS survey spectrum covering a range of 1200 eV was obtained to identify all elements and the presence of contaminants. Afterwards, a more detailed scanning of the individual elements of interest over a smaller range of 15–20 eV was undertaken to establish the different chemical states of the species and to perform a quantitative analysis. XPS photopeaks were fitted with a Shirley background, and atomic quantifications were made with the Scofield table of sensitivity factors. All peaks were calibrated against the C 1s level of adventitious carbon at a binding energy

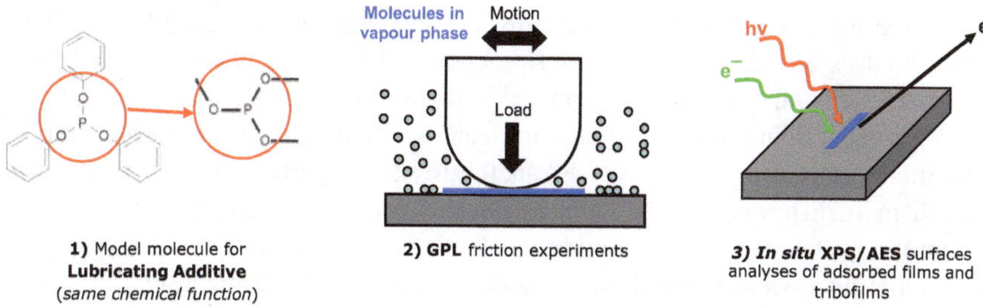

Fig. 4.2 Schematic of the experimental simulation of tribochemical reactions using GPL/in situ XPS/AES analyses

(BE) of 284.8 eV. Argon ions from a VG EXO5 ion gun were used for etching to clean the surfaces before the friction experiments and to measure XPS/AES depth profiles. In comparison with XPS, AES analysis presents a strong spatial resolution of approximately 0.1 μm (versus 100 μm for XPS), making possible accurate measurements inside the tribofilm.

Overall, GPL coupled with in situ XPS/AES surface analyses appears to be an efficient way to experimentally simulate real, complex tribochemical reactions. The operation of this approach, schematically displayed in Fig. 4.2, shows two main advantages:

- *No cleaning of the rubbed surfaces is required, avoiding any pollution and/or deterioration of the surfaces by the cleaning solvents in the case of liquid-phase lubrication;*
- *Perfect control of the chemical nature of the surfaces before and after the friction experiment (due to surface etching in the preparation chamber before the experiment) can be obtained without any issues of contamination by the surrounding air.*

This can bring new knowledge of the complex phenomena that can occur in lubricated friction experiments. Moreover, such approach appears particularly suited for combination with numerical simulations, offering unique capabilities for investigating boundary lubricated interfaces.

4.2.2 Materials

The tribochemical reactions of different kinds of lubricating additives currently used in industry were previously investigated by the authors using GPL experimental simulations. For example, phosphite, phosphate and borate additives were simulated [12–17]. The friction results obtained by GPL were corroborated by the results obtained in liquid-phase lubrication. New insights into the chemical composition of the tribofilms generated in the presence of these additives as well as more detailed information about their reaction mechanisms were obtained using GPL [13]. This chapter focuses on the investigation of tribochemical reactions of sulfur-based compounds. These compounds have been used in many applications, such as biosensors [18, 19], corrosion inhibitors [20] and lubricant additives [21–26]. Different studies in the literature have focused on N-alkanethiols and, in particular, their adsorption mechanism on different metallic surfaces, such as gold, nickel, silver and iron. However, until now, the detailed adsorption mechanism is still not fully understood [27]. The authors have shown that these compounds can adsorb on metallic surfaces to form stable self-assembled monolayers (SAMs). The structure and properties of these SAMs have been studied by different surface techniques, such as spectroscopy, microscopy and electrochemistry, to understand their formation mechanism. Self-chemisorption seems to be a particularly attractive approach in this field. The growth mechanism of SAMs on metallic surfaces strongly depends on the molecule-molecule and molecule-substrate interactions [28, 29]. It has been shown that the lateral carbon chain and deposition temperature play important roles in both the formation and the stabilization of the SAMs [30–32]. The –SH functional group promotes the adhesion of organic layers to the metal surface. The authors, by studying the mechanism of the decomposition of thiol compounds on different metallic surfaces, showed that S–H bond scission occurs upon chemisorption at low temperature, forming adsorbed alkylthiolate (R–S–Fe) [21, 33–37]. It appears that the nature of the bond formed between the adsorbate and substrate has an influence on the structure of the monolayer films formed by these species [21]. At high temperature, the S–C bond in the adsorbed thiolate breaks, and atomic sulfur remains on the surface. This thermal decomposition at high temperature results in the formation of gas products and the adsorption of alkyl groups on the metallic surfaces [35–37]. Authors have also studied the effect of an oxide layer using different oxidized metallic surfaces. They found that a thin oxide layer is sufficient to oxidize thiol to sulfonate on iron oxide surfaces [38, 39]. Different components, such as sulfinates and sulfonates, were observed to have similar behaviour on aluminium oxide [40] and titanium oxide surfaces [41]. In contrast, on clean iron substrates without an oxide layer, a chemisorptive S–Fe bond is observed between the thiol group and the substrate [20]. The adsorption kinetics of thiols on oxidized metals are quite different from those on noble metals [39]. However, the formation and properties of SAMs on gold substrates, in particular, those derived from alkanethiols, are among the most widely studied because such SAMs are easily prepared and chemically stable and present well-ordered monolayer structures [28]. Much less is known regarding the reaction between alkanethiols and steel substrates. In a previous work [42], the friction and wear behaviour of thiol and polysulfide compounds were compared using GPL. This study showed that the good friction modification and anti-wear performances of thiols under moderate pressure conditions are related to the formation of an iron sulfide-rich tribofilm. However, the key mechanism controlling its formation could not be elucidated accurately because different activation sources can operate simultaneously. Moreover, the microstructure of such a tribofilm, in particular, its stoichiometry, structure, thickness and interface with the substrate, is not known. In this paper, we investigate the adsorption and tribochemical reaction mechanisms of 1-hexanethiol on two different surfaces: (i) steel with native iron oxide layers and (ii) an ion-etched steel surface, revealing metallic iron. Moreover, we design a model GPL experiment to separate the different potential activation sources of tribofilm formation described in the literature (i.e., thermal effects, nascent surfaces, molecular shearing and exo-electron emission). For our study, we choose 1-hexanethiol (C_6H_{13}-SH), supplied by ARKEMA, which is liquid at room temperature. 1-Hexanethiol can be easily evaporated and introduced into the UHV chamber dedicated to tribometry at a limited pressure at room temperature, corresponding to the saturation vapor pressure (SVP) (Fig. 4.2). This compound was further purified by freeze–pumping–thaw cycles before being introduced into the tribometry chamber. Iron sulfide (FeS) was used as a reference for the AES surface analysis.

The friction materials, pin and flat, were made of AISI 52100 steel (composition 96.9 Fe–1.04 C–1.45 Cr–0.35 Mn–0.27 Si, in wt%), and both surfaces were polished with a 1 μm diamond paste solution to obtain a surface roughness average (Ra) of approximately 20–25 nm and then cleaned with n-heptane and 2-propanol in ultrasonic baths. The use of steel is preferred to crystalline pure iron material because the hardness of pure iron is too low to easily reach the required contact pressure in real applications. The pin has a hemispherical radius of 8 mm. The 3–4 nm-thick native oxide/hydroxide layer present at the top surface of the steel samples can be removed using a VG EXO5 Ar ion gun (ions of 3 keV) to reveal a metallic pure iron surface. Before starting the GPL experiments, the etched sample was first analysed by XPS to check its iron purity, i.e., the absence of any adventitious carbon or residual oxide layers.

4.3 Adsorption of Hexanethiol on Steel

4.3.1 Procedure for Adsorption Study

1-Hexanethiol was deposited as droplets on the steel substrate and kept in contact for 2 h. The excess liquid was removed with paper so that a very thin liquid layer remained on the surface. The sample was then introduced into the XPS analysis chamber. After XPS analysis, the sample was removed from the analysis chamber and cleaned with heptane for 30 min in an ultrasonic bath. After ultrasonic cleaning, this sample was analysed again by XPS to determine whether 1-hexanethiol chemisorption occurred on the substrate. To study adsorption on a pure iron surface, gaseous 1-hexanethiol was directly introduced in the tribometry chamber at a vapor pressure of 0.1 hPa to an ion-etched steel surface. After the 1 h adsorption experiment, the flat was transferred without exposure to air to the analytical chamber to be analysed by XPS (see Fig. 4.2).

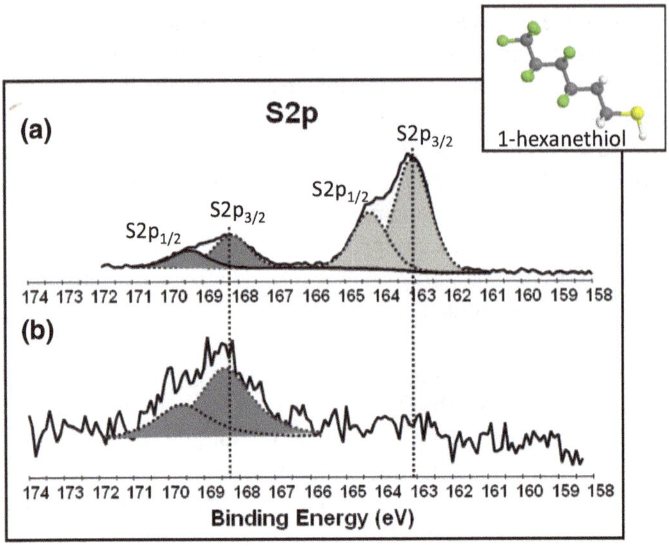

Fig. 4.3 S2p XPS spectra of liquid 1-hexanethiol deposited on the native oxide layer of steel **a** before and **b** after ultrasonic cleaning with heptane

4.3.2 Adsorption of Hexanethiol on the Native Oxide Layer of Steel

At ambient temperature, it is well known that bearing steel is covered by a 2–3 nm-thick native oxide/hydroxide layer, as revealed by XPS [43]. A survey spectrum, covering a range of 1200 eV, was first recorded to identify the elements present after the adsorption experiment in the presence of 1-hexanethiol. The results revealed the presence of sulfur, carbon, oxygen and iron on the surface. Figure 4.3 compares the S 2p XPS spectra of 1-hexanethiol covering the native oxide layer before and after ultrasonic cleaning with heptane.

Notably, the sulfur S2p photopeaks recorded before the cleaning step are the result of two contributions, the first peak at approximately 163 eV BE ($S2p_{3/2}$) corresponds to the H–S–C bond present in the 1-hexanethiol molecule. This peak completely disappears after heptane cleaning, confirming that 1-hexanethiol does not chemically adsorb as a thiolate on the native oxide layer of steel. However, another weaker photopeak at higher binding energy ($S2p_{3/2}$ core level at 168.2 eV BE) is detected before and after cleaning, although the peak is quite smaller in the latter case. According to the literature data [19, 20], this peak can be assigned to the sulfonate bond ($R–SO_3$) fixed on the oxide surface. The oxygen XPS photopeak is composed of two contributions: one located at 530 eV and attributed to iron oxide species, and another located at 532 eV and attributed to –O–C bonding. The iron $Fe2P_{3/2}$ photopeak at 710.8 eV is assigned to iron oxides, Fe_2O_3 and Fe_3O_4. To summarize, the XPS results show

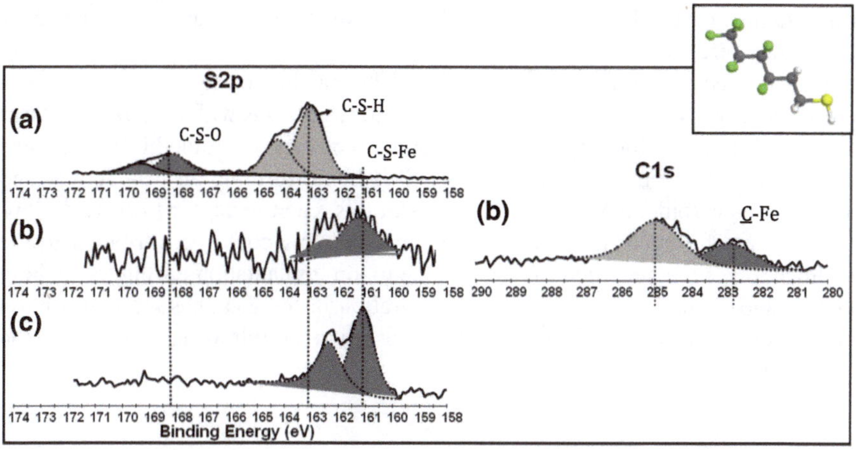

Fig. 4.4 **a** S2p XPS spectrum of liquid 1-hexanethiol covering steel, **b** S2p and C1s XPS spectra of gaseous hexanethiol adsorbed on metallic iron (etched steel), and **c** S2p XPS spectrum of iron sulfide (FeS) for reference

that n-alkanethiol molecules are weakly bound to oxygen atoms from the oxide layer by sulfur atoms, forming a few sulfonate species.

4.3.3 Adsorption of Hexanethiol on Metallic Iron (Etched Steel Surface)

1-Hexanethiol was introduced at a vapor pressure of 10^{-1} hPa in the tribometry chamber and kept in contact for 2 h with an iron surface obtained by ion-etching. At the end of the adsorption experiment, the gas was pumped down to 10^{-8} hPa, and the sample was transferred into the analytical chamber. Figure 4.4 compares the XPS results obtained after the adsorption of 1-hexanethiol on a pure iron surface (ion-etched steel surface) (Fig. 4.4b) and on the native oxide layer of steel (Fig. 4.4a). The S2p spectrum of the iron sulfide standard is also shown for reference (Fig. 4.4c).

The $S2p_{3/2}$ XPS spectrum (Fig. 4.4b) reveals a sulfur contribution of weak intensity at 161.5 eV BE, which is very close to the value reported for iron sulfide (Fe–S bond), as shown in Fig. 4.4c. This peak is easily distinguished from the R–S–H bond in the 1-hexanethiol molecule at 163 eV BE (Fig. 4.4a). The carbon C1s signal (Fig. 4.4b) is composed of a first peak at 284.8 eV BE, corresponding to the C–C/C–H bond, and a second peak at 282.8 eV BE, typical of the carbide state (C–Fe or C–Cr bond). It was observed that both the sulfur peak ($S2p_{3/2}$ at 161.5 eV BE) and the carbon peak (C1s at 284.8 eV BE) disappear from the surface after ion-etching, but the carbide peak is still visible and may originate from chromium carbide present in the AISI 52100 steel composition. The iron $Fe2p_{3/2}$ peak at 706.8 eV is attributed to metallic iron and/or iron carbide, because these species have almost the same binding energy (not shown here).

This first section clarifies the reactivity of 1-hexanethiol on steel surfaces at room temperature. We showed that 1-hexanethiol weakly interacts with iron oxide present on steel substrates by forming a sulfonate bond. In contrast, 1-hexanethiol chemisorbs on pure metallic iron by cleavage of the S–H bond, forming an iron sulfide and possibly iron carbide. However, the kinetics of this reaction appear to be very slow, as shown by the intensity of the S2p photopeak. This could be attributed to the modification of the metallic iron surface due to ion implantation during the etching process and to the lower molecular concentration in the gas phase in comparison with the liquid phase. This result will be discussed in the following section with the role of mechanically activated surfaces.

4.4 Model GPL Experiment for Understanding the Formation Mechanism of Hexanethiol-Based Tribofilm

It is still extremely challenging to provide an accurate description of the surface chemistry and interfacial properties of a tribofilm, which are relevant for understanding the functionality of lubricant additives, due to the need to clean rubbed surfaces before analysis (ex situ analyses). Part of the tribofilm may be removed by the cleaning procedure, and exposure to air may partially and/or totally oxidize its elements. This is particularly the case for adsorbed films, which are usually extremely thin. That is the reason why the ECAT system was developed, and the tribometer is schematically shown in Fig. 4.5. We illustrate the different steps used to accurately investigate both the adsorption and tribofilm formation mechanisms of additives on tribo-stressed metallic iron. First, the heptane- and propanol-cleaned steel surface is introduced in the UHV chamber, and the oxide layers are removed by Ar^+ iron etching, the metallic nature of the surface after etching can be then checked by XPS or AES (step 1 in Fig. 4.5). Second, a friction experiment under UHV (below 10^{-8} hPa) is performed to generate a nascent tribo-stressed metallic iron surface inside the wear scar (red track formed during step 2 in Fig. 4.5). Third, both the etched and tribo-stressed surfaces are exposed to gaseous hexanethiol for adsorption studies (step 3 in Fig. 4.5). Fourth, the friction experiment is performed in the presence of gaseous hexanethiol to form a tribofilm (blue track formed during step 4 in Fig. 4.5). Finally, the three different surfaces (adsorbed film on etched steel, adsorbed film on nascent tribo-stressed iron and thiol-derived tribofilm) are analysed by in situ XPS and AES analyses (step 5 in Fig. 4.5). This approach, which combines GPL and in situ surface analyses, is very efficient to study the tribochemistry mechanisms of lubricant additives and highlight the critical role of nascent metallic surfaces in the dissociation of the additives under boundary lubrication conditions.

Fig. 4.5 Model GPL experiment for elucidating the origin of tribochemical reactions of lubricating additives

4.4.1 Tribochemical Reactions of Hexanethiol on Oxidized Steel and Metallic Iron

First, the friction behaviours of oxidized steel and its metallic iron counterpart in the presence of gaseous hexanethiol were compared. For this, two-pass friction experiments were performed in the presence of 10^{-1} hPa 1-hexanethiol with metallic iron (pin and flat) surfaces (blue friction curve) and with oxidized steel (pin and flat) surfaces (green friction curve). Figure 4.6 shows the central part of the friction coefficient recorded during the back and forth motion in order to avoid the perturbing effect of changes in the sliding direction. In the case of the etched surface (Fe/Fe), the friction coefficient is initially low for a few cycles; then increases rapidly to high values near unity, indicating welding between the iron atoms from each side; and then begins to decrease until the end of the first pass. During the second (reverse) pass, the friction is much lower. This suggests that the chemisorbed layer on the nascent tribo-stressed iron flat was quickly formed during the first pass at the rear of the contact zone and is highly efficient at reducing friction during the reverse pathway. This tribolayer formed on metallic iron can be seen in the optical image of the track presented in Fig. 4.6b. The situation is very different if the contacting surfaces are oxidized (non-etched). The friction coefficient is almost stable during the first pass and increases during the second pass, reaching a value of approximately 0.6–0.7 at the end of the two passes. This result seems to indicate that no reaction occurred between oxidized steel and hexanethiol during the two friction passes. This result is in good agreement with the optical image of the track presented in Fig. 4.6c and the weak gas adsorption on oxidized steel observed previously (Fig. 4.3).

At the end of these GPL experiments, gaseous hexanethiol is pumped down, and AES analysis is performed on the different rubbed surfaces. It is important to note that the time necessary to reach a high vacuum below 10^{-8} hPa in the chamber is at least 1 h (3600 s), so some hydrocarbon adsorption from the residual gas on the steel surfaces occurs before analysis. Indeed, at a gas partial pressure of 10 nPa, the time for a monolayer to adsorb is approximately 10^4 s. Figure 4.7 compares the in situ AES spectra recorded inside the tracks formed on both the oxidized and

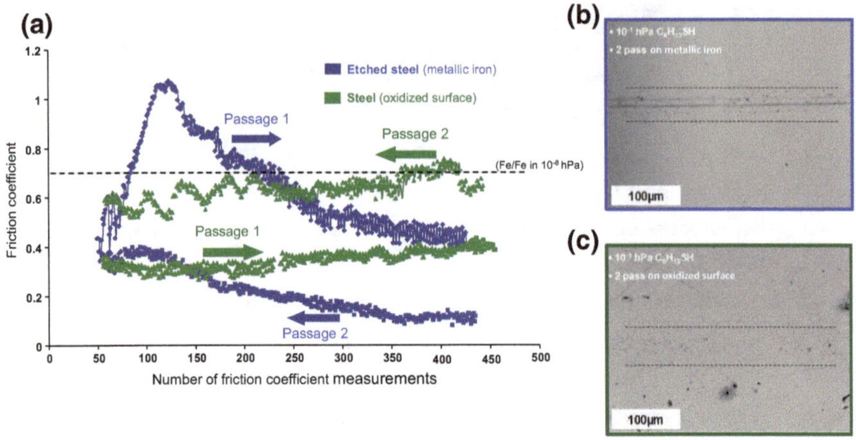

Fig. 4.6 **a** Evolution of the friction coefficient (for two passes, back and forth) obtained under GPL in the presence of 10^{-1} hPa pressure of 1-hexanethiol at room temperature. The friction curves in blue represent metallic iron surfaces (ion-etched steel), and the friction curves in green represent oxidized steel surfaces. The steady-state friction level for an etched steel surface under UHV is shown for reference. **b**, **c** Optical images of the corresponding tracks formed on the metallic iron flat and the steel flat

etched steel surfaces. In the spectrum recorded inside the tribofilm formed on etched steel, the formation of iron sulfide is clearly evidenced by the strong S_{LMM} peak at approximately 151 eV KE, confirming the decomposition of hexanethiol under shear and the release of S elements to form S–Fe bonds. It is noticed that no sulfate bonding is detected in the spectrum near 158.5 eV KE, which is in good agreement with the quite low amount of oxygen in the spectrum. For the steel surface covered by a native oxide layer, sulfur is not clearly detected in the AES spectrum. The friction reduction property (Fig. 4.6) is correlated with the quick formation of iron sulfide in the tribofilm on etched steel. These experimental results are in agreement with previous first-principles calculation results, which revealed that sulfur is very efficient at reducing the adhesion and shear strength at the iron-iron interface [17].

4.4.2 *Study of the Role of Nascent Metallic Surfaces on the Tribochemistry of Hexanethiol*

Additional AES spectra were recorded after exposure of the tribo-stressed nascent metallic surface (red) and the etched steel surface (outside the tracks) to 10^{-1} hPa hexanethiol to compare the nature of both adsorbed species and the tribofilm. Figure 4.8 compares the three different in situ AES spectra recorded. First, practically no oxygen is detected on the different surfaces (O_{KLL} at 530 eV KE), confirming that no contamination by oxygen or water from the gaseous atmosphere occurred in any

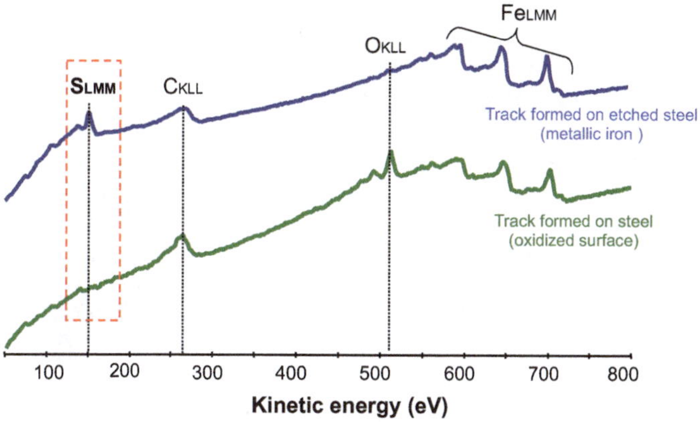

Fig. 4.7 In situ AES spectra recorded inside the tracks formed after the two-pass friction experiments with 10^{-1} hPa hexanethiol at room temperature on oxidized and etched steel surfaces

sequence (total duration of the experiment 2 h). As already observed by the authors, iron sulfide is quickly formed by tribochemical reaction (blue), whereas a very small amount of sulfur can be detected outside the tribofilm due to molecular adsorption on the etched steel. In the AES spectrum in red, corresponding to the friction track formed under UHV followed by exposure to the gas phase without friction, a small but visible signal is detectable at approximately 151 eV KE, in comparison with the spectrum registered outside the track. This peak indicates the possible formation of iron sulfide. From this result, we can conclude that the tribo-stressed iron surface, but not the ion-etched surface, chemically reacted with 1-hexanethiol. It is noticed that the mean free path of 150 eV electrons in the substrate is very small (approximately 1 nm), which is very different from the iron 700 eV electrons (mean free path of 3.5 nm). Therefore, the sulfur signal results from the first atomic layers, and the quantity of iron sulfide formed from the interaction of the molecules with the mechanically activated surface is very small, i.e., a partial monolayer.

From this model experiment, we can deduce some important features of the tribochemical reaction of 1-hexanethiol with steel surfaces:

- 1-Hexanethiol does not form iron sulfide in the presence of a native oxide layer on steel, but some sulfonate bonding was detected. This is in agreement with the chemical hardness model developed by Pearson [44], which states that a soft base (R-SH) prefers to react with a soft acid (metallic iron).
- Surprisingly, hexanethiol does not react significantly with pure metallic iron obtained by ion-etching the steel surface (free of carbon and oxygen). This is supported by the adsorption experiments shown in Fig. 4.4. This could be tentatively explained by the effect of Ar ion implantation, which can disturb the surface and remove some crystal defects. However, the presence of Ar was not detected in the AES spectra (peak at approximately 210 eV KE).

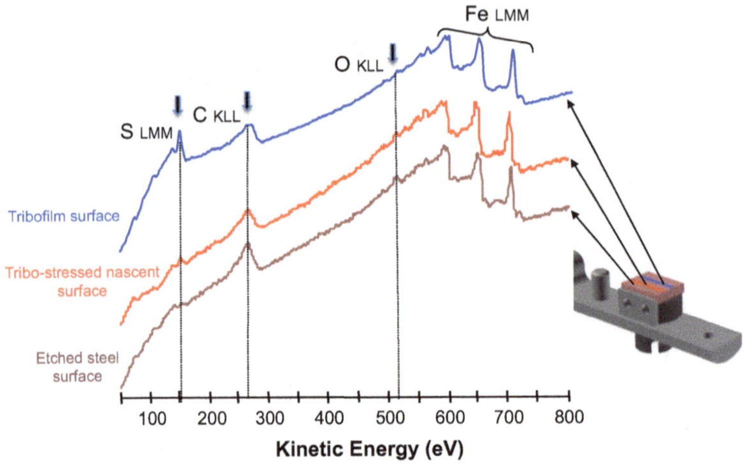

Fig. 4.8 In situ AES spectra recorded after the model GPL experiment on the tribofilm formed during the two-passes friction experiment with 10^{-1} hPa hexanethiol (blue), on the tribo-stressed nascent surface (red) and on the etched steel surface exposed to 10^{-1} hPa hexanethiol (purple)

- 1-Hexanethiol molecules dissociate on a tribo-stressed metallic iron surface (Fig. 4.8) by forming sulfide bonds (Fe–S). This dissociative reaction occurred approximately 1 h after the creation of the nascent tribo-stressed metallic iron surface. Therefore, neither the contact temperature nor electron emission is responsible for this dissociative reaction. The creation of crystal defects and dangling bonds at the metallic surface caused by high friction under UHV is certainly the origin of the activation energy of this thiol decomposition mechanism.
- 1-Hexanethiol rapidly forms a tribofilm containing iron sulfide in two passes on a metallic iron surface (Fig. 4.6). The detailed analysis of the thiol-metallic iron surface interaction as a function of the surface chemistry shows that the thiol molecules completely dissociate on the nascent surface at the contact zone exit. This confirms that the creation of the nascent metallic iron surface by mechanical action is one of the energy sources responsible for tribofilm formation. Once the iron sulfide accumulated in the tribofilm after successive passes, the friction greatly decreased. The formation of iron sulfide appeared to be widely enhanced by tribological solicitation, in particular, by shearing stress occurring at the colliding asperities.

4.5 Ex Situ Analyses of the Iron Sulfide Tribofilm Formed by Thiol

The GPL experiments coupled with in situ surface analyses showed that 1-hexanethiol efficiently reduced friction. This is correlated with the formation of iron sulfide inside the tribofilm. We highlighted the role of the active nascent metallic surface on the dissociation of adsorbed molecules and the activation of the tribochemical reaction to rapidly form a protective iron sulfide tribofilm. However, the microstructure of such tribofilm is not clearly known. Previous in situ XPS/AES chemical analyses have identified the chemistry of the top surface of the tribofilm (a few nm depth) but did not provide any information on the structure of the tribofilm and its interface with the steel substrate. Moreover, the exact stoichiometry of iron sulfide cannot be easily determined due to technical difficulties in accurately quantifying such compound and the existence of various forms of iron sulfide (FeS, FeS_2). The aim of this final section is to fully characterize the iron sulfide tribofilm formed by 1-hexanethiol on steel substrates, in particular, its stoichiometry, structure, thickness and the interfacial region with the steel substrate. A multi-technique approach is proposed, combining focused ion beam (FIB) sample preparation, transmission electron microscopy (TEM) observation, chemical analysis by energy dispersive spectroscopy (EDS) and time-of-flight secondary-ion mass spectrometry (ToF-SIMS). The studied tribofilm was obtained using GPL, taking advantage of this approach to control the contact surfaces, without the need to clean the surfaces after the friction experiment. ToF-SIMS analysis was performed on the top surface of the tribofilm, and TEM observations coupled with EDS analyses were carried out on a transversal cross section of the tribofilm prepared by FIB.

4.5.1 Formation of a 1-Hexanethiol Tribofilm by GPL

To clearly identify the nature of the tribofilm and its interfacial zone with the steel substrate, thicker tribofilms were generated by performing long-time friction experiments. The operating tribological conditions were similar to those previously described, except that 1-hexanethiol gas was introduced into the tribometry chamber of the ECAT system at a high pressure of 3 hPa and the duration of the experiment was approximately 1 h.

Figure 4.9 shows the evolution of the friction curve for the experiment in the presence of 3 hPa 1-hexanethiol gas. First, the friction coefficient slightly increased to 0.18 and then quickly decreased and stabilized at 0.15 (15th cycle). This value is in agreement with the formation of a tribofilm containing iron sulfide [42]. This tribofilm is clearly visible by optical microscopy, and some debris are present around the track, as shown in the inset of Fig. 4.9. However, 1-hexanethiol seems to have a good anti-wear property because both the width of the wear track on the steel flat (inset in Fig. 4.9) and the diameter of the wear track on the hemispherical pin almost correspond to the calculated initial Hertzian diameter, i.e., 114 mm.

4.5.2 *ToF-SIMS Chemical Analysis of the 1-Hexanethiol Tribofilm Formed by GPL*

ToF-SIMS analysis is performed on a Physical Electronics TRIFT III ToF-SIMS instrument operated with a pulsed 22 keV Au^+ ion gun (ion current of approximately 2 nA) rastered over a 100 μm × 100 μm area. No charge compensation was needed. The ion dose was kept below the static conditions limit. The data were analysed using the WinCadence software. Mass calibration was performed on the hydrocarbon secondary ions. The data were normalized to the total intensity of the secondary ions minus the ion contributions resulting from hydrogen and contaminants. The corrective eliminations were implemented because of their critical dependence on slight variations in the experimental settings (such as H ±) or because they were related to the signatures of contaminants usually observed in ToF-SIMS (such as alkalis and fatty acids).

The ToF-SIMS analysis focused specifically on the question of the exact stoichiometry of iron sulfides formed during the tribochemical reaction of 1-hexanethiol on the steel surface. For this, four different analyses were undertaken to compare the inside and outside of the tribofilm formed by hexanethiol and two references, i.e., the hexanethiol molecule and virgin steel. The standard deviations were calculated from several measurements in different areas for each case. Figure 4.10 presents the ion intensity ratios FeS^+/Fe^+ and $FeS_2{}^+/Fe^+$ recorded inside and outside the tribofilm. The chemical signatures of the steel substrate and the sulfur-containing thiol molecule could easily be identified via characteristic Fe^+ and S^- ions. In contrast,

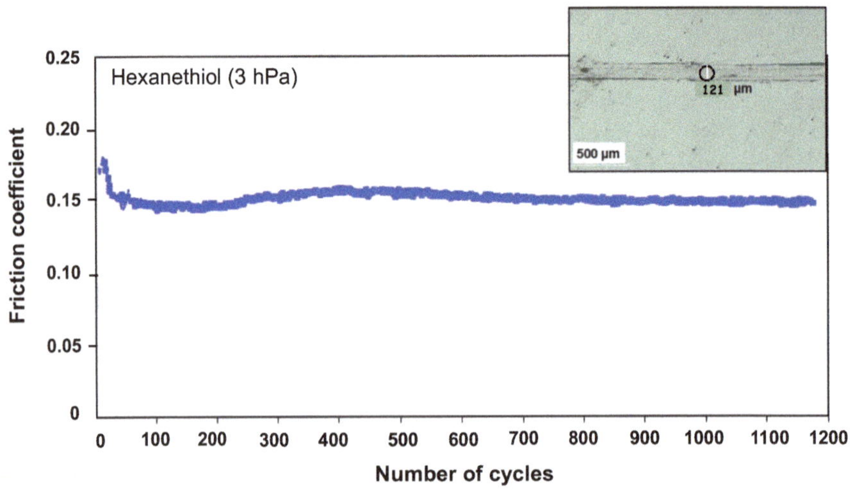

Fig. 4.9 Evolution of the friction coefficient obtained for the etched steel/etched steel friction pair under GPL in the presence of 3 hPa 1-hexanethiol versus the number of cycles at room temperature. The corresponding track formed on the etched steel flat at the end of the experiment is shown in the inset

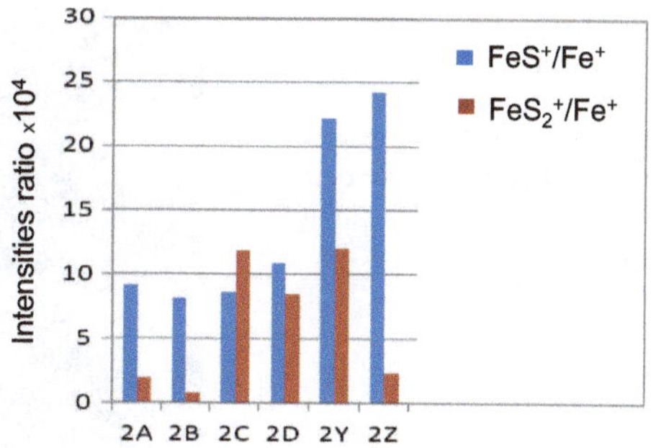

Fig. 4.10 ToF-SIMS intensity ratios FeS^+/Fe^+ and FeS_2^+/Fe^+ recorded in positive-ion mode inside (2A-2D) and outside the tribofilm (2Y-2Z) formed on steel in the presence of 3 hPa 1-hexanethiol

no specific signature for the hexanethiol molecule could be found (CxHySz⁻ ions were not detected). The normalized intensities of the Fe^+, S^+, FeS^+ and FeS_2^+ ions and the FeS^+/Fe^+ and FeS_2^+/Fe^+ ion intensity ratios were used to differentiate the spectra (so-called univariate ToF-SIMS data analysis). The results obtained from this univariate analysis (Fig. 4.10) confirmed the presence of a notable amount of these species, depending on whether the zones are contacted or not. The results seem to indicate that the tribofilm formed by hexanethiol is characterized by a higher relative FeS^+ intensity. However, this univariate data analysis made on a small number of molecular signatures is inherently incomplete and may lead to artefacts of interpretation. To confirm and complete our understanding of the nature of the iron sulfide created by the tribochemical reaction of hexanethiol on a steel surface, TEM observations combined with EDS analyses were carried out on a transversal cross section of the tribofilm.

4.5.3 TEM Observations and EDS Chemical Analyses of the 1-Hexanethiol Tribofilm Cross Section

A cross section was taken at the centre of the tribofilm formed under GPL with 3 hPa 1-hexanethiol. The cross section was thinned by FIB after deposition of a protective resin on the surface of the tribofilm. We used Ga^+ ions for nanomachining the sample. A high-resolution JEOL 2010F microscope equipped with a field-emission gun operating at 200 kV accelerating voltage was utilized to perform the TEM observations as well as the EDS analyses with a point-to-point resolution of 2 nm.

Figures 4.11 and 4.12 show a cross-sectional TEM image of the tribofilm. From Fig. 4.12 recorded with higher-resolution, it is possible to distinguish 4 zones: the steel substrate, the interfacial region, the tribofilm, and finally the Pt protective resin used for FIB preparation. Cross-sectional observations allow the measurement of the thickness and observation of the structure of the tribofilm. Even though the interface between the tribofilm and the steel substrate is not clearly marked, it appears smooth, and the thickness of the tribofilm is quite homogeneous and can be estimated to be approximately 15–30 nm. This smooth interface may be related to the fact that the

Fig. 4.11 TEM cross-sectional view of a tribofilm formed on steel under GPL in the presence of 3 hPa 1-hexanethiol at room temperature for 1180 sliding cycles

GPL friction experiment was performed directly on etched steel (metallic iron), so that the quick dissociation and reaction of thiol molecules was possible. The measured thickness is on the same order of magnitude as that of the tribofilms obtained from liquid-phase lubrication, even though in GPL, the concentration of the active element (here, sulfur) is generally lower than that in the liquid phase. Thus, the growth kinetics seem to be comparable. In fact, when molecules are introduced in vacuum, access to the surface is favoured. In the liquid phase, interaction with other additives or viscous base oils may limit access to the surface and further tribochemical reactions. This observation confirms the original hypothesis that GPL is a good way to simulate tribochemical reactions.

Further details of the tribofilm are given by the higher-resolution image in Fig. 4.13a. In addition to the quite homogeneous thickness and the defined interface, it is possible to see that the tribofilm is composed of nanocrystalline areas of approximately 3–5 nm in diameter embedded in an amorphous matrix. Chemical analysis by EDS conducted on the cross section of the tribofilm is presented in Fig. 4.13b. We carried out EDS analysis at three specific zones: the interface with the steel substrate, the centre of the tribofilm and a crystallite inside the tribofilm. Some expected contaminants and artefacts were detected in all spectra, including copper from the microscope grid and gallium from the FIB process. As expected, EDS analysis of the substrate reveals an iron-rich composition (81 at.%) with 9 and 10 at.% oxygen and carbon, respectively. The nanocrystals inside the tribofilm are mainly composed of iron and sulfur with a similar ratio of approximately 40 and 42 at.%, respectively, which can be assigned to the stoichiometry of iron sulfide (FeS). It is not unusual to find a crystalline structure for iron sulfide-based compounds because different crystalline structures can exist for these compounds. A large amount of carbon

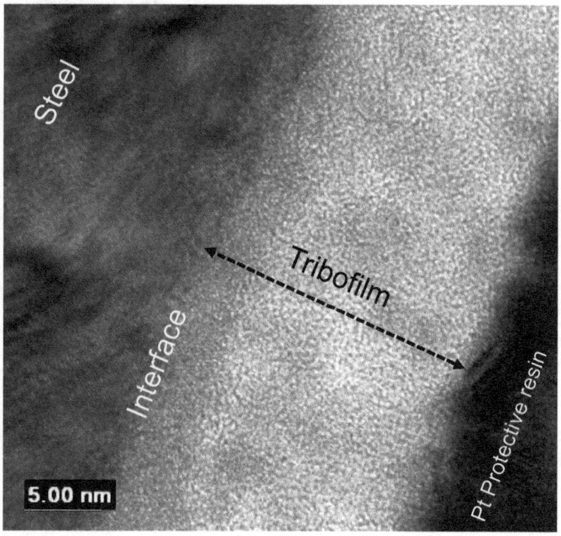

Fig. 4.12 HRTEM cross-sectional view of a tribofilm formed on steel under GPL in the presence of 3 hPa 1-hexanethiol at room temperature for 1180 sliding cycles

but little oxygen was detected inside the amorphous matrix of the tribofilm, as shown by the intermediate EDS spectrum in Fig. 4.13b. Carbon may have been introduced during tribofilm formation (via the 1-hexanethiol molecule, which contains 6 carbon atoms) or after FIB preparation (adventitious contamination of the tribofilm). The EDS line scans (Fig. 4.14) show the distribution of the main elements: platinum, iron, sulfur, carbon and oxygen. The profiles of the S and Fe elements are similar, and the elements are homogeneously distributed in the main part of the tribofilm. Near the interface with steel substrate, the amount of sulfur decreases, while that of iron increases. However, the absence of oxygen in the interfacial zone with steel may be because the GPL experiment was performed directly on etched steel (Fig. 4.14).

The TEM cross-sectional observations, EDS analyses, and ToF-SIMS analyses converge on the same conclusion about the composition of the tribofilm formed on the etched steel surface in the presence of 1-hexanethiol. This surface is mainly composed of iron sulfide with a stoichiometry close to FeS. Taking into account thermodynamics considerations, the preferential formation of iron monosulfide is not corroborated. By considering that iron sulfide formation is the result of the reaction of metallic iron Fe with atomic sulfur S after decomposition of 1-hexanethiol, all reactions in the formation of FeS_x are characterized by a negative ΔGf^0, and thus, iron sulfide forms spontaneously. However, the Gibbs free energy of formation ΔGf^0 of iron disulfide FeS_2 (-167 kJ mol^{-1}) is more favourable than that of iron monosulfide FeS (-100 kJ mol^{-1}) [45]. Consequently, the nucleation of this last phase is not the most preferential. The energy dissipated at the tribological contact allows this phase to be stabilized under friction. In addition, the thiol molecule contains only one sulfur atom, which is not considered to be very “active” since it is bonded to a carbon atom of the alkyl chain.

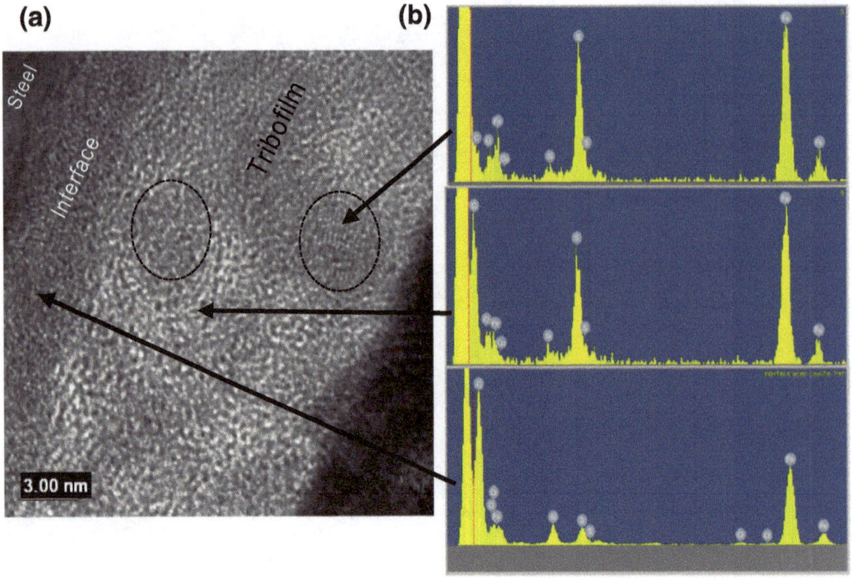

Fig. 4.13 **a** HRTEM cross-sectional view of a tribofilm formed on steel under GPL in the presence of 3 hPa 1-hexanethiol at room temperature for 1180 sliding cycles and **b** EDS spectra recorded on three different parts of the tribofilm

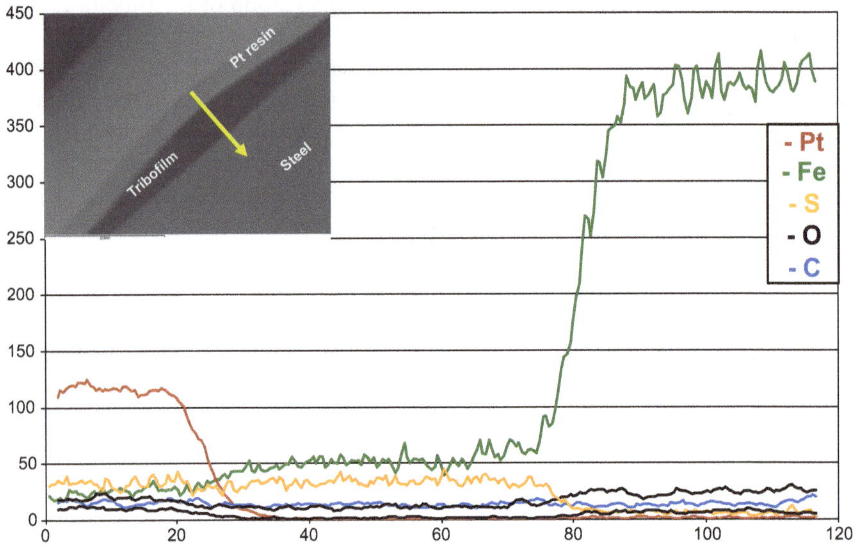

Fig. 4.14 EDS line scan analyses conducted on the cross section of the hexanethiol tribofilm (yellow arrow)

4.6 Conclusions

An approach combining gas-phase lubrication (GPL) and in situ XPS/AES analyses was found to be very efficient to investigate the tribochemistry mechanisms of lubricant additives under boundary lubrication conditions. Using a specific device called ECAT, it is possible to perform friction experiments under different gaseous atmospheres and to perfectly control the surface chemistry of the friction surfaces. In situ XPS/AES analyses allowed us to clearly detail the chemistry of the tribofilm formed by the lubricant additive under friction. This permitted us to determine the principal activation energy of the tribochemical reaction due to the model GPL experiments. Undoubtedly, we highlighted the primordial role of nascent tribo-stressed metallic surfaces obtained after shearing in the dissociation of lubricant additives under boundary lubrication conditions. The investigation into the reactivity of thiol molecules showed that they suffer dissociative adsorption on the asperities of the tribo-stressed nascent metallic surface, rapidly forming a thin protective film composed of iron sulfide. Comparison with the tribofilm obtained by tribological action under GPL showed that iron sulfide formation is largely enhanced by friction, in particular, by shearing stress. For severe collisions at the asperities, the preferential reaction between thiol species and nascent iron metal leads to the formation of an iron sulfide tribofilm, promoting the lubrication effect and efficient friction reduction. The detailed analysis of the tribofilm formed with hexanethiol using a multi-technique approach revealed that the tribofilm was mainly composed of iron and sulfur. The tribofilm displays a nanocrystalline structure made of nanograins of iron sulfide embedded in an amorphous carbon-rich matrix. The EDS analysis showed a Fe/S ratio close to unity for thiol-derived tribofilm corresponding to iron monosulfide.

Overall, this study indicates that the approach combining GPL and in situ surface analyses can be a promising way to study the potential competitive/synergic mechanisms occurring between additives in a formulated lubricant at the sliding surface.

References

1. H. Spikes, Tribol. Lett. **60**(5), 1–26 (2015)
2. F.P. Bowden, D. Tabor, *The Friction and Lubrication of Solids*, part II. Oxford University Press (1964)
3. D.H. Buckley, J. Appl. Phys. **39**, 4224 (1968)
4. A.G. Gellman, J. Vac. Sci. Tech. A **10**, 180 (1992)
5. S. Mori, *International Tribology Conference*. Yokohama (1995)
6. K. Miyoshi, F.K. Honecy, P.B. Abel, S.V. Pepper, T. Spalvins, D.R. Wheeler, STLE Trans. **36**(3), 351 (1993)
7. J. Fontaine, J.-L. Loubet, Th Le-Mogne, A. Grill, Tribol. Lett. **17**(4), 709 (2009)
8. J.L. Lauer, S.R. Dwyer, *Proceeding of the Japan International Tribology Conference*, Nagoya, vol. 989 (1990)
9. M. Boehm, J.M. Martin, C. Grossiord, Th Le Mogne, Tribol. Lett. **11**(2), 83–90 (2001)

10. M.I. De Barros-Bouchet, M.C. Righi, D. Philippon, S. Mambingo-Doumbe, T. Le Mogne, J.M. Martin, A. Bouffet, RSC Adv. **5**, 49270 (2015)
11. J.M. Martin, M.I. De Barros Bouchet, C. Matta, Q. Zhang, W. Goddard III, S. Okuda, T. Sagawa, J. Phys. Chem. C **114**, 5003 (2010)
12. D. Philippon, M.I. De Barros Bouchet, Th Le Mogne, E. Gresser, J.M. Martin. Tribology **1**(3), 113 (2007)
13. D. Philippon, M.-I. De Barros-Bouchet, Th Le Mogne, O. Lerasle, A. Bouffet, J.-M. Martin, Tribol. Int. **44**(6), 684 (2009)
14. D. Philippon, M.I. De Barros Bouchet, T. Le-Mogne, O. Lerasle, J.M. Martin, Tribol. Lett. **41**(1), 73 (2011)
15. D. Philippon, M.I. De Barros-Bouchet, Th Le-Mogne, B. Vacher, O. Lerasle, J.M. Martin, Thin Solid Films **524**(1), 191 (2012)
16. M.C. Righi, S. Loehle, M.I. De Barros-Bouchet, D. Philippon, S. Mambingo-Doumbe, T. Le Mogne, J.M. Martin, A. Bouffet, RSC Adv. **5**, 101162 (2015)
17. M.C. Righi, S. Loehle, M.I. De Barros-Bouchet, S. Mambingo-Doumbe, J.M. Martin, RSC Adv. **6**, 47753 (2016)
18. H.H. Weetall, M.J. Lee, Appl. Biochem. Biotechnol. **22**, 311 (1989)
19. S. Kulin, R. Khishore, J.B. Hubbard, Biophys. J. **83**, 1965 (2002)
20. M. Volmer-Uebing, M. Stratmann, Appl. Surf. Sci. **55**, 19 (1992)
21. B. Parker, A.J. Gellman, Surf. Sci. **292**, 223 (1993)
22. W. Davey et al., Wear **1**, 291 (1958)
23. E.S. Forbes, A.J.D. Reid, ASLE Trans. **16**, 50 (1973)
24. S. Mori, K. Hori, Y. Tamai, Junkatsu **27**, 505 (1982)
25. Z. Li, Y. Li, Y. Zhang, T. Ren, Y. Zhao, RSC Adv. **4**, 25118 (2014)
26. J. Li, B. Fan, T. Ren, Y. Zhao, Tribol. Int. **88**, 1 (2015)
27. M. Cohen-Atiya, D. Mandler, J. Electroanal. Chem. **550–551**, 267 (2003)
28. F. Schreiber, Prog. Surf. Sci. **65**, 151 (2000)
29. C. Vericat, M.E. Vela, G.A. Benitez, J.M. Gago, X. Torrelles, R.C. Salvarezza, J. Phys. Condens. Matter **18**, R867–R900 (2006)
30. N. Camillone, T. Leung, P. Schwartz, P. Eisenberger, G. Scoles, Langmuir **12**, 2737 (1996)
31. S. Xu, S.J.N. Cruchon-Dupeyart, J.C. Garno, G.Y. Liu, G.K. Jennings, T.H. Yong, P.E. Laibinis, J. Chem. Phys. **108**, 5002 (1998)
32. W. Azzam, A. Bashir, A. Terfort, T. Strunskus, C. Woll, Langmuir **22**, 3647 (2006)
33. M.R. Albert, J.P. Lu, L. Bernasek, S. Cameron, J.L. Gland, Surf. Sci. **206**, 348 (1988)
34. J.A. Syed, S.A. Sardar, S. Yagi, K. Tanaka, Surf. Sci. **556–568**, 597 (2004)
35. S.A. Sardar, J.A. Syed, E. Ikenaga, S. Yagi, T. Sekitani, S. Wada, M. Taniguchi, K. Tanaka, Nucl. Instrum. Methods Phys. Res. **199**, 240 (2003)
36. C. Kodama, T. Hayashi, H. Nozoye, Appl. Surf. Sci. **169–170**, 264 (2001)
37. D.M. Jaffey, R. Madix, Surf. Sci. **311**, 159 (1994)
38. M. Volmer, M. Stratmann, H. Viefhaus, Surf. Interface Anal. **16**, 278 (1990)
39. C. Pirlot, J. Delhalle, J.J. Pireaux, Z. Mekhalif, Surf. Coat. Technol. **138**, 166 (2001)
40. Z. Mekhalif, J. Delhalle, P. Lang, F. Garnier, J.J. Pireaux, J. Electrochem. Soc. **146**, 2913 (1999)
41. Z. Mekhalif, J. Delhalle, P. Lang, F. Garnier, J.J. Pireaux, Synth. Metals **96**, 16 (1998)
42. J. Tannous, B.M.I. de Bouchet, T. Le Mogne, P. Charles, J.M. Martin, Tribology **1**, 98 (2007)
43. A. Rossi, F.M. Piras, D. Kim, A.J. Gellman, N.D. Spencer, Tribol. Lett. **23**, 197 (2006)
44. R.G. Pearson, *Chemical Hardness*, 1st edn. (Wiley-VCH, Weinheim, 1995), p. 198
45. D.D. Wagman, W.H. Evans, V.B. Parker, R.H. Schumm, I. Halow, S.M. Bailey, K.L. Churney, R.L. Nuttall, Am. Chem. Soc. Washington, DC (1982)

Chapter 5
In-Situ Measurement of Tribochemical Processes in Ultrahigh Vacuum

Wilfred T. Tysoe

Abstract Monitoring chemical reactions occurring at a solid-solid interface is particularly challenging because of the problem of analyzing a buried interface with surface-sensitive spectroscopic techniques. This can, to some extent, be addressed if one of the contacting materials is transparent. In the case of optically opaque materials that are often of the greatest tribological interest, truly in-situ techniques are limited to the detection of gas-phase products formed by rubbing in high vacuum using a mass spectrometer, or by monitoring the contact resistance or friction coefficient variations during sliding. Optical techniques such as infrared spectroscopy can be used when one of the materials is transparent. The results of such in-situ analyses can be corroborated by using so-called pseudo in-situ techniques to analyze the surfaces after rubbing without exposing the samples to the atmosphere. Examples of such techniques are Auger spectroscopy and low-energy electron diffraction. Finally, the use of these approaches is illustrated using a simple model tribochemical reaction consisting of the gas-phase lubrication of copper by dimethyl disulfide.

5.1 Introduction

Being able to understand tribochemical processes and their reaction pathways is perhaps the most challenging of surface science problems since the reactions occur at an inaccessible, solid-solid interface. However, understanding such reaction pathways is crucial since the majority of lubricants include additive molecules that react at the sliding interface to form a film that reduces friction or wear, or both, and are used in applications as diverse as machining, in the so-called extreme pressure regime, where the interfacial temperatures are exceedingly high [1], to lubricants used in engines and machines, where the conditions are less severe. A recent study suggests that ~20% of energy used in the world is used to overcome friction and that it could

W. T. Tysoe (✉)
Department of Chemistry and Biochemistry, University of Wisconsin
Milwaukee, Milwaukee, WI, USA
e-mail: wtt@uwm.edu

M. Dienwiebel and M.-I. De Barros Bouchet (eds.), *Advanced Analytical Methods in Tribology*, Microtechnology and MEMS,
https://doi.org/10.1007/978-3-319-99897-8_5

be possible to save as much as 17.5% of the energy use in road transportation, much of it by improved lubricants [2].

The majority of the work carried out to understand the way in which lubricant additives or gas-phase lubricants operate relies on post-mortem analyses, in which the nature of the film is analysed after it has formed, to correlate the chemical composition of the film with its mechanical properties. However, while this approach is useful to understand the film properties that lead to improved tribological behavior, it generally yields few insights into the surface chemical processes that result in the formation of tribofilms. This is, in part, because chemical (including tribochemical) reactions generally consist of a sequence of elementary-step reactions that constitute an overall "reaction mechanism". Only by understanding such mechanisms for tribochemical reactions can the design of gas-phase lubricants and lubricant additives be placed on a firm scientific basis, rather than relying on trial-and-error approaches. Achieving such understanding requires that the nature of the reaction products be analysed as a function of time.

Reaction rate constants depend on temperature and are often analysed in terms of transition state theory [3]. For a simple, elementary-step reaction, $A \rightarrow B$, the transformation of A into B involves an initial increase in energy, known as the activation energy, E_{act}, before forming the product B. If this were not the case, A would spontaneously transform into B, and thus be an unstable compound. The configuration of the system at the peak of this activation barrier is known as the transition state. Transition state theory shows that the rate constant for the reaction, k, is given by: $k = A\exp\left(-\frac{E_{act}}{k_B T}\right)$, where k_B is the Boltzmann constant, T the absolute temperature in degrees Kelvin (K), and A is a pre-exponential factor, which is weakly temperature dependent, so that the overall temperature dependence of the rate constant is dominated by the exponential term, and this follows a so-called Arrhenius dependence. The derivation of this equation from transition-state theory assumes that the reactant (A) is in thermal equilibrium with the transition state. Thus, correctly modelling tribochemical reactions requires both a knowledge of the interfacial temperature during sliding and that the interface be in thermal equilibrium. This can impose restrictions on the sliding conditions since molecule dynamics simulations have suggested that the interface can deviate from thermal equilibrium during sliding [4], thereby invalidating the use of simple kinetic models. This also suggests that one mechanism by which the rates of tribochemical reactions can be accelerated is simply by frictional heating in the contact. Since the temperature appears as an exponent in the Arrhenius form of the temperature-dependence of the rate constant shown above, relatively small temperature changes can cause significant increases in chemical reaction rates.

The above considerations place quite severe strictures on carrying out-in-situ analyses of the kinetics of tribochemical reactions. In addition, depending on the contact geometry, which generally consists of a ball sliding on a flat surface, that size of the contact region may be quite small, thus often requiring the use of surface analytical techniques with good spatial resolution.

The most useful but most difficult systems to study are those which are of the greatest interest to understanding tribochemistry, namely metal and oxide surfaces since they are the most common in engineering systems. However, since they are not transparent to electrons and photons that are most commonly used as the basis for surface spectroscopic techniques, there are relatively few in-situ techniques that can be applied to such systems.

A solution to such a problem is to use contacts in which one of the materials is transparent to enable photon-in/photon-out techniques to be used for in-situ analyses of the sliding interface [5]. While this does extend the range of techniques that can be used, it does limit the tribochemical systems to ones that may not be relevant to engineering interfaces. In addition, photon-based techniques tend to be less surface sensitive that those based on electrons.

Since the range of truly in-situ techniques, even when using contacts in which one of the materials is transparent, is rather limited, we include a class of pseudo in-situ techniques. This consists of techniques that cannot be directly implemented at the solid-solid contact, but can be sequentially applied to one of the surfaces after it has been rubbed. These technique should analyze with the sample in the same location at which rubbing occurs to allow the evolution of the nature of the interface to be examined as a function of the number of times the sample has been rubbed. This definition is distinct from truly ex-situ techniques where the sample has to be removed from the tribometer in order to perform the analyses. Such ex-situ analyses can potentially suffer from the problem that the nature of the surface may be different from that at the sliding interface, either because it continues to evolve after having been rubbed or because of contamination from the ambient.

The techniques that are available for the three hierarchies of in-situ analysis are discussed in the next section.

5.2 In-Situ Analyses of Sliding Interfaces

5.2.1 In-Situ Analyses with Opaque Interfaces

Since such opaque interfaces, for example for metal-metal or ceramic-metal interfaces are opaque to both electrons and photons (light), there are no spectroscopic techniques for studying such interfaces.

5.2.1.1 Detection of Gas-Phase Products During Sliding

While the nature of the film formed on the surface is of primary tribochemical interest, the reaction of adsorbed gas-phase lubricants may evolve gas-phase products and the formation of such products can provide insights into the reactions occurring on the surface and their rates. However, in general, only a small amount of product is formed from the small contact area usually found in sliding contacts, thus requiring sensitive detection. In particular, it is not feasible to carry out such an experiment for lubricants additives in a base oil, except after a very large number of rubbing cycles, and is most informative for tribochemical reactions in high [6, 7] or ultrahigh vacuum [8], where the background pressure is low. In this case, the products are detected using a mass spectrometer that is placed in-line-of sight of and as close as possible to the contact. The experiments can be carried out either with the sample pressurized by a reactant (vapor phase lubrication) or by adsorbing a monolayer of adsorbate of the surface and measuring the evolution of gas-phase species as the surface is rubbed. In the former case, the mass spectrometer fragmentation masses of the products that are formed by rubbing must be distinct from those of the vapor-phase lubricant to allow them to be distinguished. This experiment allows the nature of the gas-phase product(s) to be identified from the relative intensities of the features at various masses.

In the second case, where a monolayer of adsorbate is rubbed, the nature of the desorbing species can similarly be identified from its mass spectrometer fragmentation pattern, and the surface reaction kinetics measured from the decay in the amount of products evolved from the surface as a function of the number of times that it has been rubbed, where the signal eventually decays to zero when all adsorbed species have completely reacted.

This is illustrated for methyl thiolate (CH_3–S) species adsorbed onto copper by dosing dimethyl disulfide, which reacts rapidly on copper by S–S bond scission at ~200 K. The resulting methyl thiolate thermally decompose to desorb hydrocarbons (methane, ethylene and ethane) at ~425 K in temperature-programmed desorption (TPD), corresponding to an activation energy of ~100 kJ/mol [9], resulting in the deposition of sulfur on the surface [10, 11]. Here the additional hydrogen required to form methane from methyl thiolate species was provided by formation of C_2 hydrocarbons.

Sliding a tip over copper covered with methyl thiolate species at a substrate temperature of ~300 K (well below the thermal decomposition temperature), causes the formation of gas-phase methane where the amount produced by each scan of the tip over the surface is displayed in Fig. 5.1. Monitoring signals at other masses confirmed methane and C_2 hydrocarbon formation, and control experiments were carried out to establish that the methane was formed by rubbing [12]. Analyses of the surface after rubbing (see Sect. 5.2) reveal that only sulfur remains on the surface after rubbing. The integrated areas of the methane signal as a function of the number of passes of the tip over the surface decrease (see inset to Fig. 5.1), eventually decreasing to zero as the surface becomes completely depleted of adsorbed methyl thiolate species. This reveals that methyl thiolate species decompose via a tribochemical (shear-induced reaction) to evolve methane and deposit sulfur on the surface. The rate of this simple reaction can be written as:

$$-\frac{d\Theta}{dt} = k\Theta^n \tag{5.1}$$

where Θ, the coverage (in monolayers) defines the proportion of the surface covered by the adsorbate, k is the rate constant of the tribochemical reaction, and n the reaction order where $n = 1$ for a first-order and $n = 2$ for a second-order reaction. Here, a

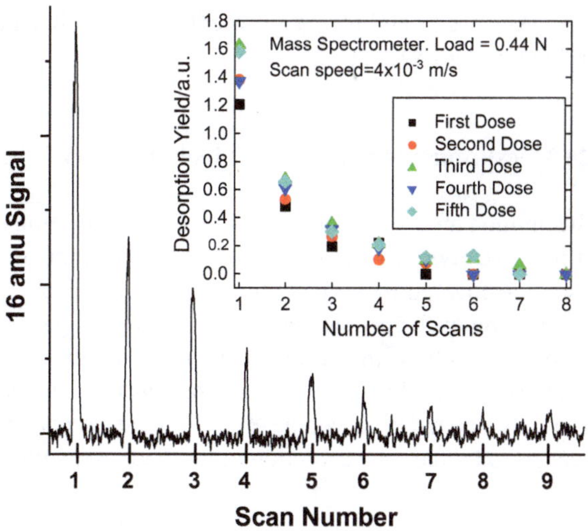

Fig. 5.1 The 16 amu (methane) signal measured by sliding a tungsten carbide tip with a copper transfer film on a copper foil as a function of the number of scans at a sliding speed of 1×10^{-3} m/s with a normal load of 0.44 N. The inset shows the desorption yield measured from the area under each methane pulse as a function of the number of scans. After collecting data for the first DMDS dose while sliding at 4×10^{-3} m/s, until no more methane was detected (■), the sample was re-saturated with DMDS and the 16 amu signal again monitored (●). This experiment was repeated for the third (▲), fourth (▼) and fifth (⬩) DMDS doses

first-order reaction would imply that the reaction occurs by a single decomposition step of an adsorbed thiolate species and a second-order reaction that it occurs via the reaction between two adsorbates and so can provide mechanistic information.

If the time that an element on the surface spends in the contact per pass is given by t_C, then the proportion of the adsorbate that decomposes during each pass can be calculated by solving (5.1) over a time t_C. The total yield of gas-phase products for each pass is simply the difference in coverages before and after the tip passes over the surface. Writing $k' = \mathrm{kt}_c$ gives the yield of gas-phase products for the pth pass, Y_p, for a first-order reaction ($n = 1$):

$$Y_p = (1 - \exp(-k'))(\exp(-k'))^{(p-1)}. \tag{5.2}$$

Similar considerations for a second-order reaction ($n = 2$) yields:

$$Y_p = \frac{1}{1 + (p-1)k'} - \frac{1}{1 + pk'}. \tag{5.3}$$

A linear plot of $\ln(Y_p)$ versus $(p - 1)$ yields a straight line indicating that shear-induced thiolate decomposition occurs via a first-order reaction, thereby allowing k' to be measured [8, 13]. Measurement of the rate constant k requires a knowledge of the contact time per pass t_C and thus a knowledge of the contact area. However, under a constant load, t_C will vary as $\frac{1}{\nu}$.

5.2.1.2 In-Situ Measurements of Coverages from Friction Forces

Perhaps the most common measurement in tribology is of the friction force or friction coefficient (friction force divided by normal load). It has been found that the friction coefficient of a surface covered by several species is proportional to the proportion of the surface covered by each species (denoted as the coverage, Θ) [14]. Thus, if during a tribochemical reaction, the surface contains a number of species, where the coverage varies with time in the contact as the reaction proceeds, then if the ith adsorbate has a time-dependent coverage $\Theta_i(t)$, with an associated characteristic friction coefficient μ_i, then the time evolution of the friction coefficient $\mu(t)$ can be written as:

$$\mu(t) = \sum_i \mu_i \Theta_i(t). \tag{5.4}$$

with the constraint that $\sum_i \Theta_i = 1$. This method requires that the characteristic friction coefficients of each species participating in the surface tribochemical reaction be known, and is generally the most useful in confirming a proposed tribological model.

However, this approach is particularly useful for analyzing simple, first-order reactions, A(ads) $\rightarrow$ B(ads) since, μ_A and μ_B can be measured before and after reaction to obtain the time-dependent coverages. In this case, (5.4) simplifies to:

$$\mu(t) = (\mu_A - \mu_B)\Theta_A(t) - \mu_B, \tag{5.5}$$

because of the constraint that the coverages sum to unity. The value of μ_B is the asymptotic friction coefficient of the sample after the reaction is complete and μ_A is the initial friction coefficient assuming that the extent of decomposition induced by sliding during the first pass is negligible. For example, for a first-order reaction, (5.5) becomes:

$$\mu(p) = (\mu_A - \mu_B)\exp\left(-k'p\right) - \mu_B. \tag{5.6}$$

This approach has recently been exploited to follow shear-induced reactions in an AFM [15].

5.2.1.3 Contact Resistance Measurements

In principle, contact resistances can provide information on the nature of the interfaces between opaque materials, and provide estimates of film thickness, t:

$$t = \frac{\rho A_C}{R}, \tag{5.7}$$

where ρ is the resistivity of the film material, R the contact resistance and A_C the real area of contact. Since the resistivity of the film material is not known, either because of its uncertain composition and non-uniformity, and the real contact area is often not well defined, even this simple interpretation makes it difficult to obtain qualitative film thickness and this approach is primarily useful for qualitative estimates of film thickness. The analysis is further complicated since the contact between real surface occurs at asperity tips that have small dimensions that may influence their conductive properties.

5.2.2 In-Situ Analyses of Transparent Interfaces

The notion that lubricating films could be interrogated with optical techniques with transparent contacts was pioneered by Cameron, who used interferometric methods to measure lubricant film thickness in-situ during sliding [16]. While such techniques have not yet been applied to sliding interfaces in ultrahigh vacuum, presumably because of the difficulty in implementing optical techniques in ultrahigh vacuum because of the potential utility of such techniques in surface analysis, the applications of such techniques are briefly discussed. The majority of optical techniques such as Raman spectroscopy have been used to study tribofilm formation in air since the Raman scattering cross section is generally too low to measure molecularly thin layers except in exceptional circumstance in which there is a surfaces enhancement in scattering intensity on rough coinage metals in a technique known a surface-enhanced Raman spectroscopy (SERS). In principle, such techniques could be applied to examining tribo-film growth in ultrahigh vacuum [17].

Infrared spectroscopy is one of the most commonly used technique for examining the surface chemistry of monolayer adsorbates primarily on metal surfaces. Since metals substrates are conductive, electric fields that are oriented parallel to the surface are screened, while electric field with components normal to the surface are enhanced [13, 18]. As a consequence, reflection-absorption infrared spectroscopy (RAIRS) is carried out on metal surfaces at grazing incident to enhance the component of the electric field normal to the surface and is sufficiently sensitive to detect submonolayer coverages of adsorbates on metals. This leads to the so-called surface selection rules in which only vibrations that have atomic motions with a component perpendicular to the surface are detected. This yields information on the orientation of the adsorbate on the surface.

Perhaps most importantly, infrared spectroscopy can provide detailed information on the chemical nature of adsorbates through characteristic "group frequencies" [19, 20]. Here, a functional group of an adsorbate is found to generally exhibit a characteristic range of frequencies that can be used to identify the nature of the adsorbate and is particularly useful for understanding surface chemistry. However,

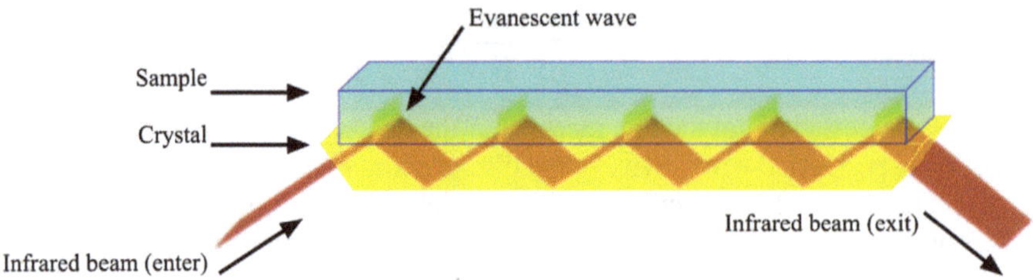

Fig. 5.2 Schematic depiction of the ATR-IR experiment

while potentially powerful for following surface tribochemical reactions, RAIRS cannot provide a truly in-situ interrogation of the sliding interface.

An alternative approach which does overcome this limitation is attenuated total internal reflection infrared spectroscopy (ATR-IR). A typical experimental geometry of the implementation of an ATR-IR experiment is illustrated in Fig. 5.2. This technique uses an ATR crystal which is fabricated from an infrared-transparent material with a high refractive index, typically germanium, or zinc selenide. The ends of the crystal have bevelled edges cut at an angle such that the infrared light impinges on the long faces of the crystal below the critical angle Θ_C defined as $\Theta_C = \frac{n_1}{n_2}$, where n_2 is the refractive index of the ATR crystal and n_1 is the refractive index of the medium outside the crystal, typically vacuum or air so that $n_1 \sim 1$. In this case, the infrared radiation repeatedly reflects off the inner faces of the crystal and exits from the opposite bevelled face, which is cut at the same angle as the opposite face. However, the infrared light does not terminate abruptly when reflecting from the inner faces of the crystal but can penetrate as an evanescent (decaying) wave some distance into the low-refractive-index region of the sample. The decay length for this evanescent wave is of the order of the wavelength of the light passing through the crystal. Thus, the spectrum of any material placed on the face of the ATR crystal that absorbs infrared radiation can be recorded.

However, since the infrared signal passes through the whole of the crystal, using a ball-on-flay geometry (where the ATR crystal comprises the flat) will result in only a relatively small signal so that the technique is generally used with a cylinder sliding on the ATR crystal to provide a wider detection area and greater sensitivity [21, 22]. A major advantage of this approach is that the ATR crystal can be coated with a thin layer of metal which, if sufficiently thin, still allows the infrared radiation to penetrate it, thus allowing tribochemical reactions to be studied in-situ for a metal-metal interface. In this configuration, the method has a transparent interface while maintaining the ability to study tribochemically realistic interfaces. While this approach has not yet been used to study tribochemical reactions in vacuum, there are no technical reasons that this should not be done.

5.2.3 *Pseudo In-Situ Analyses*

This section describes the use of what is termed "pseudo-in-situ" techniques which include surface-sensitive techniques that, while they are not able to directly probe the contacting interface while rubbing, they can be used to analyze the surface immediately after rubbing, as a function of the number of times that it is rubbed. This allows kinetic information on the nature of the surface to be obtained. In this case, it is helpful to be able to correlate the results of true in-situ analytical techniques with the pseudo-in-situ measurements. Strategies for accomplishing this will be discussed in Sect. 5.4.

Analytical techniques that are the most surface sensitive and those based on the use of the electron since the mean-free path of electrons in most materials is of the order of a few nanometers [23]. Such techniques include those that are excited by high-energy photons and where electrons are detected to render the technique surface sensitive, or those that are themselves excited by electrons. In general, to be useful in understanding tribochemistry, they should have good spatial resolution. However, photon-based techniques such as X-ray Photoelectron Spectroscopy (Chaps. 4 and 6) tend to require complex light sources that can be difficult to incorporate into the same chamber as the tribometer itself and are generally carried out by transferring the sample, in vacuo, from the chamber in which the surface is rubbed to an analysis chamber. However, they do have the advantage of being able to provide quite detailed analytical information, for example on the chemical state of the surface region.

Perhaps the simplest pseudo-in-situ analytical technique that is excited by electrons is Auger spectroscopy. This is generally excited by electrons with a few keV in energy, which can be generated using compact electron sources and can be focused to spot sizes of less than a few microns.

5.2.3.1 Surface Analysis by Auger Spectroscopy

Materials are ionized by incident high-energy electrons. In particular, high-binding energy electrons, so-called core states, can be removed if the incident electron kinetic energy is greater than the ionization energy of the electron in that state. In general, the ionization cross section, or the probability that an electron is removed, increases as the electron energy increases above the threshold ionization energy, reaching its maximum value when the incident electron kinetic energy reaches about 2–3 times the ionization energy.

The ionized atom in the material is unstable and can relax by lower-binding-energy electrons undergoing a transition onto the vacant energy level. The energy gained by such a transition can be released either by emitting a photon or another electron. Since the photons emitted during this process generally have energies in the X-ray region of the spectrum, this process is known as X-ray fluorescence. The emission of an electron during the decay process was first described by Pierre Auger and this three-electron process forms the basis for Auger stectroscopy [24].

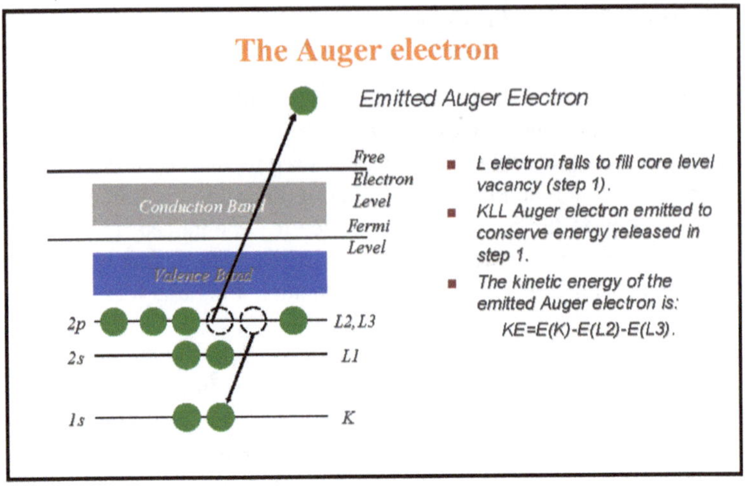

Fig. 5.3 Schematic depiction of the Auger process

The energy of the emitted Auger electron therefore depends on the energies of three electron states in the atom; the energy of the state which was initially ionized, the state from which the decaying electron originated and finally, the state from which the Auger electron was emitted (Fig. 5.3). Since the energies of the electrons in a particular atom are characteristic of that particular atom, the pattern of kinetic energies of the electrons emitted by a particular atom are characteristic of that atom so that the Auger spectrum provides detailed information on the chemical composition of the sample.

In order to collect an Auger spectrum, the sample is illuminated with high-energy electrons (with kinetic energies of a few keV), and the energies of the emitted electrons are analysed using an electron energy analyser. They are invariably electrostatic analysers that are essentially parallel plates across which a voltage is applied to create an electric field, which deflects the electrons in an arc of radius r. The magnitude electric field E across parallel plates separated by a distance d is given by $E = \frac{V}{d}$, so that the force F exerted on an electron with charge e is $F = \frac{\mathrm{eV}}{d}$. This causes the electron to move in an arc of radius r such that the centripetal force $(\frac{\mathrm{mv}^2}{r})$ balances the electrostatic force. Since the kinetic energy is given by $K = \frac{\mathrm{mv}^2}{2}$, this leads to a relationship between the kinetic energy and voltage across the plates:

$$K = \frac{\mathrm{er}}{2d}V. \tag{5.8}$$

In practice, the parallel plates are bent to have some fixed radius so that r and d are fixed and depend on the geometry of the electron energy analyser. There are several designs of electrostatic analysers, but they all operate on the same physical principle, and an example of the schematic of a typical electron energy analyser is

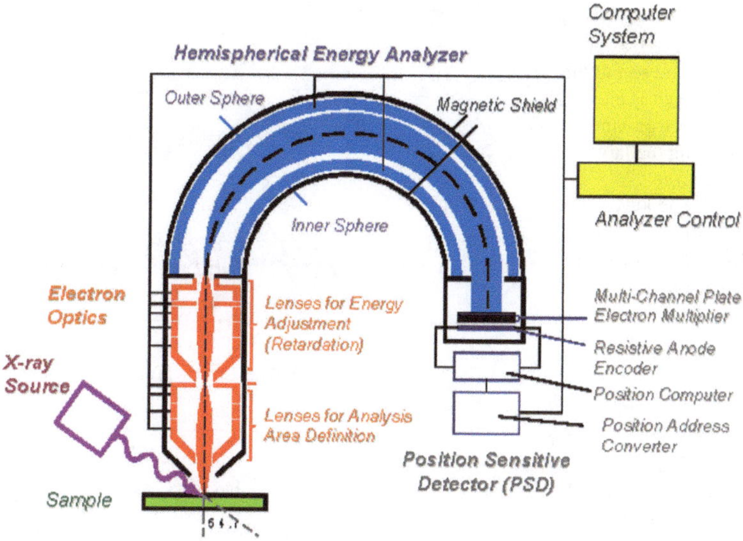

Fig. 5.4 Schematic depiction of a hemispherical electron energy electron analyzer

shown in Fig. 5.4. Here the plates have been bent into the form of two concentric hemispheres and is known as the hemispherical analyzer. Other configurations such as the cylindrical mirror analyzer and the 127° sector analyzer are used [24]. Similar types of analyzer are used for X-ray photoelectron spectroscopy. In order to provide good energy resolution, narrow slits are placed at the entrance and exit of the analyzer plates. The system shown in Fig. 5.4 includes an electron lens at the entrance of the analyzer to optimize the collection of electrons and an electron detector located after the exit slit. Based on (5.8), the spectrum is recorded by monitoring the number of electrons passing through the analyzer as a function of the voltage across the analyser plates, and this technique is used in Auger spectroscopy. It should be noted that the resolution of electrostatic analyzers also depends on the electron kinetic energy. While this is not an issue for measuring Auger spectra since the lines are generally quite broad and the measured kinetic energies relatively low, it is important in XPS where a higher resolution is required and electron analyzer is therefore configured slightly differently for XPS (see Chap. 4).

An complication with Auger spectroscopy is that, since electrons interact strongly with solids (and thus have a short mean-free path), high-energy electrons create an intense background of inelastically scatted electrons so that the peaks in the Auger spectrum are generally superimposed on a large background intensity. This is illustrated in Fig. 5.5, where the top curve shows the relatively small Auger peaks superimposed on an intense background due to the emission of inelastic electrons. Since the inelastic background intensity varies slowly with kinetic energy, while the features due to Auger emission are narrower, the first-derivative spectrum (lower curve), shows well-defined peaks with both positive and negative excursions. Note also that

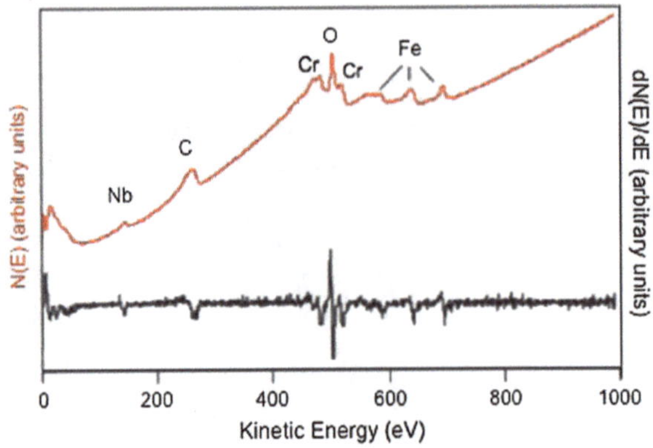

Fig. 5.5 Auger spectra of an oxidized Fe–Cr–Nb alloy. The top spectrum shows the number of counts as a function of electron kinetic energy showing the large inelastic electron background. The lower curve shows the differentiated spectrum. Reproduced with permission from [25]

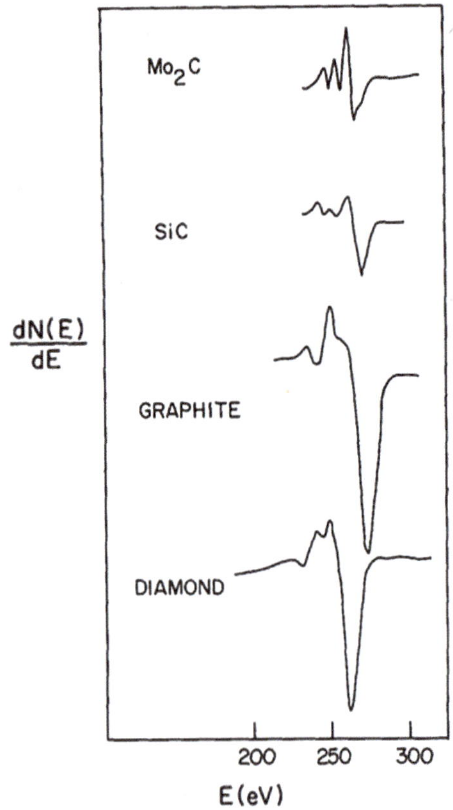

Fig. 5.6 Carbon KLL Auger spectra of various carbon-containing materials illustrating the effect of chemical environment on Auger lineshape. Reproduced with permission from [26]

the measured kinetic energies are characteristic of the elements in the sample; in this case, distinct features due to Nb, Cr and Fe are clearly evident, while the large oxygen peak indicates that the alloy is oxidized and also contains some carbon contamination.

The spectrum in Fig. 5.5 illustrates the utility of Auger spectroscopy for analyzing the elements in a sample and, because it relies on electrons, it is sensitive only to the outermost few nanometers of the sample. However, unlike XPS, it is somewhat insensitive to the chemical nature of the species near the surface. The exception is cases in which the Auger process involves the valence electrons of the element since these are strongly modified by changes in the chemical environment [26]. An example is shown in Fig. 5.6 for various carbon-containing species including carbides, graphite and diamond. Since the carbon Auger transition involves an initial ionization of the carbon 1 *s* electron (with a binding energy of ~280 eV), followed by a decay from the valence orbitals of the carbon, and the emission of an Auger electron, also from the carbon valence orbitals, significant changes occur in the lineshapes.

The composition of the surface region can also be estimated by measuring the peak-to-peak amplitudes of the features for each element in the first-derivative spectrum. The atomic composition of the sample can then be estimated using tabulated values of sensitivity factors [27].

Application of Auger Spectroscopy to Tribological Processes; Shear-Induced Loss of Sulfur

This section discusses the use of Auger spectroscopic surface analyses for studying tribochemical kinetics. The experiments are carried out in an ultrahigh vacuum (UHV) chamber, operating at a base pressure of ~1×10^{-10} Torr, where the sample can be rubbed using a tungsten carbide ball that slides on the surface with various sliding speeds and loads [8]. Sliding creates a wear track on the surface that is approximately 100 μm wide, so that the analysis spot size must be less than this to ensure that only the composition inside the wear track is analyzed. In order to carry out the experiment, a clean copper sample is rubbed ~70 times to create an initial wear track. The surface is then exposed to a model gas-phase lubricant, dimethyl disulfide (DMDS, CH_3–S–S–CH_3), which reacts by S–S bond scission to deposit methyl thiolate species (CH_3–S) on the surface. An initial Auger spectrum of the wear track is collected prior to rubbing and the peak-to-peak amplitude of the sulfur KLL Auger signal at ~160 eV kinetic energy is normalized to the peak-to-peak amplitude of the Cu LMM feature at ~920 eV. Subsequent measurements are collected in this way to eliminate the effect of variations in experimental conditions such as electron beam current.

It is a challenge in any tribological experiment to ensure that only the worn region of the sample is analyzed. In the case of the Auger spectra collected here, the spatial resolution was measured using a 100 μm diameter silver wire attached to the surface to mimic a wear track and Auger spectra collected to ensure that only silver and no copper is detected. Note that smaller electron beam spot sizes generally yield lower electron fluxes and thus less intense spectral signals so the experiment involves a compromise between surface sensitivity and spatial resolution.

In order to carry out Auger analyses sequentially on the same rubbed region of the sample, the electron gun was oriented such that the electron beam was incident on the

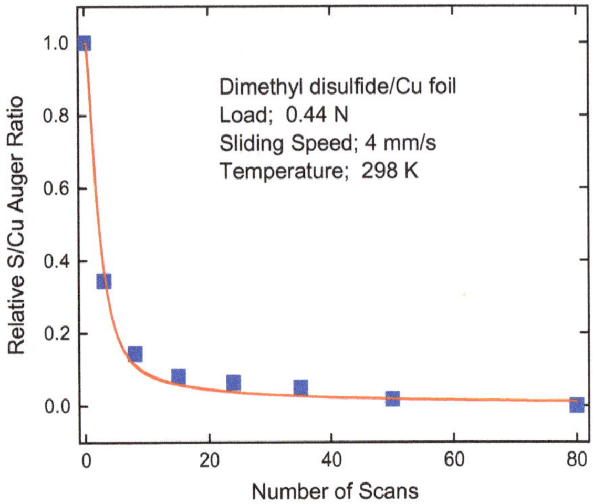

Fig. 5.7 Loss of sulfur from the surface as a function of the number of rubbing cycles following the adsorption of dimethyl disulfide on copper. Reproduce with permission from [12]

wear track to allow the sample to move repeatably between the rubbing experiments and the surface analyses. The electron gun used for these experiments also included deflection plates that allowed the electron beam to be rastered across the wear track. Scattered electrons were detected using a channeltron located inside the vacuum chamber which allowed scanning electron microscope images to be collected and the position of the wear track to be identified.

After obtaining the initial sulfur signal, the surface was rubbed at a normal load of 0.44 N using a sliding speed of 4 mm/s with the sample at room temperature. The Auger spectrum was then collected by rotating the sample towards a cylindrical mirror electrostatic electron energy analyzer (CMA) to measure the peak-to-peak ratio of the sulfur Auger signal ratioed to that of the copper substrate. The sample was then rotated back to its original position and rubbed once again to yield a plot of S/Cu Auger signal as a function of the number of passes of the ball over the surface (Fig. 5.7). This shows a rapid decrease in the amount of surface sulfur, so that, after ~80 passes over the surface, no sulfur is detected.

Because the experiment is carried out in UHV, the amount of adsorption from the gas phase when the experiment is being carried out is negligible so that the data shown in Fig. 5.7 are due to the shear-induced effects that result in the removal of sulfur from the surface. Since the data yield a time sequence of the surface coverage, the results can be analyzed to yield kinetic parameters. Accomplishing this requires a kinetic model to be developed, which, in turn, relies on being able to identify the elementary steps occurring during the tribochemical reaction. This will be discussed in greater detail in Sect. 5.3.4.

5.2.3.2 Surface Analysis by Low Energy Electron Diffraction

Low-energy electron diffraction (LEED) is a surface-sensitive electron diffraction technique that provides information on the crystal structure and order of surfaces. It is based on de Broglie's proposal that particles can have wave-like properties [28], where the wavelike properties (wavelength, λ) are related to particle properties (momentum, p) through: $\lambda = \frac{h}{p}$, where h is Planck's constant. This indicated that electrons should diffract from crystalline surfaces in a manner analogous to X-rays and this was confirmed by the first LEED experiments, although they did not call it that, by Davisson and Germer [29]. Because of the short mean-free path of electrons in solids, the diffraction pattern is sensitive to the outermost layers of the sample and this was one of the first techniques applied to the study of surfaces. The principle of the LEED experiment is relatively straightforward and consists of a monoenergetic beam of electrons with energies ~100 eV, to give de Broglie wavelengths of the order of lattice spacing in metals, impinging on a metal single crystal and observing the resulting diffractions pattern. However, as noted above, high-energy electrons produce large number of inelastically scattered electrons when interacting with surfaces and must be filtered out to allow the elastic LEED electrons to be observed. By definition, the energies of the inelastic electrons are lower than the incident beam energy and so can be filtered by a retarding grid. A typical LEED apparatus is shown in Fig. 5.8 and consists of concentric, electron-transparent grids and a fluorescent screen to detect the diffracted electrons. The sample is located at the center of curvature of the grids so that electrons diffracted from the sample impinge normal to the grids and a fluorescent screen to image the diffracted electrons. An electron gun protrudes from the center of the LEED screen and is incident onto the single crystal sample. From the de Broglie relation above, the electron wavelength can be varied merely by changing the electron kinetic energy E_p as $\lambda = \frac{h}{\sqrt{2\mathrm{mE}_p}}$ where m is the electron mass. In order to remove any inelastic electron, a retarding voltage of $+E_r$ is applied to one of the grids so that $E_r < E_p$ to allow only elastically scattered electrons to pass through the grid. Grounded grids are placed between the retarding grids and the sample, and the sample and the screen to ensure that there are field-free regions between them. The LEED pattern is imaged using a fluorescent screen using a high applied potential E_s to accelerate the electrons to the screen.

The beam is diffracted through an angle Θ given by the Bragg equation:

$$n\lambda = 2d \sin \Theta, \tag{5.9}$$

where d is the lattice spacing and n is the diffraction order and can be used to precisely measure surface lattice spacings. Since the beam energy, and hence the electron wavelength, can be easily varied, the diffraction pattern can be collected at a variety of electron wavelengths and the resulting intensity versus beam voltage data can be used to obtain detailed structural information [30].

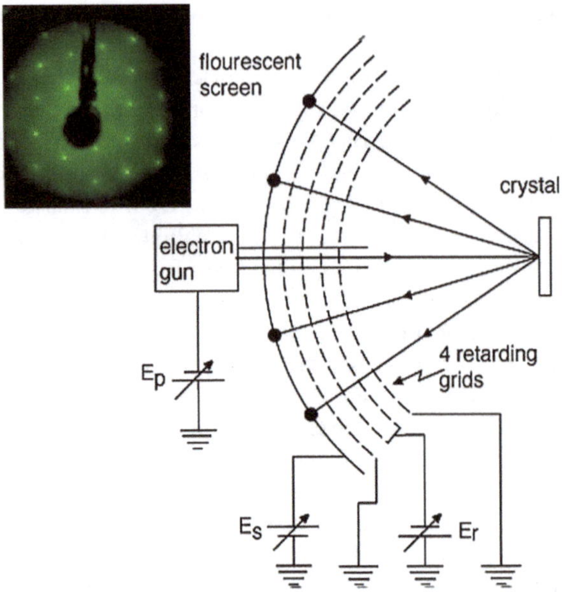

Fig. 5.8 Schematic diagram of a LEED apparatus. An example of a LEED pattern is also shown

Application of Low Energy Electron Diffraction to Material Transfer in a Contact

Buckley used LEED experiments to explore the transfer of material from the (111) surface of gold to the (111) face of copper-aluminum alloys as a function of alloy composition [31]. The experiments were carried out in a UHV chamber using various aluminium-copper alloys by bringing a clean gold sample into contact with it. The applied load was controlled by external magnets through an arm that contained the sample at the end. This arrangement allowed the adhesion force between gold and the copper-aluminum alloys to be measured after contacting them. Figure 5.9 display the adhesive force after contacting the surfaces for 10 s with a 20 mg load. The surfaces of the alloys were examined by LEED after contacting with gold where, for example, epitaxial gold films were found to be transferred to the copper-aluminum alloys with less than 0.5% aluminum.

5.3 Elementary Steps in Tribochemical Reactions

As indicated in the Introduction, fully understanding tribochemical reactions, their rates and the processes that control them require an understanding of the elementary steps in the reaction that lead to the formation of an antiwear and/or friction-reducing film, the so-called reaction pathway. A general outline of the processes that lead to the formation of a boundary film are illustrated in Fig. 5.10. The first step is necessarily

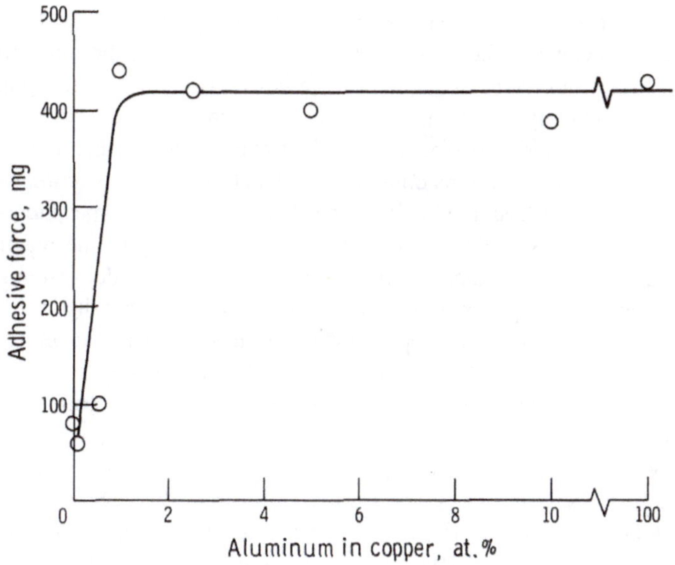

Fig. 5.9 Adhesive force of (111) gold to a (111) surface of aluminum and various aluminum-copper alloys. Initial applied load, 20 mg, Contact time 10 s. Reproduced with permission from [31]

the adsorption of the lubricant additive or gas-phase lubricant onto the surface. This chemistry will depend both on the nature of the surface and additive as well as the experimental conditions such as the temperature and concentration.

These initial thermal reactions may then be followed by sliding effects that may induce further mechano- or tribochemical reactions. These may be deleterious, as in the case of the decomposition of organic friction monolayers (comprising self-assembled monolayers of adsorbates) or beneficial when it eventually leads to the formation of a tribofilm. The final step is the formation of the tribofilm. This may, in principle, either occur by the sliding-induced products growing a film on the surface [32, 33] or by penetrating the subsurface region of the sample. The latter process will create a new surface on which the lubricant can adsorb, thus start-

Fig. 5.10 Schematic diagram indicating possible elementary steps in a tribochemical reaction to form an antiwear or friction reducing film

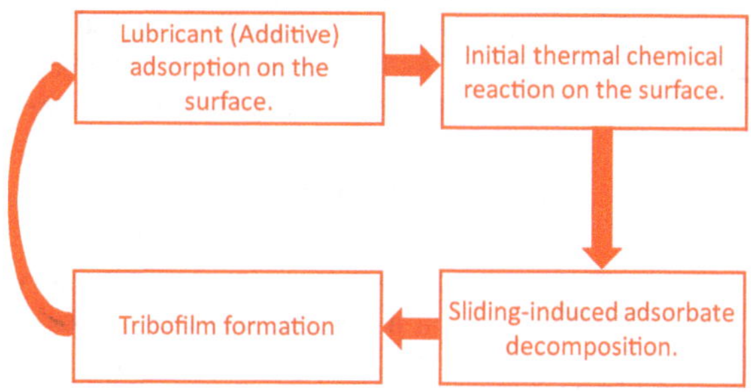

ing the second tribochemical reaction cycle. As this cycle proceeds, the nature of the surface will evolve as the tribofilm forms so that both the thermal and sliding-induced reactions and their rates may be different for each cycle, thus emphasizing the potential complexity of fully understanding tribochemical reaction pathways. In addition, many engineering surfaces themselves may be structurally and chemically rather complex, being covered by contaminants and oxides from reaction with atmospheric gases, thus emphasizing the importance of carrying out experiments in UHV on well-characterized samples. Furthermore, the complexity of the reaction cycle shown in Fig. 5.10 suggests that it is important to use simple model systems. For this purpose, we use the gas-phase lubrication of copper by dimethyl disulfide (DMDS) to illustrate the use of in-situ and "pseudo" in-situ approaches to understanding the elementary steps in a tribochemical reaction cycle and their kinetics. While this is not a commercial lubricant, it does contain the key aspects of commercial lubricant additive since it contains a S–S linkage, typical of sulfur-containing lubricant additives, as well as a hydrocarbon functionality. The discussion will address the following key steps. It is first necessary to demonstrate that DMDS does act as a gas-phase lubricant and then to identify the nature of the elementary steps in the reaction cycle shown in Fig. 5.10. The first two steps in the cycle in Fig. 5.10 involve investigations of the thermal surface chemistry and the nature and stability of the initially adsorbed species arising from exposure to DMDS and are explored using standard UHV surface science techniques. The final steps in the reaction occur at the sliding interface and illustrate the use of the in-situ techniques described above. Finally, once the nature of the elementary steps has been identified, this will allow a simple kinetic model to be constructed to establish how the surface composition evolved during rubbing.

5.3.1 Gas-Phase Lubrication of Copper by DMDS

The gas-phase lubrication of copper by DMDS is explored in UHV. As noted above, DMDS is selected because it mimics commonly used lubricant additives and sulfur-containing molecules are used as to lubricate iron [34–37]. The tribological experiments were carried out using the following protocol. First, an initial wear track was made on the clean copper substrate without DMDS dosing to obtain an initially constant friction coefficient to ensure that any subsequent changes in friction were due to DMDS dosing. As shown in Fig. 5.11, this resulted in a friction coefficient decrease from an initial value of ~0.82 to ~0.53 after rubbing 40 times, and remained essentially constant thereafter. The sample was then pressurised with 5×10^{-8} Torr of DMDS to cause a rapid initial decrease in friction, followed by a slower change for longer rubbing times, thereby showing that DMDS does act as a gas-phase lubricant. Experiments were also carried out using a model lubricant that consisted of 5% DMDS dissolved in a poly α-olefin and found a friction coefficient of ~0.05 for sliding on copper. This work established that DMDS does act as a lubricant for copper [38].

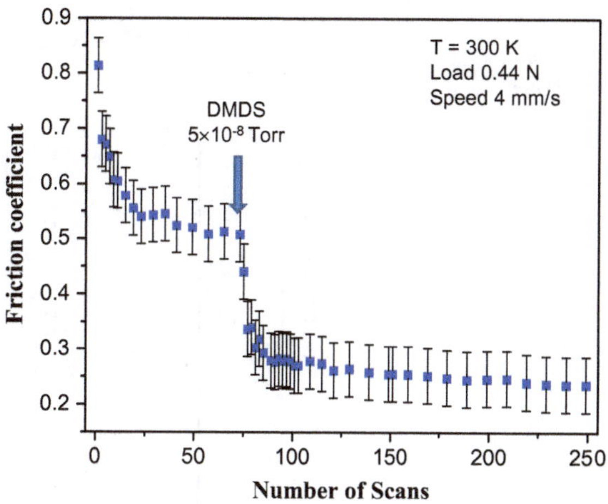

Fig. 5.11 Plot of friction coefficient versus number of cycles measured in the ultrahigh vacuum tribometer at a sliding speed of 4×10^{-3} m/s and a normal load of 0.44 N. The clean surface was initially rubbed 70 times to reach a steady-state value of friction coefficient, and then dimethyl disulfide was introduced *via* a dosing tube at a background pressure of ~5×10^{-8} Torr and the friction coefficient then recorded in the presence of gas-phase DMDS. Reproduced with permission from [39]

5.3.2 Thermal Chemistry of DMDS on Copper

The initial steps in the tribochemical reaction cycle involve the adsorption and thermal reaction of DMDS on the clean copper surface (Fig. 5.10) in UHV. Such studies are also required to explore the stability of the resulting surface species. Many of the techniques that are used to study tribological interfaces such as XPS (see previous chapter), Auger spectroscopy (Sects. 5.2.3.1), LEED (Sect. 5.2.3.3) and infrared spectroscopy (Sect. 5.2.3) are equally applicable to studying surface in UHV. An additional technique, that is not applicable to in-situ studies of the tribological interface, but is extremely useful for following surface chemical reaction pathways and their kinetics in UHV is temperature-programmed desorption (TPD). This technique involves adsorbing the molecule of interest onto a clean sample held at a sufficiently low temperature (usually ~80 K, by cooling the sample with liquid nitrogen) to prevent decomposition of the adsorbate. The sample is then placed close to and in-line-of sight of a mass spectrometer and is heated linearly as a function of time and the desorbing products monitored as a function of the sample temperature. The detection of gas-phase products is analogous to the detection of gas-phase products during sliding described in Sect. 5.2.1.1. The resulting plot of mass spectrometer signal as a function of time generally consists of peaks, where the onset temperature of the peak depends on the activation energy of the surface reaction giving rise to the peak. However, as the surface becomes depleted of reactants, the mass spectrometer signal decreases once again to produce a peaked structure.

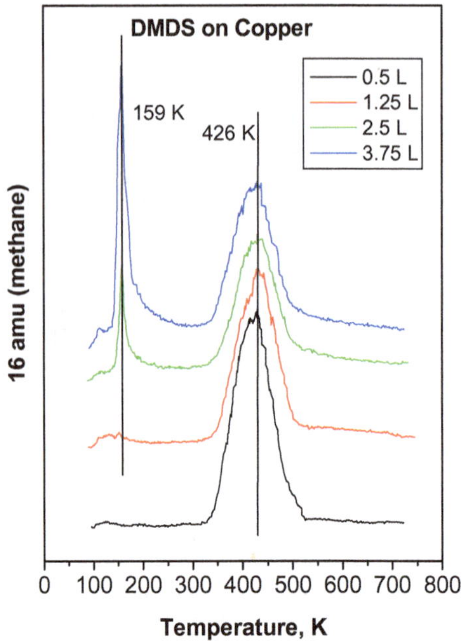

Fig. 5.12 TPD profile of DMDS adsorbed on copper. Reproduced with permission from [10]

The nature of the molecular species giving rise to a particular peak can be obtained by measuring the peak intensity at various masses and comparing with mass spectrometer ionizer fragmentation patterns of various compounds. A typical series of TPD profiles for DMDS adsorbed on copper at ~100 K are shown in Fig. 5.12 when monitoring the 16 amu mass spectrometer signal [10, 11]. The data are collected for various initial exposures of DMDS, which are indicates in Langmuirs (1 Langmuir (L) = 1×10^{-6} Torr s). This shows two distinct peaks, a sharp feature centered at ~159 K, and a broader feature at 426 K. By measuring the signals at other masses, it can be shown that the low-temperature (~179 K) peak is due to the desorption of condensed, molecular DMDS, while the higher-temperature (~426 K) is due to the reactive formation of methane due to the decomposition of adsorbed methyl thiolate (CH_3–S(ads)) species [10].

The maximum temperature of a desorption peak, T_p can be used to estimate the reaction activation energy assuming that the sample temperature varies linearly with time t:

$$T = T_0 + \beta t, \tag{5.10}$$

where T_0 is the initial temperature and β is the heating rate. For a first order reaction (5.1), and taking the amplitude of signal at a particular temperature in the TPD profile (Fig. 5.12) to be proportional to the reaction rate, r, gives:

$$-\frac{d\Theta}{dt} = r = A\exp\left(\frac{E_{act}}{k_B T}\right)\Theta, \tag{5.11}$$

where E_{act} is the activation energy for the process that gives rise to the desorption product, T is the absolute temperature and k_B the Boltzmann constant. The peak occurs when the desorption rate is the maximum. Putting $\frac{dr}{dt} = 0$ gives, for a first-order process, an equation for the activation energy in terms of T_p [9]:

$$\frac{E_{act}}{k_B T_p^2} = \left(\frac{A}{\beta}\right)\exp\left(\frac{E_{act}}{k_B T_p}\right). \tag{5.12}$$

This reveals that, for a first-order process, the value of T_p is independent of the adsorbate coverage, indicating that the methane formed in the TPD profile is due to a first-order process (methyl thiolate decomposition). In order to simplify this equation, it is found that a plot of E_{act} versus T_p is approximately linear for $10^{13} > \frac{A}{\beta} > 10^8$ and can be approximated by:

$$\frac{E_{act}}{k_B T_p} = \ln\left(\frac{AT_p}{\beta}\right) - 3.64. \tag{5.13}$$

Applying this "Redhead" equation [9] to the methane desorption peak in Fig. 5.12, and using a typical pre-exponential factor A for a first-order process of 10^{13} s^{-1}, gives an activation energy of ~100 kJ/mol. The reaction is confirmed using XPS, which shows the disappearance of carbon over a temperature range which methane desorbs and the persistence of sulfur on the surface. These results confirm that methyl thiolate species are stable on the surface at ~300 K, the temperature at which the tribological measurements are carried out.

5.3.3 *Identifying Elementary Steps During Sliding*

The subsequent sliding-induced elementary reaction steps are explored using a combination of the in-situ and pseudo in-situ techniques described above. It should be emphasized that the temperature rise for the relatively mild conditions used in the UHV tribometer is negligible [8]. The products formed during sliding were discussed in Sect. 5.2.1.1 and Fig. 5.1 shows the desorption of methane pulses as the pin slides over a methyl thiolate-covered surface. The rate of thiolate decomposition to form methane (and C_2 hydrocarbons) is accelerated by interfacial shear, since it does not occur thermally at this temperature and there are now a number of reactions for which this has been found including the decomposition of oxygen-containing adsorbates on graphene [15] and the growth of anti-wear films from ZDDP [32, 33]. However, unlike the thermal chemistry on copper, where adsorbed sulfur remains on the surface, Auger analyses of the wear track reveal that the sulfur is removed during sliding (Fig. 5.7, Sect. 5.2.2.2).

Because of the strong binding of sulfur to copper, sulfur is unlikely to desorb and no sulfur was detected in the gas phase, implying that shear causes sulfur to penetrate the subsurface region of the copper sample. This idea is tested by taking advantage of the observation made above that sulfur is thermodynamically stable on the surface of copper rather than in the bulk. That is, any subsurface sulfur is metastable so that heating the sample should cause sulfur to diffuse to the surface once again. However, the DMDS-dosed copped sample is covered by methyl thiolate species which must be removed for this experiment to be carried out. The results of this experiment are displayed in Fig. 5.13, which display the profiles of the sulfur coverage across a wear track measured using small-spot-size Auger spectroscopy. The center of the wear track is located at 300 μm. The baselines of the scans shown in Fig. 5.13a have as S/Cu Auger ratio of ~0.42, corresponding to the sulfur in a saturated monolayer of methyl thiolate species, while in Fig. 5.13b, the baseline has a S/Cu ratio of zero. Profile I shows a scan across a previously formed wear track for a copper surface that has been saturated with methyl thiolate species. It seems that slightly more methyl thiolate species can adsorb in the wear track, possibly because it is somewhat rougher than the surface outside the wear track. Profile II shows the effect of rubbing the surface 50 times, confirming that sulfur has been removed in the rubbed region as found in the data plotted in Fig. 5.7. This sulfur is proposed to have penetrated the bulk of the sample. In order to test this, the surface sulfur was carefully removed by Argon ion bombardment and the resulting profile of the sulfur Auger signal is shown in scan III; all of the surface sulfur has been removed. Profile IV shows the result of heating to ~780 K where now sulfur is detected on the surface. This is proposed to have diffused from the subsurface region and is wider than the original wear track presumably due to lateral diffusion of sulfur on the copper surface. This effect provides a shear-induced pathway for the low-temperature formation of a tribofilm and formally results in an oxidation of the copper by sulfur.

A possible explanation for this shear-induced surface-to-bulk transport come from molecular dynamics simulations [40–43] which suggest that shear can induced mixing of the surface regions of soft metals during sliding. Analytical models of this effect [44] show that the surface-to-bulk transport depends on the strain-rate sensitivity of the material. For copper, which has a low value of the strain-rate sensitivity [45–52], this theory predicts that the distance that the sulfur on the surface penetrates the bulk is proportional to the number of times that the surface has been rubbed and this dependence has recently been verified for subsurface sulfur in copper [53]. The elementary steps for the tribochemical reaction of DMDS with copper are summarized in Fig. 5.14.

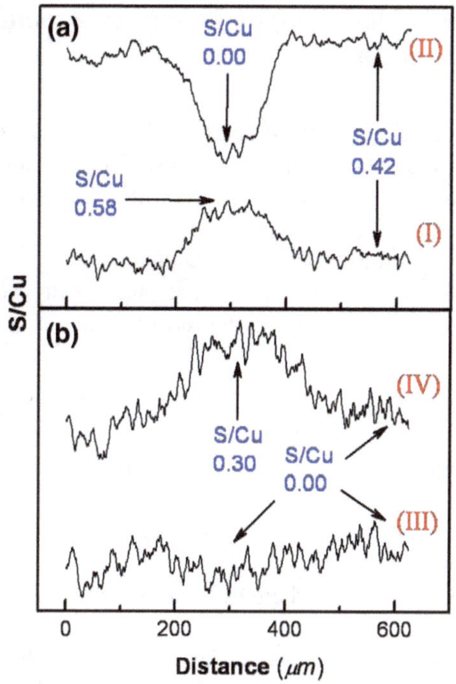

Fig. 5.13 Plots of sulfur peak-to-peak LMM Auger intensity (152 eV kinetic energy) ratioed to the peak-to-peak intensity of the Cu LMM Auger transition (920 eV kinetic energy) as a function of position through the wear track, which is centered at ~300 μm. **a** Shows the effect of rubbing, where profile I plots the sulfur coverage after the sample was dosed with DMDS after rubbing the clean surface, and profile II shows the result of subsequently rubbing the surface 50 times under a normal load of 0.44 N at a sliding speed of 4×10^{-3} m/s. **b** Shows that all of the sulfur is removed from the surface by argon ion bombardment (profile III), but that subsequently heating the sample to ~780 K causes sulfur to segregate to the surface only in the region of the wear track (profile IV). Reproduced with permission from [12]

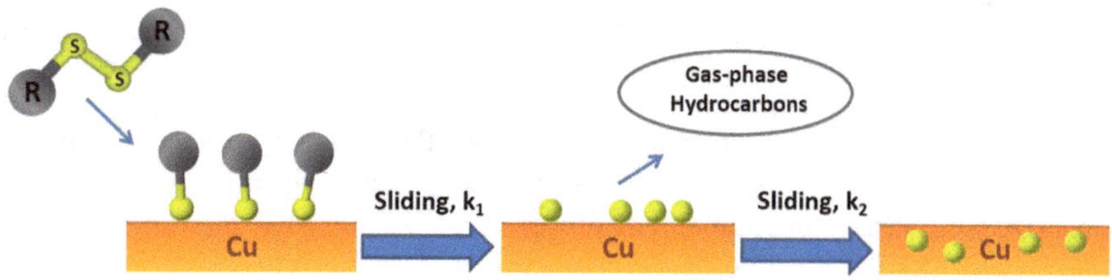

Fig. 5.14 A schematic of the elementary steps in the tribochemistry of DMDS on copper

5.3.4 Microkinetic Modelling of Tribofilm Formation for DMDS on Copper

Having defined the elementary steps in the tribochemical reaction pathway sets the stage for developing microkinetic models to describe the film formation reaction rates based on the reaction pathway (Fig. 5.14), which consists of a sequence of two shear-induced reactions: $RS_{(ads)} \rightarrow S_{(ads)} \rightarrow S_{(subsurface)}$, where the rate constant for the first reaction is k_1 and for the second is denoted as k_2.

The data in Fig. 5.1 for the formation of gas-phase hydrocarbons indicated that methyl thiolate decomposes via a first-order reaction (Sect. 5.2.1):

$$-\frac{d\Theta_{th}}{dt} = k_1\Theta_{th}, \tag{5.14}$$

where Θ_{th} is the thiolate coverage. Writing $k_1' = k_1 t_C$ and integrating (5.14) gives:

$$\Theta_{th} = \exp\left(-k_1' p\right), \tag{5.15}$$

where p is the number of passes, where the method for obtaining k_1' is discussed in Sect. 5.2.1.

Denoting the coverage of adsorbed sulfur, $S_{(ads)}$ formed by this reaction as Θ_S then, from the sequential reaction shown in Fig. 5.14, adsorbed sulfur is formed by alkyl thiolate decomposition and removed by surface-to-bulk transport:

$$\frac{d\Theta_S}{dt} = k_1\Theta_{th} - k_2\Theta_S, \tag{5.16}$$

and assuming that surface-to-bulk transport occurs via a first-order process as will be discussed in greater detail below, substituting for the alkyl thiolate coverage Θ_{th} from (5.15) and solving gives:

$$\Theta_S(p) = \frac{k_1{}'}{k_2{}' - k_1{}'}\left(e^{-k_1' p} - e^{-k_2' p}\right), \tag{5.17}$$

where $k_2' = k_2 t_C$. The rate constant k_2' is that for the transport of adsorbed sulfur into the subsurface of copper where the film penetrates a distance $d(p)$ into the subsurface region as a function of the number of scans as: $d(p) = d_p p^x$, where d_p is the distance that the sulfur penetrates the subsurface region per pass [44] and it has been shown the $x \sim 1$ [53]. If the initial film thickness at $t = 0$ is d_0 (corresponding to the thickness of the original saturated sulfur overlayer), then it can be shown that $k_2' = \frac{d_p}{t_C}$ [54].

Since the Auger analyses of the surface during rubbing (Fig. 5.7) measure the sulfur from both adsorbed sulfur and alkyl thiolate species, the coverage of all sulfur-containing species is calculated as $\Theta_S(\text{tot}) = \Theta_S + \Theta_{th}$, and this gives:

$$\Theta_S(\text{tot}) = \frac{k_2'}{k_2' - k_1'} \mathrm{e}^{-k_1' p} - \frac{k_1'}{k_2' - k_1'} \mathrm{e}^{-k_2' p}. \tag{5.18}$$

Finally, the coverage of the clean surface, Θ_{clean} is calculated from:

$$\Theta_{\text{clean}} = 1 - \Theta_S - \Theta_{\text{th}}. \tag{5.19}$$

However, surface spectroscopies, including Auger spectroscopy, are sensitive not only to adsorbates but also include signals from the sub-surface region [24, 55]. It has been previously shown that the Auger signal due to subsurface sulfur varies as $1/p$ [53] so that the Auger signal contribution arising from subsurface sulfur is taken to be $\propto \frac{\Theta_{\text{subsurface}}}{p+1}$ where $\Theta_{\text{subsurface}}$ is the subsurface sulfur coverage and $p + 1$ is used to avoid a singularity at $p = 0$. Therefore the sulfur Auger signal $I_{\text{Auger}}(p)$ is given by:

$$I_{\text{Auger}}(p) \propto I_{\text{Auger}}(S) + I_{\text{Auger}}(\text{th}) + \frac{\alpha \Theta_{\text{subsurface}}}{p + 1}, \tag{5.20}$$

where α is a constant. Since $\Theta_{\text{subsurface}} = 1 - \Theta_{\text{th}} - \Theta_S$, this yields:

$$I_{\text{Auger}}(p) = \left(\frac{k_2'}{k_2' - k_1'} \mathrm{e}^{-k_1' p} - \frac{k_1'}{k_2' - k_1'} \mathrm{e}^{-k_2' p} \right) \left(1 - \frac{\alpha}{p + 1} \right) + \left(\frac{\alpha}{p + 1} \right), \tag{5.21}$$

which is normalized to the initial Auger signal of the clean surface.

The above kinetic equations are used to analyze the shear-induced elementary reaction steps of alkyl thiolates on copper (Fig. 5.14). The rate of decomposition of methyl thiolate species, analyzed as described in Sect. 5.2.1.1 gives $k_1' = 0.63 \pm 0.03$. In order to obtain k_2', the sulfur Auger data (Fig. 5.7) are fit to (5.21) to yield values of $\alpha = 0.9 \pm 0.1$ and $k_2' = 2.5 \pm 0.5$. The line through the experimental data in Fig. 5.14 is a fit to this equation.

Since these kinetic parameters derive from a pseudo in-situ technique, Auger spectroscopy, the results are verified using an in-situ method measuring the friction coefficient as a function of number of rubbing cycles (Fig. 5.15). Adapting (5.4) (Sect. 5.2.1.2) to the reaction sequence shown in Fig. 5.14 gives an equation for the friction coefficient $\mu(p)$ as a function of the number of passes p:

$$\mu(p) = \mu_{\text{th}} \Theta_{\text{th}}(p) + \mu_S \Theta_S(p) + \mu_{\text{clean}}(1 - \Theta_{\text{th}}(p) - \Theta_S(p)), \tag{5.22}$$

where μ_{th}, μ_S and μ_{clean} are the characteristic friction coefficients of the thiolate-covered, sulfur-covered and clean surfaces, respectively. Here, the rate constants were constrained to be in the range determined above and μ_{clean} was allowed to vary between 0.45 and 0.55 found for sliding on clean copper. The fit to the data is shown as a solid line in Fig. 5.15 and yields $\mu_{\text{th}} = 0.07 \pm 0.02$, $\mu_S = 0.39 \pm 0.06$, and $\mu_{\text{clean}} = 0.47 \pm 0.05$. The agreement between experiment and theory suggests that the value of k_2' measured from the Auger data is reasonable. The resulting plot of

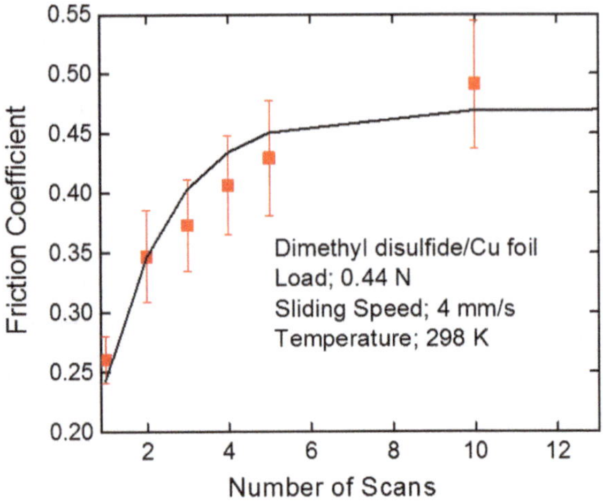

Fig. 5.15 Plot of the friction coefficient as a function of the number of scans while rubbing a methyl-thiolate-covered copper surface at a normal load of 0.44 N and a sliding speed of 4 mm/s. The line shows a fit to the data. Reproduced with permission from [54]

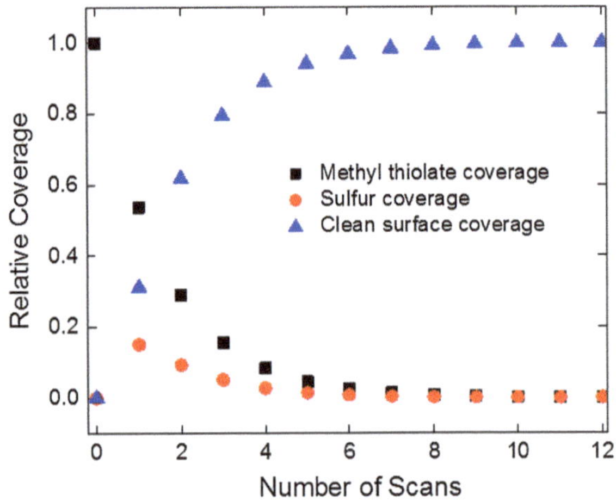

Fig. 5.16 Plot of the calculated coverages of adsorbed methyl thiolate species (■), adsorbed sulfur (●) and the clean copper surface (▲) as a function of the number of scans over a methyl-thiolate covered surface with a normal load of 0.44 N and a sliding speed of 4 mm/s. Reproduced with permission from [54]

the adsorbate coverages as a function of the number of scans for an initially methyl thiolate-saturated surface, at a load of 0.44 N and a sliding speed of 4 mm/s, is shown in Fig. 5.16. This shows that a clean surface is produced after ~8 scans, since the clean surface coverage (▲) becomes almost unity after this number of scans, thereby allowing the tribochemical cycle to be repeated to eventually form a sulfur-containing boundary film. In addition, as a consequence of k_2' being much larger than k_1', the sulfur is transported into the subsurface region almost as quickly as it is formed on the surface resulting in the adsorbed sulfur coverage always remaining less than 0.2 monolayers (●).

5.4 Conclusions

This Chapter describes a number of in-situ and pseudo in-situ techniques that can be used to explore the elementary steps in tribochemical reactions to form friction-reducing or antiwear films under UHV conditions. The application of some of these techniques is illustrated using a simple model system of DMDS reacting with copper to form a sulfur-containing tribofilm. The results of these experiments allow the elementary steps in the reaction pathway to be identified. In particular, two shear-induced processes were found. The first consisted of an increase in the rate of decomposition of methyl thiolate species into adsorbed sulfur and gas-phase hydrocarbons. Here, the shear-induced reaction pathway was identical to that found for the thermal reaction. This behaviour can be rationalized using model that was first applied to cell adhesion [56] but has its roots in earlier models proposed by Prandl and Eyring [57] and has also been shown to apply to adsorbate decomposition measured in the AFM [8]. This approach of in-situ measurements of tribochemical reaction pathways on model systems will lead to a more detailed fundamental understanding of the way in which interfacial shear accelerates the rates of surface chemical reactions and lead to the formation of tribofilms.

The second shear-induced step in the reaction pathway consisted of the surface-to-bulk transport of sulfur to initiate the formation of a tribofilm. As shown in Fig. 5.16, this results in the regeneration of a clean surface to allow methyl thiolate species to once again adsorb onto copper to provide a tribochemical reaction cycle (Fig. 5.10). However, alternative film formation processes have also been observed with ZDDP, where the growing film forms on top of the substrate [32, 33].

Finally, while this Chapter has focussed on the film formation chemistry because of its relevance to tribology, the reaction shown in Fig. 5.14 could equally as well be viewed as a tribocatalytic reaction in which a combination of surface reactivity and interfacial shear combine to accelerate the rate of DMDS decomposition into small hydrocarbons and form a copper sulfide film.

References

1. T.J. Blunt, P.V. Kotvis, W.T. Tysoe, Determination of interfacial temperatures under extreme pressure conditions. Tribol. Lett. **2**(3), 221–230 (1996)
2. K. Holmberg, P. Andersson, A. Erdemir, Global energy consumption due to friction in passenger cars. Tribol. Int. **47**, 221–234
3. K.J. Laidler, *Chemical kinetics* (McGraw-Hill, New York, 1965)
4. O.A. Mazyar, H. Xie, W.L. Hase, Nonequilibrium energy dissipation at the interface of sliding model hydroxylated Î±-alumina surfaces. J Chem. Phys. **122**(9), 094713 (2005)
5. W.G. Sawyer, K.J. Wahl, Accessing inaccessible interfaces: in situ approaches to materials tribology. MRS Bulletin **33**(12), 1145–1150 (2008)
6. S. Mori, W. Morales, Tribological reactions of perfluoroalkyl polyether oils with stainless steel under ultrahigh vacuum conditions at room temperature. Wear **132**(1), 111–121 (1989). https://doi.org/10.1016/0043-1648(89)90206-8

7. I. Minami, T. Kubo, S. Fujiwara, Y. Ogasawara, H. Nanao, S. Mori, Investigation of tribochemistry by means of stable isotopic tracers: TOF-SIMS analysis of langmuir-Blodgett films and examination of their tribological properties. Tribol. Lett. **20**(3–4), 287–297 (2005). https://doi.org/10.1007/s11249-005-9068-2
8. H.L. Adams, M.T. Garvey, U.S. Ramasamy, Z. Ye, A. Martini, W.T. Tysoe, Shear-induced mechanochemistry: Pushing molecules around. J. Phys. Chem. C **119**(13), 7115–7123 (2015). https://doi.org/10.1021/jp5121146
9. P.A. Redhead, Thermal desorption of gases. Vacuum **12**, 9 (1962)
10. O.J. Furlong, B.P. Miller, Z. Li, J. Walker, L. Burkholder, W.T. Tysoe, The surface chemistry of dimethyl disulfide on copper. Langmuir **26**(21), 16375–16380 (2010). https://doi.org/10.1021/la101769y
11. O. Furlong, B. Miller, Z. Li, W.T. Tysoe, The surface chemistry of diethyl disulfide on copper. Surf. Sci. **605**(5–6), 606–611 (2011). https://doi.org/10.1016/j.susc.2010.12.026
12. O. Furlong, B. Miller, W. Tysoe, Shear-induced surface-to-bulk transport at room temperature in a sliding metal-metal interface. Tribol. Lett. **41**(1), 257–261 (2011). https://doi.org/10.1007/s11249-010-9711-4
13. R.G. Greenler, Reflection method for obtaining the infrared spectrum of a thin layer on a metal surface. J. Chem. Phys. **50**(5), 1963–1968 (1969). https://doi.org/10.1063/1.1671315
14. G. Wu, F. Gao, M. Kaltchev, J. Gutow, J.K. Mowlem, W.C. Schramm, P.V. Kotvis, W.T. Tysoe, An investigation of the tribological properties of thin KCl films on iron in ultrahigh vacuum: modeling the extreme-pressure lubricating interface. Wear **252**(7–8), 595–606 (2002)
15. J.R. Felts, A.J. Oyer, S.C. Hernández, K.E. Whitener Jr., J.T. Robinson, S.G. Walton, P.E. Sheehan, Direct mechanochemical cleavage of functional groups from graphene. Nat. Commun. **6**, 6467 (2015). https://doi.org/10.1038/ncomms7467
16. A. Cameron, R. Gohar, Theoretical and experimental studies of the oil film in lubricated point contact. Proc. Royal Soc. A Mathematical, Phys. Eng. Sci. **291**(1427), 520–536 (1966)
17. P.L. Stiles, J.A. Dieringer, N.C. Shah, R.P. Van Duyne, Surface-enhanced Raman spectroscopy. Annu. Rev. Anal. Chem. **1**(1), 601–626 (2008)
18. R.G. Greenler, Infrared study of adsorbed molecules on metal surfaces by reflection techniques. J. Chem. Phys. **44**(1), 310–315 (1966). https://doi.org/10.1063/1.1726462
19. K. Nakamoto, *Infrared and Raman Spectra of Inorganic and Coordination Compounds* (Wiley, New York, 1978)
20. N.B. Colthup, L.H. Daly, S.E. Wiberley, *Introduction to Infrared and Raman Spectroscopy*, 2nd edn. (Academic Press, New York, 1975)
21. F.M. Piras, A. Rossi, N.D. Spencer, Growth of tribological films: a in situ characterization based on attenuated total reflection infrared spectroscopy. Langmuir **18**(17), 6606–6613 (2002)
22. F. Mangolini, A. Rossi, N.D. Spencer, Chemical reactivity of triphenyl phosphorothionate (TPPT) with Iron: an ATR/FT-IR and XPS investigation. J. Phys. Chem. C **115**(4), 1339–1354 (2010)
23. E.H. Sondheimer, The mean free path of electrons in metals. Adv. Phys. **50**(6), 499–537 (2001). https://doi.org/10.1080/00018730110102187
24. B. Briggs, M.P. Seah, *Practical Surface Analysis: Auger and X-ray Photoelectron Spectroscopy* (Wiley, New York, 1996)
25. J.C. Vickerman, *Surface Analysis: The Principal Techniques* (Wiley, Chichester, 1997)
26. T.W. Haas, J.T. Grant, G.J. Dooley III, Chemical effects in auger electron spectroscopy. J. Appl. Phys. **43**(4), 1853–1860 (1972)
27. S. Mroczkowski, D. Lichtman, Calculated Auger yields and sensitivity factors for KLL–NOO transitions with 1–10 kV primary beams. J. Vac. Sci. Technol. A: Vac. Surf. Films **3**(4), 1860–1865 (1985)
28. L.D. Broglie, A tentative theory of light quanta. Philos. Mag. Ser. 6 **47**(278), 446–458 (1924). https://doi.org/10.1080/14786442408634378
29. C. Davisson, L.H. Germer, Diffraction of electrons by a crystal of nickel. Phys. Rev. **30**(6), 705–740 (1927)

30. J.B. Pendry, *Low Energy Electron Diffraction : The Theory and Its Application to Determination of Surface Structure* (Academic Press, London, 1974)
31. D.H. Buckley, A LEED Study of the Adhesion of Gold to Copper and Copper-Aluminum Alloys. NASA Technical Report NASA-TN-D-5351 (1969)
32. N.N. Gosvami, J.A. Bares, F. Mangolini, A.R. Konicek, D.G. Yablon, R.W. Carpick, Mechanisms of antiwear tribofilm growth revealed in situ by single-asperity sliding contacts. Science (2015). https://doi.org/10.1126/science.1258788
33. J. Zhang, H. Spikes, On the Mechanism of ZDDP antiwear film formation. Tribol. Lett. **63**(2), 1–15 (2016). https://doi.org/10.1007/s11249-016-0706-7
34. W. Davey, E.D. Edwards, The extreme-pressure lubricating properties of some sulphides and disulphides, in mineral oil, as assessed by the four-ball machine. Wear **1**(4), 291–304 (1958). https://doi.org/10.1016/0043-1648(58)90002-4
35. E.S. Forbes, The load-carrying action of organo-sulphur compounds—A review. Wear **15**(2), 87–96 (1970). https://doi.org/10.1016/0043-1648(70)90002-5
36. M. Kaltchev, P.V. Kotvis, T.J. Blunt, J. Lara, W.T. Tysoe, A molecular beam study of the tribological chemistry of dialkyl disulfides. Tribol. Lett. **10**(1), 45–50 (2001). https://doi.org/10.1023/a:1009020725936
37. J. Lara, T. Blunt, P. Kotvis, A. Riga, W.T. Tysoe, Surface chemistry and extreme-pressure lubricant properties of dimethyl disulfide. J. Phys. Chem. B **102**(10), 1703–1709 (1998). https://doi.org/10.1021/jp980238y
38. O.J. Furlong, B.P. Miller, P. Kotvis, W.T. Tysoe, Low-temperature, Shear-induced tribofilm formation from dimethyl disulfide on copper. ACS Appl. Mater. Interfaces **3**(3), 795–800 (2011)
39. O.J. Furlong, B.P. Miller, P. Kotvis, W.T. Tysoe, Low-temperature, shear-induced tribofilm formation from dimethyl disulfide on copper. ACS Appl. Mater. Interfaces **3**(3), 795–800 (2011). https://doi.org/10.1021/am101149p
40. D.A. Rigney, Transfer, mixing and associated chemical and mechanical processes during the sliding of ductile materials. Wear **245**(1–2), 1–9 (2000). https://doi.org/10.1016/s0043-1648(00)00460-9
41. X.-Y. Fu, D.A. Rigney, M.L. Falk, Sliding and deformation of metallic glass: Experiments and MD simulations. J. Non-Cryst. Solids **317**(1–2), 206–214 (2003). https://doi.org/10.1016/s0022-3093(02)01999-3
42. H.J. Kim, W.K. Kim, M.L. Falk, D.A. Rigney, MD simulations of microstructure evolution during high-velocity sliding between crystalline materials. Tribol. Lett. **28**(3), 299–306 (2007). https://doi.org/10.1007/s11249-007-9273-2
43. A. Emge, S. Karthikeyan, H.J. Kim, D.A. Rigney, The effect of sliding velocity on the tribological behavior of copper. Wear **263**, 614–618 (2007). https://doi.org/10.1016/j.wear.2007.01.095
44. S. Karthikeyan, H.J. Kim, D.A. Rigney, Velocity and strain-rate profiles in materials subjected to unlubricated sliding. Phys. Rev. Lett. **95**(10), 106001 (2005). https://doi.org/10.1103/physrevlett.95.106001
45. A. Mishra, M. Martin, N.N. Thadhani, B.K. Kad, E.A. Kenik, M.A. Meyers, High-strain-rate response of ultra-fine-grained copper. Acta Mater. **56**(12), 2770–2783 (2008). https://doi.org/10.1016/j.actamat.2008.02.023
46. M.A. Meyers, A. Mishra, D.J. Benson, Mechanical properties of nanocrystalline materials. Prog. Mater. Sci. **51**(4), 427–556 (2006). https://doi.org/10.1016/j.pmatsci.2005.08.003
47. A. Mishra, B.K. Kad, F. Gregori, M.A. Meyers, Microstructural evolution in copper subjected to severe plastic deformation: Experiments and analysis. Acta Mater. **55**(1), 13–28 (2007). https://doi.org/10.1016/j.actamat.2006.07.008
48. T. Zhu, J. Li, A. Samanta, H.G. Kim, S. Suresh, Interfacial plasticity governs strain rate sensitivity and ductility in nanostructured metals. Proc. Natl. Acad. Sci. **104**(9), 3031–3036 (2007). https://doi.org/10.1073/pnas.0611097104
49. R. Schwaiger, B. Moser, M. Dao, N. Chollacoop, S. Suresh, Some critical experiments on the strain-rate sensitivity of nanocrystalline nickel. Acta Mater. **51**(17), 5159–5172 (2003). https://doi.org/10.1016/s1359-6454(03)00365-3

50. Y.F. Shen, L. Lu, M. Dao, S. Suresh, Strain rate sensitivity of Cu with nanoscale twins. Scripta Mater. **55**(4), 319–322 (2006). https://doi.org/10.1016/j.scriptamat.2006.04.046
51. H.W. Höppel, J. May, M. Göken, Enhanced strength and ductility in ultrafine-grained aluminium produced by accumulative roll bonding. Adv. Eng. Mater. **6**(9), 781–784 (2004). https://doi.org/10.1002/adem.200306582
52. G.T. Gray III, T.C. Lowe, C.M. Cady, R.Z. Valiev, I.V. Aleksandrov, Influence of strain rate and temperature on the mechanical response of ultrafine-grained Cu, Ni, and Al–4Cu–0.5Zr. Nanostruct. Mater. **9**(1–8), 477–480 (1997). https://doi.org/10.1016/s0965-9773(97)00104-9
53. B. Miller, O. Furlong, W. Tysoe, The kinetics of shear-induced boundary film formation from dimethyl disulfide on copper. Tribol. Lett. **49**(1), 39–46 (2013). https://doi.org/10.1007/s11249-012-0040-7
54. H. Adams, B.P. Miller, P.V. Kotvis, O.J. Furlong, A. Martini, W.T. Tysoe, In situ measurements of boundary film formation pathways and kinetics: Dimethyl and diethyl disulfide on copper. Tribol. Lett. **62**(1), 1–9 (2016). https://doi.org/10.1007/s11249-016-0664-0
55. P.J. Cumpson, Angle-resolved XPS and AES: Depth-resolution limits and a general comparison of properties of depth-profile reconstruction methods. J. Electron Spectrosc. Relat. Phenom. **73**(1), 25–52 (1995). https://doi.org/10.1016/0368-2048(94)02270-4
56. G. Bell, Models for the specific adhesion of cells to cells. Science **200**(4342), 618–627 (1978). https://doi.org/10.1126/science.347575
57. H. Spikes, W. Tysoe, On the commonality between theoretical models for fluid and solid friction, wear and tribochemistry. Tribol. Lett. **59**(1), 1–14 (2015). https://doi.org/10.1007/s11249-015-0544-z

Chapter 6
Tribochemistry and Morphology of P-Based Antiwear Films

A. Dorgham, A. Neville and A. Morina

6.1 Introduction

Zincdialkyldithiophosphate (ZDDP) is one of the most widely used additives, whether in oils or greases. ZDDP is multifunctional as it can work as antiwear, antioxidant and anticorrosion additive [1]. It exhibits these beneficial functionalities by its ability to decompose under rubbing, heating or possibly high pressure to form a protective film. In the case of this film is formed during heating without rubbing it is called a thermal antiwear film whereas in the case of rubbing, the film is called a tribofilm [2]. The antiwear mechanism of this tribofilm originates from its capability to separate the contacting surface [3–7], to digest the abrasive wear debris [8–10] and to decompose peroxides and peroxy-radicals [11–13].

Despite the numerous benefits of using the antiwear ZDDP additive, it has disadvantages as well. Firstly, ZDDP increases micropitting of the steel surface and thus decreases its bearing life [14–17]. Secondly, it can poison the catalyst in the cars equipped with a catalytic converter and therefore degrades the converter's capability to reduce harmful emissions in the exhaust gas [18, 19]. This led to an ever expanding strict rules on the allowed concentrations of ZDDP, phosphorus and zinc in the oil [20, 21]. The trend of these rules suggests that the ZDDP should be replaced. One of the available options is to use neutral or acidic ashless dialkyldithiophosphate (DDP). The ashless additives can provide better [22, 23], comparable [24] or worse

A. Dorgham · A. Neville · A. Morina (✉)
School of Mechanical Engineering, Institute of Functional Surfaces, University of Leeds, Leeds LS2 9JT, UK
e-mail: a.morina@leeds.ac.uk

A. Dorgham
e-mail: A.Dorgham@leeds.ac.uk

A. Neville
e-mail: a.neville@leeds.ac.uk

M. Dienwiebel and M.-I. De Barros Bouchet (eds.), *Advanced Analytical Methods in Tribology*, Microtechnology and MEMS,
https://doi.org/10.1007/978-3-319-99897-8_6

[25, 26] antiwear properties than the ZDDP, which seem to depend greatly on the operating conditions, properties of contacting surfaces and chemistries of base oil and additives.

Replacing the ZDDP completely is not an easy task due to the lack of complete understanding of the complex pathways involved in its decomposition reactions and the possible interactions with the other additives in the oil. This statement is also true for the case of the DDP, which shares in general the same complexity. Nonetheless, despite these difficulties, a myriad of experimental works were carried out over the last 70 years in order to understand the ZDDP decomposition reactions. These works have been discussed in different reviews [21, 27, 28]. The review of Barnes et al. [27] discussed the role of the ZDDP in the oil and its functions as an antioxidant and antiwear additive. In addition, the review covered the interactions between the ZDDP and various other components that can be present in the oil. The ZDDP review of Spikes [21] presented a historical overview of the additive starting from its inception until its current use. The review focused on the reaction mechanisms of the ZDDP in the oil and the composition of the formed protective tribofilm and its formation and removal kinetics. It also examined the contribution of different experimental techniques to our current understanding of the ZDDP tribochemistry. Finally, the review of Nicholls et al. [28] focused on the decomposition reactions of the ZDDP in the absence and presence of other components in the oil, the composition of the formed tribofilms and their mechanical properties. This review is the last one focusing on the ZDDP and was presented more than a decade ago.

It should be noted that the early works in the field of tribology focused on using a single surface characterisation technique such as XPS, FTIR and EDX to study the complex tribochemistry of the ZDDP and DDP tribofilms. However, every technique, despite its numerous advantages, has certain limitations as summarised in Table 6.1, which are mainly related to the experimental atmosphere; whether ambient (A), high vacuum (HV) or ultra-high vacuum (UHV); detection sensitivity, chemical and structural information capability, lateral resolution and sampling depth. Therefore, as more and more of these techniques become available added to the insufficient information gathered from a single technique only, the recent tribological studies have started employing what is commonly known as the multi-technique approach. In this approach, observations and evidence are collected from various experimental techniques to help understand not only the composition but also the tribological and mechanical properties of ZDDP and DDP tribofilms.

Despite the extensive research in the tribological and mechanical properties of the ZDDP and DDP tribofilms and their compositions, the link between these properties is still not clear. This review aims at discussing these properties and highlighting the different links between them in order to provide a clearer picture on their tribochemistry in the base oil and the properties of the formed antiwear films. As the main body of the literature is related to the ZDDP and due to the large similarities between the ZDDP and DDP additives, the focus of this review will be mainly on the ZDDP while making a direct comparison to the DDP whenever appropriate.

Table 6.1 Comparison between the surface characterisation techniques

Method	Atmosphere[a]	Sensitivity	Chemical information	Structure	Lateral resolution	Detected depth
XPS	UHV	1%	Yes	No	10 μm	3–8 nm
AES	UHV	1%	Yes	No	20 nm	3–8 nm
XAS	UHV-A	50 ppm	Yes	Yes	50 nm-mm	1 nm—bulk
EDX	HV	0.1%	No	No	1 μm	1 μm
Raman	A	0.1 at.%	Yes	Yes	1 μm	0.2–10 μm
FTIR	A	1 ppm	Yes	Yes	10–100 μm	20 Å–1 μm
SIMS	HV	<1 ppm	No	No	60 μm	0–10 μm
XRD	A	1%	No	Yes	30 μm	0.1–10 μm
XRF	A	1 ppb	No	No	1 cm	Bulk
AFM	A	NA	No	Yes	<0.1 Å	0.1 Å

[a]Experimental atmosphere whether ambient (A), high vacuum (HV) or ultra-high vacuum (UHV)

6.2 Chemical Nature of ZDDP and DDP

Most of the added ZDDP to the base oil is either neutral, basic or hybrid of both [21, 27, 29, 30], which are depicted in Fig. 6.1. Similarly, the DDP can be either neutral, acidic or a mixture of both, as depicted in Fig. 6.2. The neutral ZDDP usually exists in equilibrium state between the monomeric structure ($Zn[PS_2(RO)_2]_2$) shown in Fig. 6.1(I) and the dimeric one ($Zn_2[PS_2(RO)_2]_4$) shown in Fig. 6.1(II) [31]. On the other hand, the basic ZDDP is stable and has a structure with the chemical formula ($Zn_4[PS_2(RO)_2]_6O$) [29, 30, 32, 33], which is depicted in Fig. 6.1(III). Within these types of ZDDP and DDP additives, new categories can be defined as primary, secondary, tertiary, etc., depending on the number of carbon atoms, i.e. one, two or three respectively, that are attached to the carbon atom of the alkyl (R) group [32]. These functional groups predetermine the thermal stability of the additive, which can be arranged as follows [30]: aryl > primary alkyl > secondary alkyl. It is interesting to note that the least thermally stable additive provides the best wear protection. Therefore, according to the alkyl functional group (R), the wear protection performance can be arranged as follows [21]:

secondary alkyl > primary alkyl > aryl

This can be attributed to the fact that primarily the decomposition process of the additive is thermally activated, which will be discussed in more detail in the subsequent sections.

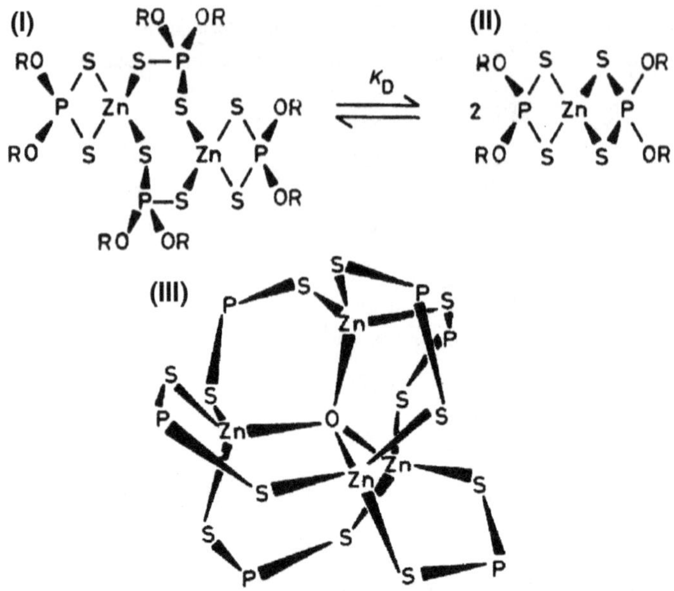

Fig. 6.1 The different structures of ZDDP: (I) neutral dimeric in equilibrium with (II) neutral monomeric; (III) basic ZDDP. Reprinted from Harrison and Kikabhai [31]

Fig. 6.2 The different structures of DDP: (I) neutral and (II) acidic DDP. Reprinted from Kim et al. [34]

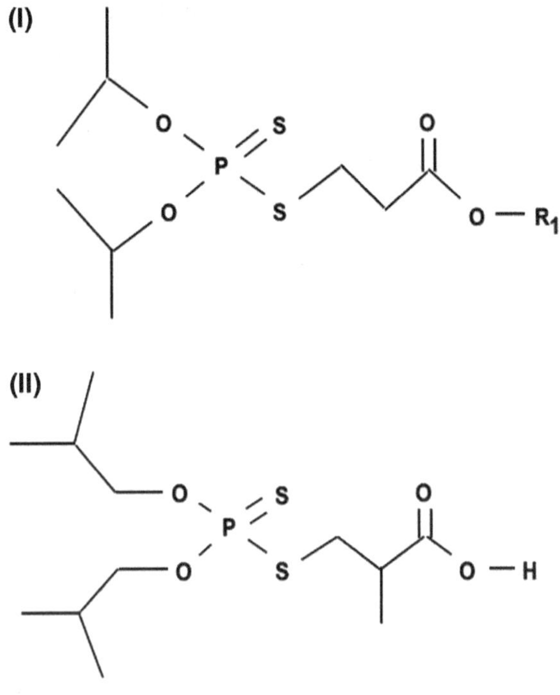

6.3 Reactions of ZDDP and DDP

Different theories were proposed to explain the reactions of P-based additives, such as ZDDP and DDP, in lubricating oils, which can be grouped into three main categories [21]: (i) ligand exchange, (ii) peroxides and peroxy-radicals decomposition and (iii) thermal, oxidative or hydrolytic degradation. These reactions can have very complex pathways due to the fact that the commercial oils containing ZDDP or DDP may as well contain some impurities, other additives, detergents or dispersant that might alter these paths and ultimately alter the precursors and the final reaction products forming the tribo- or thermal films [35].

To avoid such complexities, the three reactions mentioned earlier will be discussed in detail in the following sections assuming that only the P-based additive is present in the oil.

6.3.1 *Ligand Exchange*

During the ligand exchange reaction, the zinc cations (Zn^{2+}) in the ZDDP or any of its decomposition products can be exchanged by another cation. For example, the Zn^{2+} in the monomeric form of the neutral ZDDP can undergo a ligand exchange reaction with another metal ion (M^{2+}), as follows [21]:

$$Zn\left[PS_2(RO)_2\right]_2 + M^{2+} \rightarrow M\left[PS_2(RO)_2\right]_2 + Zn^{2+} \quad (6.1)$$

The relative order of the ability of one cation to replace another one is as follows [36]:

$$Pd^{2+} > Au^{3+} > Cu^{+} > Cu^{2+} > Fe^{3+} > Pb^{2+} > Zn^{2+} \quad (6.2)$$

This means that Zn^{2+} can be replaced easily by Fe^{3+}, which can be released from the metal surface as a result of wear of parts of the contacting surfaces. This reaction is very important as the nature of the metal cations in the substrate or the metal dialkyl dithiophosphate molecules can alter the thermal decomposition process by changing the decomposition temperature and kinetics and possibly the final composition of the formed tribofilm [37].

6.3.2 *Decomposition of Peroxides and Peroxy-Radicals*

Peroxides and peroxy-radicals can oxidise the steel surface as follows [11]:

$$ROO^{\cdot} + Fe \rightarrow RO^{\cdot} + FeO \quad (6.3)$$

$$2ROOH + Fe \rightarrow 2RO^{\cdot} + 2OH^{-} + Fe^{2+} \quad (6.4)$$

$$ROOH + Fe^{2+} \rightarrow RO^{\cdot} + OH^{-} + Fe^{3+} \quad (6.5)$$

Several studies [11, 12] found that in the presence of hydrogen peroxide, the oil that has antiwear additives, such as ZDDP, results in less wear than the oil without the additive. This suggested that one of the different antiwear mechanisms to protect the contacting surfaces is by decomposing the peroxides and hence terminating the oxidation chain reaction. However, other authors noted that when the antiwear additive acts as an antioxidant additive it cannot completely protect the surface from wear [38]. This indicates that the additive is used in decomposing the peroxides instead of forming a protective film. Therefore, there is a need to tailor the proactive tribological considerations based on not only the general application but also the operating conditions while inspecting the compatibility between the additive and contacting surfaces [38].

6.3.3 Formation of Tribo- and Thermal Films

The currently accepted view on the formation of tribo- and thermal films regards the decomposition of P-based antiwear additives as a thermally-activated and stress-assisted reaction [40–42], which can be catalysed by either heat or mechanical action in the form of rubbing or shear at the interface of the contacting bodies. Nevertheless, the classical view of this decomposition reaction considers it as a chemical reaction that can be either thermal, oxidative, hydrolytic or hybrid. These mechanisms will be discussed in detail in the following sections while highlighting the effect of rubbing on the reaction kinetics and the final decomposition products.

6.3.3.1 Thermal Decomposition

Numerous studies suggested that the decomposition of the ZDDP or DDP additive occurs thermally [35, 36, 43–45], which means that the extent and rate of the decomposition process are temperature dependent. The high temperature is especially needed at the early stage of the decomposition process as reported by Jones and Coy [46]. This was based on the observation that after the initial decomposition of the ZDDP at high temperature, the decomposition continued at a high rate even at low temperature as indicated by the continuous formation of white deposition in the oil. This suggested that the reaction can be multistage where intermediates must be formed before the final products. The initial stage of this multistage reaction can occur at elevated temperature in solution [36, 40, 47–49] or on the steel surface, which was argued to be essential for this process [4, 50–52].

In the case that the steel surface is not required, it was suggested that the formation of the protective film at high temperature can occur due to an in situ deposition process on the steel surface by one or more of the following mechanisms [49]:

1. Polymerisation of small molecules to form a complex large molecule on the steel surface.
2. Isomerisation of one or more of the additive molecules to other molecules deposited on the steel surface as a protective layer.
3. Decomposition of the additive molecules and the deposition of the resulting products on the steel surface.
4. Chemical reaction between two or more intermediate molecules to form the surface layer.

On the other hand, Coy and Jones [50] suggested that the availability of the steel surface can play a vital role in the thermal decomposition process. A reaction with the steel surface can transform the organic phosphates into inorganic ones, which can subsequently polymerise to a range of polyphosphates [46].

Jones and Coy [46] explained these reactions based on Pearson's hard and soft acids and bases (HSAB) principle [53–57]. The principle suggests stable pathways for any chemical reaction according to the general observation that hard acids prefer to form bonds with hard bases whereas soft acids favour forming stable compounds with soft bases. Following the hardness classification of the acids and bases of the most relevant compounds to ZDDP and DDP additives, which are summarised in Table 6.2, Jones and Coy [46] proposed that the following chain of reactions can take place during the thermal decomposition of the ZDDP:

1. Migration of the soft acid alkyl from the hard base oxygen atoms to the soft base sulphur atoms of thiophosphoryl (P=S)
2. Formation of phosphorus acid as a result of the elimination of the thioalkyl (_SR) functions.
3. Formation of phosphates P_O_P as a result of the nucleophilic reaction of one hard acid tetravalent phosphorus O_P in one short phosphate segment with a hard base oxygen bonded to another phosphate segment O_P.

Table 6.2 Classification of the hard and soft acids and bases relevant to ZDDP [55]

Type	Hard	Soft	Borderline
Acid	Tetravalent phosphorus –P–	Tetravalent carbon (CH_4)	Ferrous Fe^{2+}, Zn^{2+}
Base	H_2O, OH–, O_2– Phosphoryl (P=O), PO^{-3}_{4},	Thiophosphoryl (P=S) Thiolate (RS^-)	Sulphite (SO^{-2}_{3})
	ROH, RO–, SO^{-2}_{4}	Phosphine (PR_3)	

4. Formation of zinc mercaptide $Zn(SR)_2$ as intermediate, which upon reaction with an alkylating agent forms dialkyl sulphide whereas upon reacting with mercaptide forms dithiophosphate, trithiophosphate and finally tetrathiophosphate.
5. Formation of a mixture of zinc thiophosphate, zinc pyro-thiophosphate and zinc polypyro-thiophosphate as a final deposition.

Although the predictions of Jones and Coy [46] do not entirely match the current consensus that the final decomposition product is a mixture of zinc pyro, poly- and meta-phosphate glass, the way they utilised the HSAB principle to explain the thermal decomposition reaction was revolutionary at the time. They were one of the first to explore the thermal decomposition mechanism not as a black box but as a tractable chain of chemical reactions in which several intermediates can be identified. It took nearly two decades until Martin et al. [8] explored the HSAB in more detail to explain various important parts of the ZDDP reactions using the chemical hardness approach while matching the predictions with the experimental findings. This will be discussed in the following section.

6.3.3.2 Thermo-Oxidative Decomposition

A number of studies proposed that the decomposition reaction of the P-based additives is a thermo-oxidative in nature. Willermet et al. [58] suggested that the antiwear additive decomposition takes an oxidative pathway in the case of an equimolar quantity of free radicals is present in the oil, otherwise a thermally controlled decomposition occurred. However, the authors pointed out that in some localised areas of a thermally controlled decomposition, products of a thermooxidative controlled decomposition coexisted as well. Based on this observation, they concluded that apart from the availability of free radicals, stress plays a controlling role in determining the predominant decomposition pathway. To test the hypothesis of the thermo-oxidative controlled decomposition, which is controlled by the availability of O_2, Willermet et al. [58] conducted their tribological experiments in air and argon atmospheres. In the two cases, they did not observe any differences in the tribofilm composition. This should have necessary ruled out the oxidative decomposition mechanism. However, the authors argued that the experimental conditions somehow did not allow the oxidative decomposition to compete well with the thermal one. This apparent discrepancy can be related to the role of the ZDDP as an antiwear additive and oxidation inhibitor [21, 27], i.e. helps decompose the peroxy-radicals [59, 60].

Yin et al. [1] proposed a thermo-oxidative mechanism for the ZDDP tribofilm growth starting with the strong chemisorption of the ZDDP to the oxide layer on the steel surface followed by the fast formation of long polyphosphate chains and the slow formation of short polyphosphate chains, which can be summarised as follows:

- Step 1: Physisorption or chemisorption

$$Zn\left[(RO)_2PS_2\right]_2(\text{solution}) \rightarrow Zn[(RO_2)PS_2]_2(\text{adsorbed}) \quad (6.6)$$

- Step 2: Formation of long polyphosphate chains

$$Zn\left[(RO)_2PS_2\right]_2 + O_2(\text{or ROOH}) \rightarrow Zn(PO_3)_2 + \text{sulphur species} \tag{6.7}$$

- Step 3: Formation of short polyphosphate chains

$$Zn(PO_3)_2 + FeO \rightarrow FeZnP_2O_7 \tag{6.8a}$$

or

$$Zn(PO_3)_2 + 2FeO \rightarrow Fe_2Zn(PO_4)_2 \tag{6.8b}$$

Following Yin et al. [1], Najman et al. [25] proposed similar steps for the decomposition of the DDP additive, as follows:

- Step 1: Physisorption or chemisorption

$$(RO)_3PS_2(\text{solution}) \rightarrow (RO)_3PS_2(\text{adsorbed}) \tag{6.9}$$

- Step 2: Formation of long polyphosphate chains

$$(RO)_3PS_2 + O_2 + FeO \rightarrow Fe(PO_3)_2 + \text{sulphur species} \tag{6.10}$$

- Step 3: Formation of short polyphosphate chains

$$Fe(PO_3)_2 + FeO \rightarrow Fe_2P_2O_7 \tag{6.11}$$

- Step 4: Formation of sulphate species

$$S + \frac{3}{2}O_2 + FeO \rightarrow FeSO_4 \tag{6.12}$$

Similar to the above reactions, Martin et al. [8, 9, 61] suggested that the depolymerisation of the initially formed long phosphate chains of the ZDDP tribofilm into shorter ones is a result of their reaction with iron oxides on the steel surface, as follows:

$$5Zn(PO_2)_2 + Fe_2O_3 \overset{\theta,\tau,P}{\rightarrow} Fe_2Zn_3P_{10}O_{31} + 2ZnO \tag{6.13}$$

which can be accompanied by other intermediate reactions, such as:

$$4Zn_6(P_{10}O_{29}S_2) + 10Fe_2O_3 \rightarrow 10Fe_2Zn(P_2O_7)_2 + 8ZnS + 6ZnO \tag{6.14}$$

These proposed reactions were justified based on Pearson's principle of hard and soft acids and bases (HSAB) [53]. According to this principle, the harder acid Fe^{3+} than Zn^{2+} will react preferentially with the hard base tetravalent phosphorus. Thus, under high temperature, shear and pressure, iron can be easily digested by the zinc phosphate glass to form mixed iron and zinc phosphates. However, under severe conditions of shear, the authors suggested that the digestion of iron can convert the long zinc phosphate chains completely into short iron phosphate chains according to the following reaction:

$$Zn(PO_3)_2 + Fe_2O_3 \xrightarrow{\theta,\tau,P} 2FePO_4 + ZnO \tag{6.15}$$

As these reactions need iron oxide and thus wear to commence near the metal surface, the tribofilm is expected to have a uniform structure of zinc phosphate free of iron in the case of mild wear conditions whereas it should have a layered structure in the case of severe wear. In this structure, the short chains of zinc and iron phosphate or iron phosphate are formed on the metal surface whereas the long chains of zinc phosphate are continuously formed away from the metal surface.

Fuller et al. [62] suggested a thermo-oxidative decomposition mechanism similar to one proposed by Yin et al. [1] but with two main modifications. The first one concerns the additive adsorption [step 1 in (6.6)]. It was suggested that when the antiwear additive such as the ZDDP adsorbs to the steel surface, it undergoes a transformation into a rearranged linkage isomer in which the alkyl groups have migrated from O to S atoms. This alkylation reaction is based on the mechanism proposed by Jones and Coy [46], which was discussed in the previous section. However, Fuller et al. [62] further suggested that in this linkage isomer all the sulphurs are partially or totally replaced by oxygen. The other modification concerns the formation of the short phosphate chains [step 3 in (6.8a) and (6.8b)]. Instead of being a product of the reaction of the long phosphate chains with iron oxide, the short phosphate chains can also be formed as a product of the reaction between the long phosphate chains and water, which increases with increasing temperature. These steps can be summarised for the case of the ZDDP additive as follows:

- Step 1: Physisorption or chemisorption of ZDDP to the metal surface

$$Zn\left[(RO)_2PS_2\right]_2(\text{solution}) \rightarrow Zn\left[(RO)_2PS_2\right]_2(\text{ZDDP adsorbed}) \tag{6.16}$$

- Step 2: Decomposition of ZDDP to LI-ZDDP

$$Zn\left[(RO)_2PS_2\right]_2(\text{solution}) \rightarrow Zn[O_2P(SR)_2]_2(\text{LI-ZDDP insolution}) \tag{6.17}$$

- Step 3: Physisorption or chemisorption of any LI-ZDDP in oil to the metal surface

$$Zn\left[O_2P(SR)_2\right]_2(\text{solution}) \rightarrow Zn\left[O_2P(SR)_2\right]_2(\text{LI-ZDDP adsorbed}) \tag{6.18}$$

- Step 4: Thermal-oxidation of ZDDP and LI-ZDDP to form long polyphosphate chains

$$Zn\left[O_2P(SR)_2\right]_2 + O_2(\text{or ROOH}) \rightarrow Zn(PO_3)_2 + \text{sulphur species} \tag{6.19}$$

- Step 5: Hydrolysis of the long polyphosphate chains to form short polyphosphate chains

$$7Zn(PO_3)_2 + 6H_2O \rightarrow Zn_7(P_5O_{16})_2 + 4H_3PO_4 \tag{6.20}$$

$$2Zn(PO_3)_2 + 3H_2O \rightarrow Zn_2P_2O_7 + 2H_3PO_4 \tag{6.21}$$

Bell et al. [6] suggested that the adsorption step described earlier occurs only on localised areas of the wear scar where asperities are rubbing against each other. This heterogeneity in the adsorption was suggested to be responsible for the noticed heterogeneity of the tribofilm thickness and composition. In addition, the authors suggested that after the adsorption of the rearranged linkage isomer (step 3) and before the formation of the long polyphosphate chains (step 4), sulphide products can react with the steel surface to form iron sulphide (FeS).

6.3.4 Hydrolytic Decomposition

Few studies suggested that the decomposition of P-based additives such as the ZDDP is hydrolytic in nature, i.e. catalysed by water. In order to provide evidence for this mechanism, Spedding and Watkins [35] showed that in the absence of water, e.g. by heating the sample up to 100–170 °C in order to evaporate all water as well as by flooding the sample with dry nitrogen, the decomposition reaction was suppressed. In contrast, when the sample was flooded by watersaturated nitrogen, a rapid decomposition rate was observed. Therefore, they proposed that the following hydrolysis reaction chain can take place:

1. alcohol formation

$$RO{-}P{-} + H_2O \rightarrow ROH + HO{-}P{-} \tag{6.22}$$

2. alcohol dehydration to olefin

$$ROH \xrightarrow{H^+} R' = CH_2 + H_2O \tag{6.23}$$

   The sum of the above two reactions demonstrates the catalytic action of water in the overall reaction, as follows:

$$RO{-}P \xrightarrow{H_2O} R' = CH_2 + HO{-}P{-} \tag{6.24}$$

3. polyphosphate formation

$$-P{-}OH + HO{-}P \rightarrow -P{-}O{-}P{-} + H_2O \tag{6.25}$$

4. hydrogen sulphide formation due to the consumption of water by the sulphur species

$$-P-_S + H_2O \rightarrow -P-_O + H_2S \quad (6.26)$$

It should be noted that when Spedding and Watkins [35] conducted tests at 200 °C, at which most of the water should evaporate, the decomposition proceeded without any significant reduction in the reaction rate. Willermet et al. [13] argued that even lower temperatures than 200 °C could not slow down the reaction rate. This either disproves the hydrolytic decomposition mechanism or simply suggests that more complex reaction pathways or multiple mechanisms, e.g. thermo-hydrolytic or thermo-oxidative-hydrolytic, can take place at the same time.

6.4 Composition of Antiwear Films

The tribo- and thermal films of P-based additives such as the ZDDP and DDP have a dynamic tribochemical nature as their compositions change continuously during the decomposition process [1, 63–66]. Nevertheless, the typical decomposition products of such additives are some volatile products, e.g. olefins and mercaptans, and some non-volatile products [45]. The non-volatile products can be oil soluble, e.g. organic sulphur-phosphorus compounds, or can be oil insoluble, e.g. phosphate polymers, zinc or iron thio- and poly-phosphate [43, 45, 46]. This indicates that the formed antiwear films, in general, consist mainly of Zn or Fe, P, S, O and C. This was confirmed using electron probe micro analysis (EPMA) [6, 67], scanning electron microscopy (SEM) [5, 6, 43, 67–70], Auger electron spectroscopy (AES) [5, 58, 67–69, 71], X-ray fluorescence (XRF) spectroscopy [3, 4, 44, 72, 73], X-ray photoelectron spectroscopy (XPS) [3, 6, 51, 52, 58, 69, 71, 74–77], X-ray absorption near edge spectroscopy (XANES) [1, 62, 63, 65, 75, 78–80] and secondary ion beam spectroscopy (SIMS) [3, 6, 35]. Willermet et al. [69] found that the carbon can only be detected in the thinnest parts of the tribofilm. This suggested that this carbon was originated from the environment or steel substrate but not a real part of the tribofilm. Ancillary experiments conducted by Lindsay et al. [71] supported the same conclusion that carbon existed as a result of adventitious sources, i.e. contaminants. Furthermore, the results of Martin et al. [81] showed that no carbon existed in the samples under study, which confirms its adventitious nature.

It should be noted that although the composition of the tribo- and thermal films are similar they are not completely identical. Kasrai et al. [75] found differences in the ratio of the elements forming the thermal film and tribofilm. Hence, other factors than the high temperature, e.g. reaction with the steel surface, rubbing and high contact pressure, can also play a role in determining the final decomposition products [50, 75].

In the subsequent sections, the decomposition species of the ZDDP and DDP additives will be discussed in detail. Phosphorus, zinc and iron species will be examined first followed by sulphur species. The role of the operating conditions on the formation of these species will be discussed as well.

6.4.1 Phosphorus, Zinc and Iron Species

The solid precipitates of the decomposition reaction of the ZDDP or DDP additive is a complex mixture of zinc or iron polyphosphates depending on the available cations. DDP forms, in general, tribofilms of short chains of iron polyphosphate [25, 34, 66, 82, 83] with a minor concentration of sulphur species consisting mainly of iron sulphate under high contact pressure [84, 85] and iron sulphides, e.g. FeS or FeS_2, under low pressure [66, 84]. On the other hand, the ZDDP forms tribo- and thermal films of zinc phosphate of different chain lengths and a small content of sulphur in the reduced form of sulphides [46]. The chain length is typically a complex function of the operating conditions, e.g. load and temperature, as well as the type of additives, dispersant and contaminants such as water that can be present in the oil. For instance, in the case of thermal films, increasing the temperature can lead to the formation of longer chains of polyphosphates [35]. However, in the case of tribofilms formed under rubbing, increasing the temperature can lead to the formation of short chain pyrophosphate [45]. This apparent controversy can be attributed to the observation that regardless of the oil temperature short phosphate chains are likely to be formed near the steel surface as a result of the depolymerisation reaction occurring to the long phosphate chains [8, 9, 86–89]. The depolymerisation can occur due to the high shear stress at the asperity-asperity contacts, which can possibly cleave the long phosphate chains into shorter ones, and to wear that can remove the weakly adhered long phosphate chains from the surface [89]. In addition, other studies [8, 9, 86] suggested that in the presence of iron oxide the long phosphate chains are depolymerised into short ones of mixed Fe–Zn or Fe phosphates, as discussed in Sect. 10.3.3. The depolymerisation effect is significant especially in the case of ZDDP tribofilms consisting initially of long to medium phosphate chains, whereas it might not affect the short chains of iron phosphate predominantly formed in the case of the DDP additive [84].

The ratios between metal cations (M^+) and phosphorus (P) [90] on the one hand, and between oxygen (O) and phosphorus (P) [58] on the other hand, are useful indicators for the length of the phosphate chains. For instance, in the case of the ZDDP, assuming that the zinc phosphates have a general formula of $x(\mathrm{ZnO}).(1-x)\mathrm{P_2O_5}$, then the chain length n can be related to the ratio of the mole fraction of ZnO, x, to the one of P_2O_5, $1-x$, as follows [91]:

$$\frac{x}{1-x} = \frac{n+2}{n} \tag{6.27}$$

Alternatively, if the zinc phosphate is assumed to have the general formula $Zn_{n+2}[P_nO_{3n+1}]_2$, for odd n and $Zn_{(n+2)/2}P_nO_{3n+1}$ for even n, then the chain length can be related to ratio of the atomic concentration of phosphorus to the one of oxygen, as follows [9]:

$$\frac{P}{O} = \frac{n}{n+1} \tag{6.28}$$

These two ratios above appeared to give comparable results for the chain length of phosphates [92]. Alternatively, another option to quantify the length of the phosphate chains would be the intensity ratio of the bridging oxygen (BO) to non-bridging oxygen (NBO) [86, 92–97], as follows:

$$\frac{B}{NBO} = \frac{1}{2}\frac{n-1}{n+1} \tag{6.29}$$

Based on (6.27), the phosphate chains seem to change from metaphosphate ($n \rightarrow \infty$) at $x = 0.5$, to orthophosphate ($n \rightarrow 1$) at $x = 0.75$. As the chain length changes drastically over a small concentration range of ZnO, or FeO and Fe_2O_3 in the case of DDP, one should examine the presence of zinc and iron more closely. Increasing the metal oxides content of these cations in the phosphate can increase the fragmentation of the long chains and thus depolymerises them into shorter chains [97–99].

6.4.2 *Sulphur Species*

There is no clear consensus on the evolution of the sulphur species present in the tribo- and thermal films. For instance, Bird and Galvin [51] suggested that the thermal film contains sulphur in the form of sulphate and free sulphur whereas the tribofilm contains large patchy areas of sulphide and a small amount of sulphate. Zhang et al. [26] showed that for the ZDDP tribofilms, the evolution of P/S ratio was nearly constant over time, whereas it showed a gradual increase in the case of DDP tribofilms. Kim et al. [34] observed that for the ZDDP the sulphate concentration increases with heating while sulphide decreases possibly due to its oxidation to sulphate. For one of the tested DDPs, both sulphides and sulphates were detected whereas for the other no sulphides were observed but only sulphates with a similar behaviour to the ZDDP. Similarly, Najman et al. [85] showed that the DDP reacts rapidly with the substrate covered with oxides, which leads to the oxidation of the sulphur species into iron sulphate. This was confirmed for both neutral and acidic DDPs [100]. Other studies [25, 84] showed that under high contact pressure, DDP tribo- or thermal films contain Fe sulphate near the steel surface whereas under less harsh conditions initially mixed iron sulphide, as FeS and FeS_2, and sulphate, as $FeSO_4$, are formed where the sulphides can oxidise over rubbing time yielding primarily sulphates at the end.

The role of the operating conditions can explain some of the conflicting results. For instance, Zhang et al. [66] reported that the DDP additive forms mainly FeS whereas the ZDDP forms FeS in the early stage near the metal surface and ZnS in the later stages. The observed sulphides in the case of DDP, which is in contrast to the previously discussed studies suggesting sulphates, can be related to the different operating conditions used while generating the various tribofilms.

The complex decomposition reaction of the antiwear additives might have several intermediates. One of these intermediates, which might be formed initially in the solution even at low temperature and be deposited on the metal surface, can be a sulphur-rich thiophosphate [43]. Subsequently, when the temperature of the oil is raised or the local temperature at the asperity-asperity contacts of the rubbing surfaces increases due to frictional heating, the reaction between the thiophosphate deposit and rubbing surfaces becomes possible to occur.

The amount of the deposits of the sulphur-rich thiophosphate and its rate depend closely on the thermal stability of the additive. In the case of the ZDDP, Spedding and Watkins [35] found that by increasing the temperature, more of the sulphur is consumed. In contrast, Kim et al. [34] found that for the ZDDP the progression of heating makes the thermal films richer in sulphur whereas for the DDP it makes the films richer in phosphorus. Despite the trend, this indicates that the local temperature during the tribofilm formation can be inferred from the local composition. This is possible by examining the local ratio of zinc to sulphur or phosphorus to sulphur, which indicates the minimum temperature attained during the formation of that part of the tribofilm [43].

Several other studies [3, 5, 6, 25, 26, 66, 82, 85, 101] suggested that the decomposition products can be a result of a direct reaction between the additive and contacting surfaces instead of being formed in the oil and deposited on the surface. This starts with the adsorption of the additive to the steel surface with the maximum coverage occurs when the additive molecules are flat on the surface, which means that the sulphur atoms lay near the surface [102]. Several authors [51, 103–105] have already found that the amount of sulphur chemisorption products are higher on the steel surface. Bell et al. [6] suggested that immediately after the adsorption of the ZDDP to the steel surface, sulphide products from the decomposition of the ZDDP react with the steel surface to form iron sulphides. In agreement with these results, Loeser et al. [76] also found that the amount of sulphur is localised in the areas of high pressure, i.e. asperity-asperity contacts. These various reports indeed corroborate the observation of Dacre and Bovington [102] that initially the four sulphur atoms should be near the steel surface.

Watkins [3] suggested that FeS reacts with the oxide layer to form a eutectic system, which its phase and potential diagrams are depicted in Fig. 6.3. This system has a melting temperature of 900 °C and was postulated to form a viscous film at the contacting surfaces under extreme conditions. The results of Glaeser et al. [5] supported the idea of the formation of iron-oxide-sulphide complex. On the other hand, Barcroft et al. [43] could not detect any complex mixture of zinc oxide, phosphate, and sulphide as the one proposed by Watkins [3]. Nevertheless, they suggested that this system can still be formed at the asperity contacts where temperature can be high. Bell et al. [6] suggested that the iron sulphide and iron oxide can enter the phosphate layer as cations or fragments from the worn surface. They also suggested that the replenishment of sulphur and oxide to the layer on the steel surface occurs continuously either by entrainment, mixing, or diffusion. Similar to these findings,

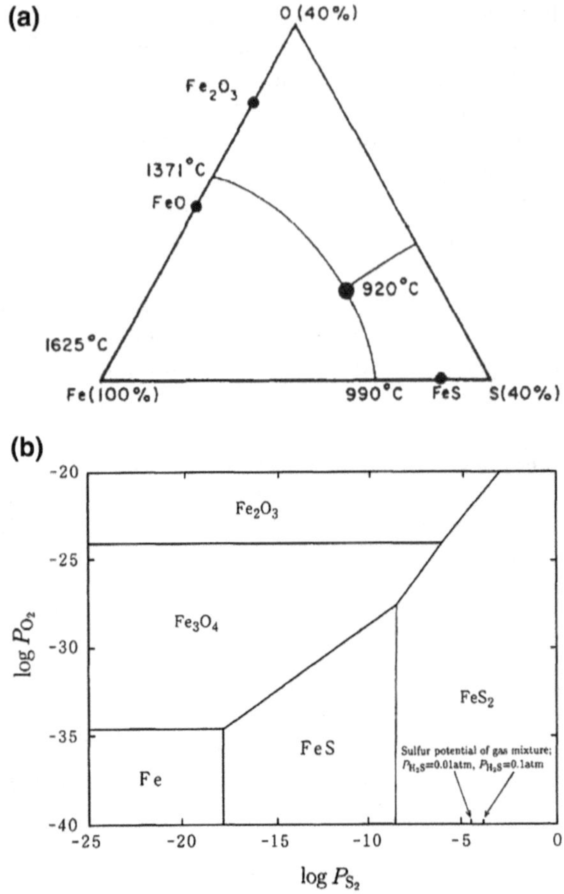

Fig. 6.3 The Fe_O_S ternary system that can be formed on the steel surface under rubbing. **a** Is its Phase diagram. Reprinted from Watkins [3]. And **b** is its potential diagram at 673 K. Reprinted from Watanabe et al. [106]

Glaeser et al. [5] suggested that the iron sulphide forms a chemisorbed layer of iron-sulphide-iron oxide that prevents direct contact between asperities.

Nevertheless, it should be noted that other studies [51, 58, 68, 71] showed that iron is absent in the formed tribofilm, i.e. iron detected was mainly iron oxide rather than iron sulphide. These results led to the conclusion that the tribofilm is deposited over the substrate rather than being a result of a chemical reaction with the rubbing surface. This also suggests that the formation of the tribofilm occurs through a thermal route rather than an oxidative one.

6.4.3 Factors Affecting the Tribofilms' Composition and Formation

6.4.3.1 Material of Counterbodies

As the additive molecules adsorb to the substrate before decomposing to form a tenacious tribofilm, it is expected that the type of substrate, e.g. steel, DLC-coated, ceramic and etc., should have an important effect on the tribofilm formation and its tenacity. There are two views related to the effect of substrate. The first considers the substrate effect to be mainly related to its mechanical properties such as hardness. On the other hand, the second view considers that the substrate effect is related to a more complex chemo-mechanical rather than mechanical properties of the rubbing interfaces, such as the chemical, electronic and structural properties of the substrate. These two views are discussed in the following sections with much of the emphasis is put on the second.

6.4.3.2 Mechanical Properties

Several studies investigated the effect of the mechanical properties of the substrate, especially hardness and elastic modulus, on the tribofilm formation. The overall conclusion is that the tribofilm formation, friction and wear are affected by not only the mechanical properties of the substrate but also the compatibility between the substrate and additive [38]. For instance, Li et al. [107] used XANES to investigate the effect of steel hardness on the composition of the formed ZDDP antiwear films. The results indicated that longer phosphate chains are formed on the hard substrates. The results also indicated that the softer the substrate the larger the surface roughness, tribofilm thickness, wear scar width and friction. Furthermore, more uniform but thinner tribofilms were formed on the hard substrate but less uniform but thick tribofilms were formed on the soft ones. FIB/SEM was used to examine the elements present in the formed tribofilms and to assess the substrate damage based on the dimensions of the wear scar. It was found that mainly sulphides and few sulphates were formed on the hard substrates whereas the soft ones contained more sulphates and fewer sulphides.

Sheasby et al. [108] also studied the effect of steel hardness on the formation of protective tribofilm and wear performance. The study found that the wear performance was enhanced with increasing the substrate hardness, although wear was in general small, even for the soft substrates. The results showed that in several occasions, medium softened substrates resulted in less wear than harder substrates. They argued that mechanical mixing can occur between the formed tribofilm and the soft substrate, which can result in improving the overall mechanical properties of the interface and thus better resisting wear. However, the results of Vengudusamy et al. [109], using six different types of DLCs of hardness ranged from about 760 to 6800 HV, suggested that there is no direct correlation between wear resistance of the contacting surfaces and the surface hardness. This further highlights the significance of tailoring the tribological and mechanical properties of the interface by taking into account the compatibility between the substrate and used additives [38].

6.4.3.3 Chemo-Mechanical Properties

Apart from hardness, the substrate chemical properties are expected to affect the decomposition of the antiwear additives and the formation of their protective tribofilms. This effect is manifested in the fundamental question of whether there is a substantive requirement for the presence of metal cations, e.g. Fe, W and Ti, for the tribofilm to be formed or such metallic cations are dispensable. There is a definite consensus that the ZDDP molecules can adsorb to steel surfaces and decompose to form protective tribofilms of excellent tenacity [21, 27, 28]. Furthermore, other studies showed that ZDDP tribofilms can also be formed on surfaces other than steel. For instance, Zhang and Spikes [41] were able to generate ZDDP tribofilms on a WC substrate. The results showed that the rate of formation ranges from 0.2 to 0.7 nm/min depending on the interfacial shear stresses. Similarly, Gosvami et al. [40] using elaborate in situ AFM tribotests showed that the ZDDP can adsorb and decompose to form tribofilms on both Fe-coated and uncoated Si substrates with a similar rate depending on the temperature and contact pressure. It is not clear whether the ability to form tribofilms in these cases was due to the presence of W and Si in particular or due to the operating conditions. This can be better understood by studying some cases of coated surfaces, e.g. DLC, both non-doped and doped with various metallic cations. For such surfaces, the exact chemo-mechanical nature of the coating is expected to play a vital role in the adsorption and decomposition of the P-based additives.

The wear and friction performance of DLC coatings and the properties of any formed tribofilms on them were reviewed extensively in the literature [38, 110–114]. Several previous studies reported that the P-based additives such as ZDDP can react and form protective tribofilms on DLC coatings even without containing any doped metallic cations [109, 115–120]. In contrast, other studies found that no tribofilms can be formed on non-doped DLC without metallic cations [121–126]. We will review some of these reports and try to identify the reasons behind this apparent discrepancy. The available literature will be divided into three main themes: (i) formation of tribofilms on DLC, (ii) structure of the formed tribofilms on DLC and (iii) effect of the formed tribofilms on friction and wear.

(i) **Formation of tribofilms on DLC coatings**

Regarding the formation of tribofilms on DLC coatings, it is interesting to note that the literature is somehow divided between two contrasting views. The first suggests that no tribofilm formation is possible on non-doped DLC whereas the other suggests that the formation is possible on any type of DLC coatings.

In support of the first view, Haque et al. [121] found, using XPS, that no ZDDP-derived tribofilm was formed on a-C:H DLC even in the case of one of the counterbodies is made of cast iron. The role of iron cations was suggested to be deactivated by the presence of a carbon transfer layer, which might be graphitic, on the cast iron counterbody. This is also in line with the results of Bouchet et al. [125] regarding the interaction of the ZDDP and MoDTC additives with hydrogen-free DLC (a-C), hydrogenated (a-C:H) and Ti-doped (Ti–C:H) DLC coating s. The XPS and TEM/EELS analysis revealed that no P-derived tribofilm was formed and no iron was present, which indicated that the pin surface was protected by a transfer layer from the DLC. Similarly, Podgornik et al. [122] investigated the effect of coating one contacting surface or both with WC doped hydrogenated DLC coatings (Me–C:H) in the presence of ZDDP antiwear additive. Based on the SEM/EDX analysis, the study found that no P-based antiwear tribofilm was formed on the DLC coating. Furthermore, Podgornik et al. [124, 127] using SEM/EDX confirmed that for (a-C:H) DLC no tribofilm was formed for both S-based and P-based additives. The results also suggested that probably W from the coating can combine with S to form a protective layer. Furthermore, Ban et al. [126] tested the reaction of Si-doped (a-Si:H) and non-doped (a-C:H) DLC coatings with ZDDP. Using XPS, no tribofilm was detected on the (a-C:H) DLC coating whereas a P-based tribofilm was formed on the Si-doped DLC. Kalin et al. [128] also found that for the self-mated (DLC/DLC) nondoped and doped Ti-, W-, and Si-DLC coatings, the EDX and FTIR results did not provide evidence to support a reaction between the DLC coatings and the extreme pressure (EP) additives. In another study, Kalin et al. [118] found that the tribo- and thermal films can be formed on the Ti-doped DLC, which was found to have 10 times higher activity than the W-doped (P/S > 25 folds). They also found some reactivity with a-C:H DLC but much slower. They concluded that the metal doped DLCs behave more like metal steel that catalyses the decomposition of the additive and the formation of tribo- or thermal films.

In contrast to the previous view, Akbari et al. [120], using ATR-FTIR and XPS, found that the thermal films can be formed equally on bare steel surfaces and Si-doped (a-Si:H) and non-doped (a-C:H) DLC coatings. The thickness of the thermal films as estimated from XPS sputtering was thicker in the case of bare steel as compared to the coated ones, which suggested that steel can still catalyses the thermal decomposition of the ZDDP even without rubbing. Apart from this, the decomposition products of the ZDDP does not seem to be affected by the substrate as the short chain pyrophosphate and zinc oxide was detected on all the tested surfaces. Nonetheless, ATR-FTIR showed that organic sulphides (R–S) were present on all the surfaces but sulfhydryl (R–SH) groups were detected on the steel surface only. These species were confirmed by their XPS analysis except for the H–S, which could not be resolved. Nevertheless, it should be noted that these results apply only to thermal films, which could indicate that in the case of tribofilms any formed species on the DLC coatings, especially if non-doped, are of low tenacity and thus under rubbing can be easily removed. This can partly explain the discrepancy in the literature regarding whether a tribofilm can be formed on non-doped DLC coatings or not. However, other studies, e.g. see [109, 115, 116], suggested that tribofilms can indeed be formed on the non-doped DLCs. However, despite the assertion, the data provided do not seem to be conclusive for the a-C:H non-doped DLC, as will be discussed in the next paragraph describing the structure of what was perceived to be a tribofilm on DLC coatings.

(ii) **Structure of the tribofilms formed on DLC coatings**

The studies that showed a formation of P-based tribofilms on DLC coatings indicated unique tribofilms' structures with some similarities to the ones formed on bare steel surfaces. For instance, Vengudusamy et al. [109] analysed the ZDDP tribofilms possibly formed on six different types of DLCs using various experimental techniques including AFM, SLIM, ToF-SEM and EDX. The DLCs had a hardness ranging from about 760 to 6800 HV. The structure of the tribofilm was analysed using the AFM. The study found that a tribofilm with a pad-like structure similar to one formed on steel surfaces can only be observed in the case of a DLC containing metallic elements, e.g. W-doped DLC. For the other types of DLC, tribofilms of minuscule amount with scattered patches were formed instead. The thickness of the pads formed on the W-DLC was <30 nm, which was much larger than the one formed on the other types of DLC. This may suggest that the metallic cations can catalyse the decomposition of the additive and the formation of a protective tribofilm. SLIM interferometry images after different rubbing times showed an interesting behaviour that the tribofilms formed on DLCDLC contacting surfaces are less tenacious than the ones formed on metal surfaces, which was also reported by other studies as well [115, 116]. The low tenacity was evidenced based on the decrease in the concentrations of the decomposition products over rubbing cycles. The ToF-SIMS analysis of Equey et al. [115] and the XPS and TEM/EELS analysis of Haque et al. [121] showed that no iron is present in the formed tribofilms on DLC surfaces, which suggests that the low tenacity to the surface can be related to the absence of mixed oxide/sulphide base layer.

In agreement with the previous discussion, Equey et al. [115] using AFM found that the tribofilm seemed to lack the pad-like structure observed in the tribofilms formed on steel surfaces but instead it appeared to have a similar structure like the one of DLC coating, i.e. rounded nodular structure. The AFM images indicated that the tribofilm does not cover the metal surface completely but forms more like scattered islands of thickness up to 100 nm. In a different study, Equey et al. [116] tested the decomposition of ZDDP, butylated triphenyl phosphorothionate (b-TPPT) and amine phosphate (AP) additives on (a-C:H) DLC coating. Using ToF-SIMS and AFM, it was found that the ZDDP and b-TPPT can form thin tribofilms, although AFM was unable to quantify their thicknesses as they seemed to be within the roughness of the DLC coating. This is in line with the conclusions of Topolovec et al. [117] regarding (a-C:H) DLC and Cr-doped non-hydrogenated and graphitic DLC coatings.

The above discussion indicates that the main factor behind the wide disparity between the observations of the tribofilm formation on DLC coatings is the tribofilm tenacity. Any formed tribofilm does not seem to strongly adsorb to the DLC surface, especially if non-doped, leading to its effortless removal under rubbing once formed. Another explanation was provided by Kalin and Vižintin [129] who suggested the presence of a different thermal activation barrier for the additive reaction with doped as opposed to nondoped DLC, which is also different from the one of steel. The difference was attributed to the low thermal conductivity of the non-doped DLC in comparison to steel or metal doped-DLC. This can lead to higher contact temperature that helps in decomposing the additive and forming a tribofilm.

(iii) **Effect of the formed tribofilms on friction and wear**

Apart from the additive decomposition to form a tribofilm on the DLC surface, its role in friction and wear is contentious. For instance, Haque et al. [130] suggested that the decomposition of the ZDDP additive seems to be essential in protecting the surface from polishing wear. This was based on the observation that outside the boundary lubrication regime where the conditions might not be suitable for the decomposition of the additive, the coating surface undergoes sp^3 to sp^2 conversion (possibly graphitisation), which reduced friction and the DLC hardness leading to more wear. Yang et al. [131] showed that the extent of this is additive dependent but could not establish a direct relation between the friction reduction and sp^2/sp^3 ratio in the coating or between the hardness reduction and wear volume. Contrary to these results, Vengudusamy et al. [109] suggested that no graphitisation takes place because of rubbing and that for most of the tested DLCs including a similar a-C:H DLC coating used by Haque et al. [130], wear was less in the absence of ZDDP than in its presence. In contrast, the results of Bouchet et al. [125] suggested that ZDDP additives can improve wear properties of the hydrogen-free DLC (a-C), hydrogenated (a-C:H) and Ti-doped (Ti–C:H) DLC coatings. Similarly, Equey et al. [115] tested (a-C:H) DLC coating and found that although wear was insignificant in the presence or absence of the ZDDP, its absence caused some abrasive wear scars to appear on the surface of the coating.

Kalin and Vižintin [129] investigated the interactions occurring between the contacting surfaces when only one of them is coated, i.e. DLC versus steel. They studied a-C:H, Ti–C:H, a-C:H/a-C:H–W multilayer and a-C:H/a-Si:O single layer when interacting with antiwear additive (mixture of amine phosphates) and EP additive (dialkyl dithiophosphate). Wear was found to be dependent on the coating and additive types. Metal doped DLCs and steel were found to have similar tribological behaviour except for steel/W-DLC, which shows large friction and wear. The EDX surface analysis shows no P derived tribofilm was formed on the DLC coatings. In the case of non-doped DLC, the additive presence seems to increase wear whereas the opposite was found in the case of doped DLCs. They related this to the formation of a soft tribofilm on the metal surface or metal doped DLC coating instead of a transfer layer from the coating. The soft tribofilm can decrease wear of the substrate more than in the case where the additive is not present. In line with these results, Podgornik et al. [122] tested the effect of coating one contacting surface or both with WC doped hydrogenated DLC coatings (Me–C:H) in the presence of ZDDP antiwear additive and found that although no P-based antiwear tribofilm was formed on the DLC coating, the additive type seems to have a great effect on the running-in period. Interestingly, the results showed that the DLC/iron combination gives the best tribological performance in terms of low friction and wear, which is in line with other studies [123]. The EDX analysis revealed that in this case a mixed material from the DLC coating and decomposition material from the additive can form a protective tribofilm on the uncoated or exposed steel surfaces, which seems to exhibit a superior protective properties than the individual components. Donnet and Grill [132] indicated that the low shear stress of the carbon-rich transfer layer formed on the DLC is responsible for the improved tribological properties in comparison to steel surfaces. Ban et al. [126] tested the reaction of Si-doped (a-Si:H) and non-doped (a-C:H) DLC coatings with ZDDP. Using XPS, a P-based tribofilm was formed on the Si-doped DLC, which seemed to have low shear stress that helped lower friction force. In contrast, Kalin et al. [128] found that for the self-mated (DLC/DLC) non-doped and doped Ti-, W-, and Si-DLC coatings, the use of EP additives mitigated wear but friction relatively increased. The increases in friction suggested a formation of a high shear strength interface, despite the observed reduction in surface roughness. However, the EDX and FTIR results did not provide evidence to support a reaction between the DLC coatings and additives.

Based on the above discussion, any tribofilm formed on the DLC coating can generally serve two purposes [119]. First, it can form a lubricating layer that helps separating the contacting surfaces. Second, the formed layer is soft and thus can reduce the peak shear stresses and strains at the surface of the coating [133, 134]. This can result in possibly graphitisation suppression and wear reduction [119] though the effect on friction can be higher or lower depending on the type of the formed interface, i.e. friction increases in the case of ZDDP tribofilms but decreases in the case of carbon-rich layers.

6.4.3.4 Type of Base Oil

P-based additives, such as ZDDP and DDP, can be represented typically as a polar moiety, attached to P or S atom, and a non-polar tail, which is typically an alkyl moiety [135]. The affinity of such a molecule to the steel surface originates from the molecules' polarity due to the electron charge difference, i.e. asymmetric charge distribution, between the different parts of the molecule due to the difference in electronegativity between the bonded atoms [135]. Tomala et al. [136] used the AFM to study the effect of base oil polarity on the decomposition of the ZDDP. The study found that in the case of non-polar oils, the tribofilms were formed faster, thicker and caused less surface roughening than the ones formed in the case of polar oils. The explanation behind this is that polar base oils, as expected, have a large affinity to the steel surface and thus compete with the additive and hinder its accessibility to the metal surface. On the other hand, non-polar oils was found to improve the additive accessibility to the steel surface, which results in accelerated formation of thicker tribofilms than in the case of polar oils [135, 137]. The composition of the tribofilms formed in polar and non-polar oils were similar consisting mainly of phosphate but a difference was observed in the formed sulphur species. For the ZDDP in polar oils, sulphide species were formed as opposed to the mixture of sulphide and sulphate species in the case of non-polar oils [135].

Kar et al. [138] used TEM, SEM, XPS, AFM and microhardness tests to investigate the effect of base oil polarity on the decomposition of additives. The results showed that polar oils provided smoother tribofilms than the ones formed in non-polar oils. The results indicated that the coverage of polar oils on the metal surface enhanced the effective lubrication and protected the metal surface. In contrast, other studies [139, 140] found that the P-containing additives reduce wear more efficiently when used in non-polar oils due to the ease of access of the additive to the steel surface and thus the better formation of the protective antiwear tribofilm. Suarez et al. [137] used SLIM, SEM-EDX XPS, AFM and nanoindentation to study the effect of oil polarity on the interplay between the tribological properties and composition of the formed ZDDP antiwear tribofilms. The study showed that the tribofilm thickness is not the only parameter determining the wear and friction performance but the composition can play a vital role as well. The results showed that although the polar base oil suppresses the formation of thick tribofilms and decreases their formation rate, it reduces friction more efficiently than the thicker tribofilms formed in the non-polar oils. The tribofilm structure in the polar oil seemed smoother and more homogeneous and continuous than the one formed in non-polar oils. This smoother tribofilm provided better wear properties, which was proposed to be related to the harder tribofilm formed in the case of polar oils as opposed to the softer tribofilm in the case of non-polar oil. The study suggested that this softer tribofilm leads to a larger contact area and thus larger wear. However, larger contact area should necessarily mean lower contact pressure and thus wear is expected to be less. This discrepancy cannot be explained based on hardness point of view only. In addition, measuring the hardness of a tribofilm of thickness less than 40 nm with the NI technique using an indentation depth of 20 nm, will inevitably include a significant effect from the hard substrate underneath.

6.4.3.5 Duration of Rubbing and Heating

The duration of rubbing and heating can greatly affect the composition of the formed tribofilms and therefore their antiwear properties. Yin et al. [1] used XANES to study the decomposition species of the ZDDP after different rubbing times, i.e. 5 min, 30 min, 6 h and 12 h. They also investigated the effect of rubbing a mature film formed after 30 min in base oil for another 5.5 h to examine the durability of the tribofilm and the effect of rubbing without additives. The results indicated that the tribofilms formed after short rubbing times exhibit different fingerprints than the ones formed after longer rubbing times, which suggested that different species are formed initially before the phosphate chains. They also found that initially a large concentration of unreacted ZDDP adsorbs to the metal surface, which can coexist with the reacted ZDDP. Rubbing the ZDDP tribofilm in base oil led to the depolymerisation of the long chains into shorter ones.

Similarly, Gosvami et al. [40] used the in situ AFM liquid cell shown in Fig. 6.4 to study the evolution of ZDDP tribofilms over time. The study showed that initially the tribofilm volume increased linearly over rubbing time. However, after a certain threshold the volume started to have exponential growth. The EDX and Auger electron spectroscopy (AES) analysis of the tribofilm revealed that it consisted mainly of Zn, P and S. Furthermore, the distribution of Fe was uniform inside and outside

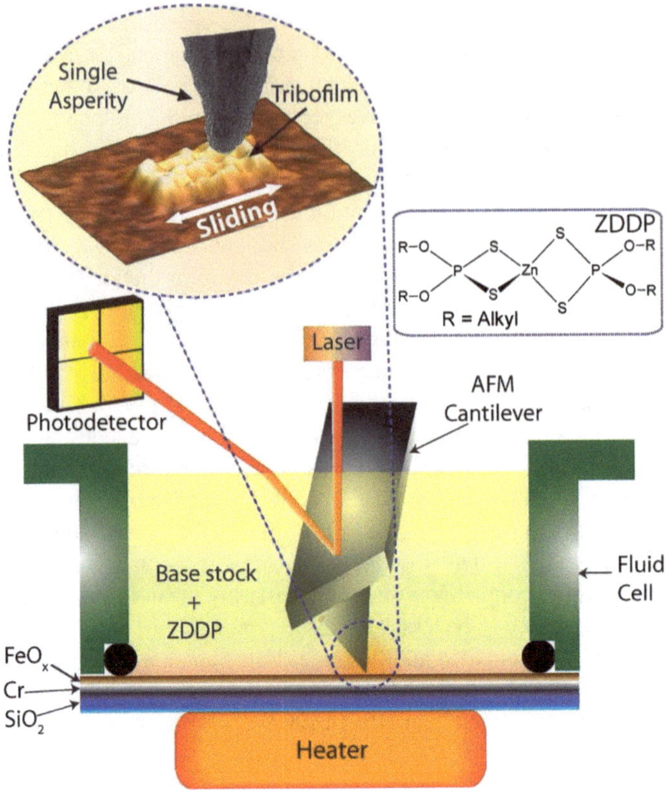

Fig. 6.4 In-situ AFM liquid cell used by Gosvami et al. [40]

the wear scar indicating that the tribofilm does not have any significant amount of Fe, if any.

Different other studies also used in situ XANES to follow the change in composition of ZDDP thermal films over heating time. For instance, Dorgham et al. [141] used the in situ XANES tribotester shown in Fig. 6.5 to follow the change in composition of ZDDP tribo- and thermal films over time. The results showed that the formation rate of the tribofilm is much faster than the thermal one but the two eventually have a similar composition. It was also found that the decomposition of ZDDP starts with the formation of sulphate species, which over time is reduced into sulphide along with the formation of zinc phosphate. In agreement with these results, Morina et al. [64] using the in situ XANES heating cell shown in Fig. 6.6 showed that in the beginning of the thermal decomposition process, sulphide and sulphate species are formed. In order to have a surface sensitive signal, the in situ experiments were performed in the total external reflectance mode by tilting the samples to an angle less than the glancing angle. The XANES results showed that in the beginning of the thermal decomposition reaction, sulphide and sulphate species are formed. The

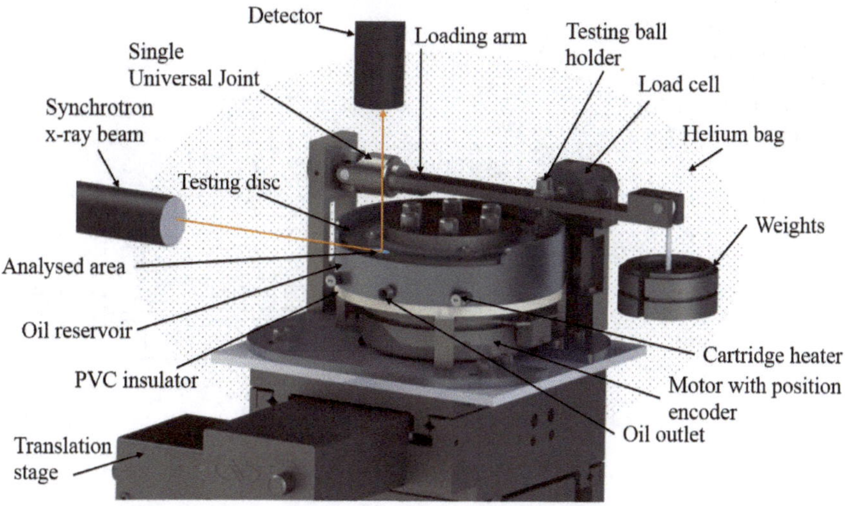

Fig. 6.5 Schematic of the assembly of the tribological apparatus used by Dorgham et al. [141] for in-situ XAS experiments

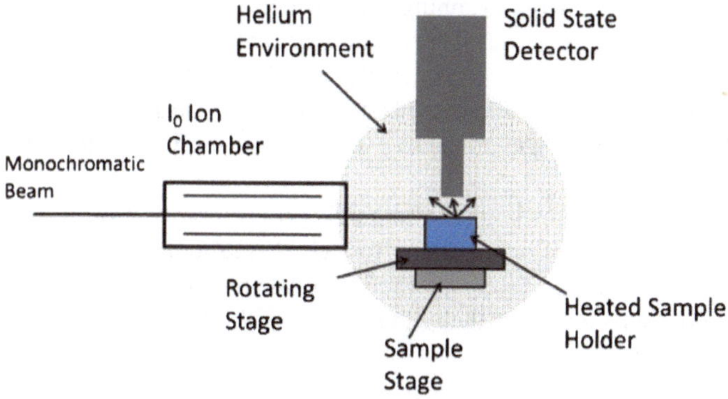

Fig. 6.6 Schematic of the assembly of the thermal liquid cell used by Morina et al. [64] in-situ XAS experiments

sulphate formation was related to the effect of relative humidity [64, 143, 144], i.e. water contamination in the oil. The presence of sulphate species was also detected by Ferrari et al. [142, 145] who used the heating cell shown in Fig. 6.7 to follow the ZDDP thermal films over heating time.

For the DDP antiwear additives, Kim et al. [35] tested two types of DDPs and compared them with ZDDP. They performed these experiments at 170 °C at which the additive decomposes in the oil and produces solid precipitates, which are subsequently deposited on steel coupons underneath. The results indicated that the progression of heating makes the ZDDP thermal films richer in sulphur whereas the DDP films richer in phosphorus. Zhang et al. [26] found that the evolution of P/S

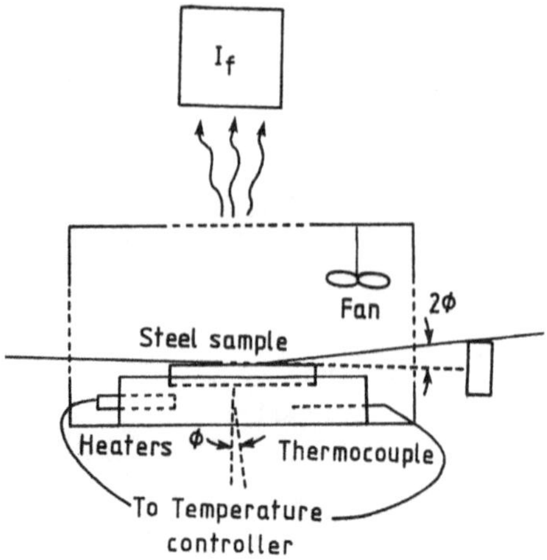

Fig. 6.7 Schematic of the assembly of the thermal liquid cell used by Ferrari et al. [142] for in-situ XAS experiments

ratio was nearly constant over time for the ZDDP tribofilms, whereas it showed a gradual increase for the case of DDP tribofilms. This is despite the general increase in the tribofilm thickness over rubbing time, which indicates that the composition evolves mainly during the initial stage of rubbing [25]. Najman et al. [25] suggested that initially the DDP additive reacts rapidly with the substrate covered with oxides, which leads to the oxidation of the sulphur species into iron sulphate. These species do not help in protecting the contacting surfaces but only does the subsequently formed phosphate [85].

The DDP additive forms Fe phosphate of short chains whereas the ZDDP forms initially zinc phosphate of short chains that grow into longer ones away from the metal surfaces whereas the layers near the substrate remain of shorter chains due to the presence of Fe cations near the steel substrate [25, 66].

6.4.3.6 Concentration of Additive

The concentration of the ZDDP affects its adsorption and coverage on the metal surface [105, 146], i.e. the larger the concentration of the ZDDP, the larger its adsorption and coverage. Yin et al. [1] used XANES to study the effect of the ZDDP concentration, i.e. 0.25%, 0.5%, 1.0% and 2.0 wt%, on the decomposition species. The results showed that a low concentration of ZDDP can lead to the formation of short phosphate chains whereas a high concentration leads to the formation of long chains in addition to the short ones and unreacted ZDDP. Furthermore, Tomala et al. [136] found that the larger the additive concentration (i.e. 2 and 5%) the thicker the tribofilm and the larger its roughness. Similarly, Ghanbarzadeh et al. [147] indicated that increasing the additive concentration increases the formation rate and terminal film thickness and decreases the average wear depth. Thus, the additive concentration evidently has a strong effect on the decomposition kinetics. This is further supported by the study of Akbari et al. [120] of the effect of ZDDP concentration, i.e. 1%, 2.0% and 20 wt%, on the composition of thermal films formed on bare steel, Si-doped (a-Si:H) and non-doped (a-C:H) DLC coatings. The results showed that the thickness of the thermal film increases linearly with the concentration with a rate about 2.75 nm/min on bare steel and 1.5 nm/min on a-C:H DLC coating.

6.4.3.7 Temperature and Load

Temperature and load are expected to affect not only the composition of the formed tribofilm but also its structure as well as its tribological and mechanical properties. Zhang and Spikes [40] showed using the AFM that at low temperature of 60 °C the tribofilm morphology was patchy of roughness similar to the large initial roughness of the substrate. As temperature increased to 100 and 120 °C, the tribofilm thickness increased and a more pad-like structure was formed with wide pads elongated in the direction of rubbing. Similarly, Yin et al. [1] showed that the high temperature accelerates the decomposition of the unreacted ZDDP and helps increase the length of the formed polyphosphate chains. However, the study found that after a certain threshold, the high temperature can be detrimental to the chains length. Short chains were formed at 200 °C as opposed to the long chains formed at 100 and 150 °C. The study also found that high temperatures promote sulphate species formation. At 100 °C, sulphide species were detected whereas sulphate species were found at 200 °C. Other studies observed a similar effect for the load [84]. For instance, under high contact pressure, the DDP additive forms mainly sulphate, e.g. $FeSO_4$, whereas under low pressure mainly disulphide (FeS_2). Contact pressure did not seem to affect the already short chains composing the DDP tribofilms but reduced the chain length of the ZDDP tribofilms from long to medium.

Similar to load, temperature can notably increase the formation rate and terminal film thickness of the formed tribofilms but decrease the average wear depth [147]. The tribofilm thickness during the early stages of rubbing, e.g. running-in period, and not only the steady state thickness, seems to be responsible for the antiwear protection [147]. Similarly, Parsaeian et al. [148, 149] found that by reducing the temperature from 100 to 80 °C, the tribofilm thickness decreases accompanied by a conspicuous increase in wear.

The effect of temperature on the mechanical properties of postmortem tribofilms was studied by Pereira et al. [150] using nanoindentation. The tribofilms consisted mainly of polyphosphate of medium chain length and sulphides as indicated by the XANES analysis. The tribofilm thickness measurement based on FIB-SEM cross-section suggested a thickness of 180 ± 60 nm, which is consistent with the XANES P k-edge estimation of 105 nm. The evolution of the indentation modulus over temperature, i.e. from 25 to 200 °C, showed that up to 200 °C the modulus was nearly constant at around 100 GPa but dropped to about 70 GPa at 200 °C. The decrease in the modulus might be responsible for the good antiwear properties of the ZDDP tribofilm, i.e. the low modulus means a compliant sacrificial tribofilm that can be easily worn instead of the substrate.

The combined effect of load and temperature on the evolution of ZDDP tribofilms was investigated by Gosvami et al. [40] using in situ AFM tribotests. The study showed that temperature exponentially increases the growth rate of the tribofilm. Similar to temperature, load ranging from 2 to 7 GPa, appears to increase the growth rate of the tribofilm exponentially until reaching steady state. Therefore, the study concluded that load and temperature have the same catalytic effect on the tribofilm formation. This is in agreement with the results of Yin et al. [1], which found that load has the same effect as temperature, i.e. accelerating the decomposition of the unreacted ZDDP and the formation of long phosphate chains. The study also found that load does not affect the sulphur species type or concentration, i.e. sulphides were the only sulphur species to be formed and their intensity did not change much with load. Zhang and Spikes [41] showed using SLIM that in the EHL regime the larger the load (50–75 N) the faster the tribofilm formation and thicker the terminal thickness of the formed tribofilms on WC. The formation rate ranges from 0.2 to 0.7 nm/min depending on the maximum shear stress (220–250 MPa) at the edge of the tribofilm. Similarly, Tomala et al. [136] found that the larger the load, ranging from 1.3 to 2.4 GPa, the thicker the tribofilm and the larger its roughness.

The effect of load on the ZDDP decomposition was also studied by Ji et al. [151] using XANES and AFM. The results indicated that the formed tribofilms consisted of polyphosphate of thickness between 10 and 100 nm. The results also showed that the larger the load the thicker the tribofilm. Additionally, based on the correlation between the friction force behaviour and electrical contact resistance measurements, they concluded that the ZDDP decomposition and tribofilm formation undergo three different stages:

1. Induction period, which increases with load and generally decreases with sliding speed. During this period, ZDDP molecules adsorb to the substrate and friction increases.
2. Tribofilm growth period during which tribofilm thickness increases and friction stays constant.
3. Tribofilm growth and removal period during which tribofilm thickness and friction stay constant.

The effect of load and temperature on DDP additives seems to be similar to the one of ZDDP. It was reported that the higher the temperature the thicker the thermal film [82]. Furthermore, the tribofilm thickness was found to be at least twice the thickness of the thermal film [82] and in general increases gradually over rubbing time [25], which suggests that load and temperature can both help accelerate the decomposition reaction.

In summary, the above discussion suggests a similar effect of temperature and contact pressure on the growth rate of ZDDP and DDP tribofilms, which indicates they have a catalytic effect on the additive decomposition process. Thus, it can be stated that the decomposition of P-based additives is a thermally and mechanically assisted reaction, which can be activated by the availability of either shear or heat.

6.4.3.8 Sliding and Rolling Speeds

It is understood that the ratio of sliding to rolling speeds can affect the contact severity and thus the lubrication regime. To further understand this effect on the decomposition of the ZDDP, Suarez et al. [135] followed the ZDDP tribofilm formation under different levels of slide-to-roll ratio (SRR), i.e. 0, 2, 5 and 10%. The study found that the larger the SRR, the faster the tribofilm formation. However, the study showed that the exact sliding percentage, as long as >0, does not have a significant effect on the limiting tribofilm thickness. On the other hand, the SRR appeared to have a different effect on the topography of the tribofilm depending on the base oil. In the case of polar oil, the tribofilm appeared homogeneous throughout the different levels of the SRR. When nonpolar oil was used, low SRR generated smooth tribofilms whereas large SRR ($\geq$5%) generated a rougher pad-like structure. That is the larger the SRR is, the rougher the tribofilm topography becomes and the more the structure appears patchy.

The effect of sliding speed on the ZDDP decomposition was also investigated by Ji et al. [151] using XANES and AFM. The results showed that lower velocities and higher load increase the tribofilm thickness whereas higher speeds and lower loads decrease the tribofilm thickness. In terms of friction relation to velocity, there was no simple relation. Under all the tested loads there seems to be a certain threshold of sliding speed below which the friction appears to decrease with increasing the sliding speeds whereas the opposite trend was observed above this threshold.

The effect of high SRR, i.e. 50–230%, on the ZDDP tribofilm formation was examined by Shimizu and Spikes [152] using SLIM, EDX and AFM. The study found that in the mixed sliding and rolling condition, the formation rate of the tribofilm is less sensitive to the exact SRR, which indicates that the rubbing time is far more important than the sliding distance.

6.5 Mechanical Properties of Antiwear Films

6.5.1 *Structure*

The first report on the structure of the P-based antiwear films goes back to the study of Bird and Galvin [51] who suggested that the ZDDP tribofilm is polymeric and has a patchy structure. The authors conjectured that the first layer of this structure is a discontinuous sulphide, possibly zinc sulphide, or sulphate layer covering the metal surface. This layer in turn is covered with islands of unknown compounds containing Zn, P, and S. They also found that the ratio of Zn:P:S changes considerably along and across the wear scar. This heterogeneity can be related to the different local conditions, e.g. temperature, pressure and load. For instance, Sheasby et al. [67] and Palacios [7] observed that at low temperatures, e.g. 35 °C, a thin but uniform film was formed as compared to a thick but less uniform film at higher temperatures, e.g.

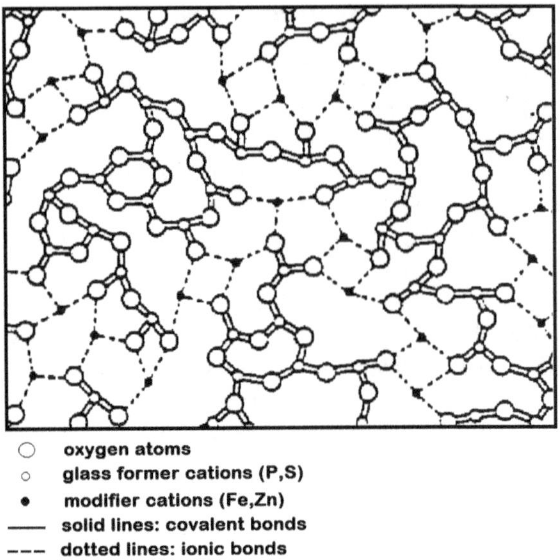

Fig. 6.8 Arrangements of atoms in the amorphous tribofilm. Reprinted from Martin et al. [81]

150 °C. The tribofilm formed at high temperature was patchy in nature suggesting that it was only formed at the asperity-asperity contacts where temperature was high. In addition, the thickness was not found uniform amongst the different patches. Martin et al. [81] used X-ray absorption fine structure (XAFS) spectroscopy to show that the ZDDP wear particles have a continuously random amorphous structure, as depicted in Fig. 6.8. This structure was hypothesised to be formed due to friction- and shear-induced atomic-scale mixing processes at the interface [81] in addition to the digestion of the iron oxide by the phosphate layers near the metal surface [8]. Furthermore, it could be formed due to thermal effects by the interface quenching [81].

Later on, the tribofilm was identified of having a multilayer structure [1, 6, 9, 153–155]. The layers close the substrate appear to be solid and adhere strongly to the rubbed surface. However, the outer layers are viscous or semi-solid of hydrocarbons and organic radicals, which adhere weakly to the lower layers and hence they can be removed easily by rinsing with a solvent [6, 153, 155]. The concentrations of these products decrease along the depth whereas the concentrations of iron, iron oxide and iron sulphide increase. Above the iron sulphide/oxide layer, a layer of amorphous polyphosphate glass was identified, which can contain zinc oxide and zinc sulphide, as depicted in Fig. 6.9. The mechanical and physical properties of this amorphous 3D network are predetermined by the Zn^{2+} cations and organic radicals. However, the layers near the metal surface are mainly affected by Fe^{2+} and Fe^{3+} cations. Bell et al. [6] reported that the detected phosphate was similar to P_2O_5 glass. Therefore, they proposed that the ZDDP tribofilm consists mainly of glassy polyphosphate that has different physical and mechanical properties along the depth of the tribofilm. In addition, they noticed that the molar ratio of M_2O/P_2O_5 increased from 1.2 at the

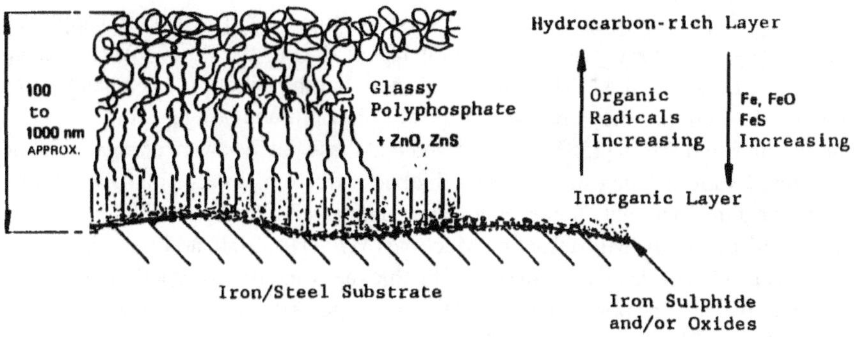

Fig. 6.9 Schematic of the structural model of the ZDDP tribofilm elucidating its multilayer characteristic. Reprinted from Bell et al. [6]

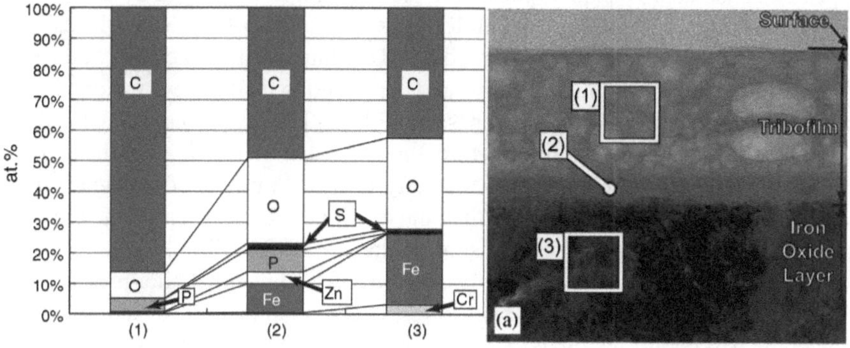

Fig. 6.10 TEM Cross section image of the ZDDP Tribofilm layers and EDX semi-quantitative analysis of the elements constituting the different layers. Reprinted from Ito et al. [103]

outer surface to about 2 in the bulk of the tribofilm. This suggested that the outer layers have longer chain length than those in the bulk. This also indicated that the outer layers have less concentration of zinc. Furthermore, recent experiments conducted by Ito et al. [103, 104] showed that the concentration of zinc is much higher near the metal surface, as shown in Fig. 6.10. This indicates that the cation exchange reaction between zinc and iron at high temperature should have occurred during the initial stage of the ZDDP decomposition. This also suggests that the presence of zinc near the metal surface plays a significant role in the observed short phosphate chains near the substrate.

Yin et al. [1] noticed that the XANES results of the Total Electron Yield (TEY), which is surface sensitive, and Fluorescence Yield (FY), which is bulk sensitive, were different. This suggested that the tribofilm has a layered structure consisting of a layer of short phosphate chains laid with a layer of long phosphate chains. Bec et al. [155] used the Surface-Force Apparatus (SFA) to determine the structure of the ZDDP tribofilm. The authors reported a full schematic of the ZDDP tribofilm before and after washing with solvents as shown in Fig. 6.11. The tribofilm appeared to be patchy and every patch has several layers. Using the AFM, Warren et al. [156] and Graham et al. [157] further confirmed that the tribofilm has a patchy-like structure. In addition, they observed stripes of discontinuous ridges along the wear scar. Pidduck and Smith [158] also observed that these patches were elongated in the direction of sliding. The estimated film thickness was in the range of 100–140 nm, which is in agreement with the recent measurements using the Spacer Layer Imaging Method (SLIM) of the Mini-Traction Machine (MTM) [159]. In addition, the results of Bec et al. [155] indicated that the tribofilm thickness and friction coefficient change substantially from one position to another. This suggests that these parameters should be considered as local properties rather than as averaged values.

Aktary et al. [160] followed the topography evolution of ZDDP thermal films using Atomic Force Microscopy (AFM). During the first two hours, the structure appeared as isolated islands of phosphate precipitates, as confirmed by the Fourier Transform Infrared spectroscopy (FTIR). However, after 3–6 h, these islands coalesced and the thermal film became smooth and continuous. The thickness of the film increased linearly with immersion time and continued to increase even after 6 h during which it reached an average thickness of 420 nm. The results indicated that there is some correlation between the structure, thickness and morphology of the thermal film. The FTIR results showed that the early formed thin film of isolated islands consists of shorter phosphate chains than the mature thick and continuous film. The follow-up study of Aktary et al. [70] showed that the ZDDP tribofilm evolves in the same way as the thermal film. The only difference found was that after long rubbing time, the continuous film disintegrates to form very small pads. The results of Canning et al. [161] confirmed that the tribofilm is largely heterogeneous in the lateral direction. The tribofilm was found to have ridges of mainly long phosphate chains and troughs of mainly short phosphate chains. Same conclusions were also reached by Nicholls et al. [20, 28, 80, 162] using X-ray Photoemission Electron Microscopy (XPEEM) and AFM. The results of these studies showed that the ZDDP tribofilm is also heterogeneous along the depth. The large pads of the tribofilm were found to have a multilayer system where long polyphosphate chains lay on the top of short polyphosphate chains. Based on these results amongst others, Spikes [21] proposed that the ZDDP tribofilm has a structure similar to one depicted in Fig. 6.12. This structure is in line with the findings of the previous studies that the tribofilm consists of multilayer structure, which is largely heterogeneous in the lateral direction and along its depth. This is also similar to the structure of DDP tribofilms, which initially appears less patchy than the ones of ZDDP but eventually the two provide a similar uniform pad-like structure [25, 26]. Distinctively though, the DDP additive forms Fe phosphate of predominantly short chains as opposed to the initially zinc phosphate of short chains that grow into longer ones away from the metal surface in the case of ZDDP [25, 66]. Furthermore, the formation of iron sulphides or sulphates in the DDP tribofilms near the metal surface depends greatly on the operating conditions [84], as discussed in detail in the previous sections.

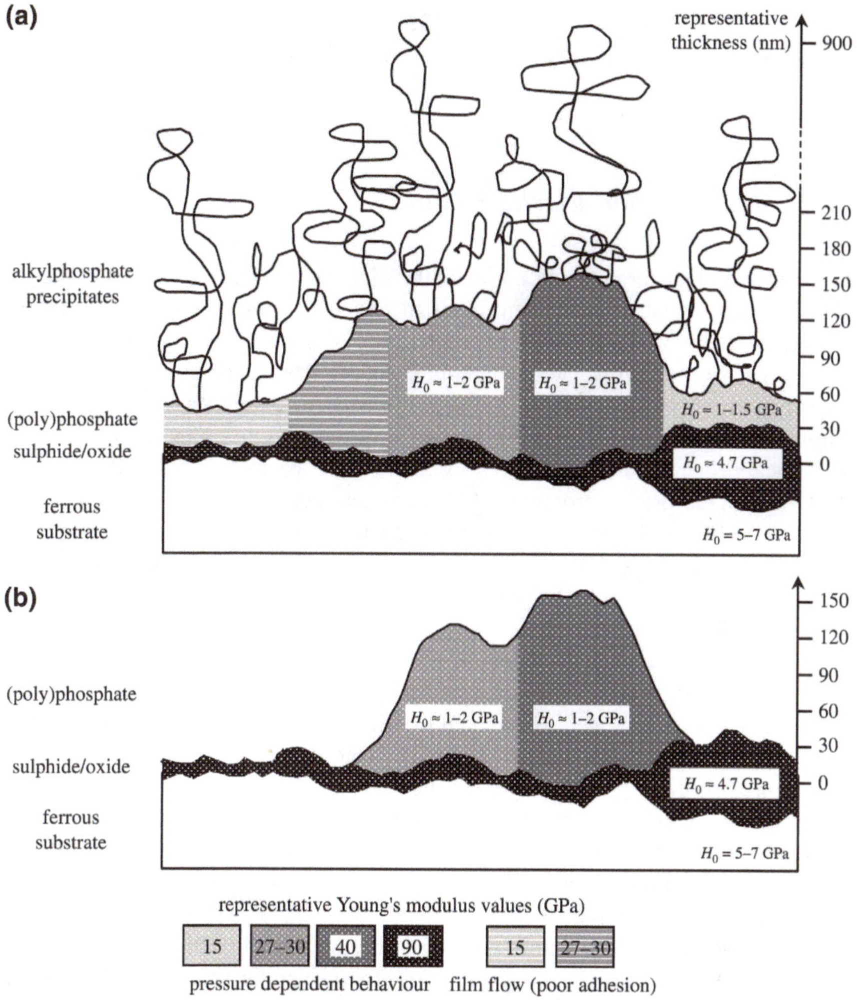

Fig. 6.11 Schematic of the multilayer structure of the ZDDP tribofilm **a** before and **b** after washing with solvent. Reprinted from Bec et al. [155]

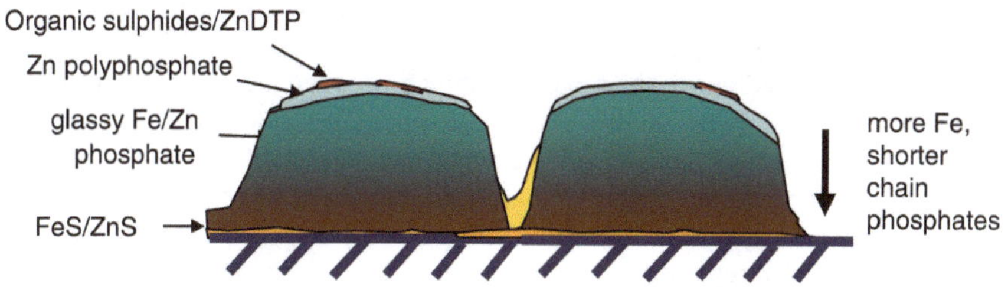

Fig. 6.12 Schematic of the patchy structure of the ZDDP tribofilm. Reprinted from Spikes [21]

The structures proposed by Bec et al. [155], Martin et al. [9] and Spikes [21] seem to be the most accepted picture of the structure of P-based antiwear tribofilms up to date.

6.5.2 Hardness and Elastic Modulus

Due to the largely heterogeneous nature of the ZDDP or DDP antiwear films, whether tribofilms or thermal films, probing their mechanical properties is a complex and intricate task. The main difficulties are attributed to the error brought by averaging over a large area and propagated by the different uncertainties associated with every experimental technique. This necessitates the need to analyse a large area of the sample at a high lateral resolution and at different length scales. Measuring the mechanical properties at a small length scale became feasible with the advent of the AFM, interfacial force microscope (IFM) and nanoindentation.

Many studies have already used the AFM to study the mechanical properties of the ZDDP tribofilms [70, 77, 156, 157, 160, 163–166] as well as tribofilms of other additives [167, 168]. This enormous work concluded that the layers close to the metal surface are most likely elastic solid whereas the outer layers are probably viscous. This indicates that the layers in between can have a combination of the two. Furthermore, Warren et al. [156] observed stripes of discontinuous ridges along the wear scar. These stripes were speculated to originate from repetitive sliding, which suggested that the ridges were the only part of the tribofilm that carry the load. In agreement with these results, the experiments of Graham et al. [157] showed that these ridges were elongated in the sliding direction. Aktary et al. [70] attributed this heterogeneity in the tribofilm growth to the local variations in the contact pressure between the contacting surfaces. However, as the tribofilm is not heterogeneous only in the lateral direction but also along its depth, this indicates that the heterogeneity of the tribofilm is not only due to the local variations in the contact pressure [77]. Warren et al. [156] proposed three possibilities for the patchy structure of the P-based tribofilms, which are as follows:

1. Ridges are formed due to the high contact pressure between the asperities whereas the troughs are formed due to the three body interaction between the wear debris and the contacting surfaces.
2. Ridges are formed due to the repeated rubbing between the asperity contacts at the same average locations whereas the troughs are formed due to occasional contact between the asperities at the troughs regions.
3. Ridges are formed due to the repeated rubbing between the asperity contacts whereas the troughs are originated from the flow of some material from the ridges to troughs' regions.

The authors suggested that the way the ridges and troughs appeared indicated that the third mechanism is likely to be the case. Other possibilities to account for the heterogeneity in the tribofilm growth have also been proposed recently by Gosvami et al. [40] to be due to the following:

1. Heterogeneous random nucleation sites due to the surface roughness or any surface defects.
2. Instabilities in the decomposition and growth mechanisms due to any variations in the operational conditions.

These possibilities are plausible and can be combined with the ones proposed by Warren et al. [156] to give a wider understanding.

To quantify the mechanical properties of the ZDDP or DDP tribofilms, several studies attempting at measuring the elastic modulus using nanoindentation, AFM and SFA. Table 6.3 summarises the indentation modulus that was reported for the ZDDP and the measurement conditions. Aktary et al. [70] measured the elastic modulus of the ZDDP tribofilm after different rubbing times and found that the modulus and hardness of the tribofilm are independent of the rubbing time. They explained this result by pointing out that the majority of the bulk material of the tribofilm at any time consists mainly of short polyphosphate chains. Therefore, the nanomechanical properties are predetermined by the properties of this bulk material. However, Nicolls et al. [162] reported that the large pads of the tribofilm consist mainly of long polyphosphate chains whereas the small pads and troughs consist of short chains. Graham et al. [157] showed that the pads have a great elastic response. The measured indentation modulus was about 180–250 GPa at the centre of the pad compared with 50–110 GPa at the edge. Nicolls et al. [80] found that the indentation modulus was 120 GPa at the centre of the pad compared with 90 GPa at the edge. In a later study, Nicolls et al. [162] reported an indentation modulus of 80 GPa for the large pads but could not measure the modulus of the small pads due to the effect of the substrate on the indentation measurements of these thin regions.

This wide range of values of the reported indentation modulus could be related to the extent of the additive decomposition to form polyphosphates [20] in addition to the intrinsic uncertainties associated with probing such thin heterogeneous films using any indentation technique.

In contrast to the aforementioned studies, which suggested that the tribofilm is highly elastic, Aktary et al. [70] found that the ZDDP tribofilm exhibits high plasticity. They suggested that this is in line with the sacrificial nature of the tribofilm, which is continuously formed and removed at the interface. Conversely, Warren et al. [156] showed that the ridges of the tribofilm exhibit a great capacity of elastic deformation, which was also confirmed by the results of Graham et al. [157] and Ye et al. [163] showing that the ridges of the ZDDP tribofilm resist plastic deformation to a great extent.

Ye et al. [77, 163, 164] studied various tribofilms formed in oils containing ZDDP and ZDDP/MoDTC additives. In the two cases, the results showed a gradual increase of hardness and modulus over depth until they reach the ones of the substrate. This was considered as evidence that the tribofilm consists of a multilayer system of a hard layer covered by a softer one. Similarly, Bec et al. [155] showed that the ZDDP tribofilm resists indentation by increasing hardness and elastic modulus with the indentation

Table 6.3 Elastic modulus and hardness of the ZDDP thermal and tribo-film

Material	Elastic modulus (GPa)	Method	Lubricant	Test conditions	References
5 min tribofilm	88.5 ± 23.7	Nanoindentation	1.49 wt%	Plint	[70]
10 min tribofilm	92.8 ± 18.6		ZDDP	100 C	
40 min tribofilm	88.6 ± 29.9		In MCT-10	225 N	
1 h tribofilm	88.9 ± 12.1			25 Hz	
2 h Tribofilm	96.1 ± 25.7				
3 h thermal film	34.8 ± 9.7				
1 h tribofilm	87.8 ± 3.9	Nanoindentation	1.2 wt%	Plint	[80]
Edge of pad			ZDDP	100 C	
1 h tribofilm	119.5 ± 5.8		In MCT-10	220 N	
Centre of pad				25 Hz	
Sulphide oxide	90	SFA		Amsler	[155]
$t \leq 80$ nm				100 C	
Polyphosphate	15			400 N	
$20 < t < 30$ nm				5 h	
Polyphosphate	27–30				
$70 < t < 100$ nm Polyphosphate $t > 140$ nm	40				
6 h tribofilm	25	IFM	1.2 wt%	Plint	[156]
Trough			ZDDP	100 C	
6 h tribofilm	81		In paraffinic	220 N	
Ridge			Base oil	25 Hz	
1 h thermal film	36 ± 9			200 C	
Large pads	209 ± 38	IFM	5 mM/kg	Plint	[157]
Ridges			ZDDP	100 C	
Large pads	87 ± 23		In paraffinic	225 N	
Troughs			Base oil	25 Hz	
Small pads	74 ± 20				
Off pads	37 ± 7.3				
1 h tribofilm	85.1 ± 11.1	Nanoindentation	1.2 wt%	Plint	[169]
		And IFM	ZDDP	100 C	
			In MCT-10	220 N 25 Hz	
1 h tribofilm	80.5 ± 4.5	Nanoindentation	1.2 wt%	Plint	[80]

(continued)

Table 6.3 (continued)

Material	Elastic modulus (GPa)	Method	Lubricant	Test conditions	References
Ridges			ZDDP	100 C	
1 h tribofilm	30		In MCT-10	220 N	
Troughs				25 Hz	
10 min tribofilm	78–110 ± 8	Nanoindentation	1.2 wt%	Plint	[162]
		And IFM	ZDDP	60 C	
1 h tribofilm	97–110 ± 8		In MCT-10	60 N 25 Hz	
1 h tribofilm	80–215	Nanoindentation		Pin-on-	[163]
	Depends on		ZDDP in	Disc	
	Contact		5 W–30 SG	490 N	
	Depth		Engine oil	0.03 m/s	

depth. Therefore, they suggested that the heterogeneity of the mechanical properties of the tribofilm along its depth is due to work hardening but not a real intrinsic property. However, in contrast to these results, the loading-unloading experiments of Warren et al. [156] using AFM indicated that the tribofilm ridges have two layers of which the one that is more compliant lay beneath the stiffer surface layer. The base layer seemed to exhibit a significant tenacity, which was manifested in the apparent adhesion during the retraction of the AFM tip from the sample. These features were neither observed in the troughs of the tribofilm nor in the thermal film. The mechanical properties of the trough region of the tribofilm seemed identical to the ones of the thermal film. Based on these results, Warren et al. [156] suggested that the thermal film is formed first as a precursor to the tribofilm. In line with these results, Kim et al. [22] showed that the surface layer of the ZDDP tribofilms, away from the substrate, is harder than the bulk layers. In contrast, DDP tribofilms showed a more compliant tribofilm without the presence of a hard surface layer. They suggested that the hard crust protects against wear while the compliant bulk helps dissipate energy. This is despite the fact that they observed that the more compliant but thicker DDP tribofilm showed a much better tribological performance than the one of ZDDP. To resolve this discrepancy, they suggested that the effective coverage of the DDP tribofilm is higher than the ZDDP one leading to better protection against wear.

Based on the discussion above, it can be concluded that the ZDDP and DDP antiwear films have rich mechanical properties. These properties can be affected by the decomposition mechanism, load and temperature. In addition, the mechanical properties might also be affected by the rheological properties of the tribofilm. For instance, in case of the tribofilm is viscoelastic then its mechanical response can look similar to the plastic behaviour of a compliant material [170]. Moreover, a viscoelastic tribofilm may suggest that its mechanical properties are rate dependent.

Hence, the history and rate of measurement can play a major role in the measured properties. Therefore, studying the rheological properties of the antiwear tribofilm is necessary to give insight into its mechanical as well as tribological properties. These rheological properties will be discussed in detail in the subsequent section after the tenacity and durability of the tribofilm is reviewed.

6.5.3 *Tenacity and Durability*

In order to study the tenacity and durability of ZDDP tribofilms, Bancroft et al. [2] examined the effect of the ZDDP concentration in oil on the tribofilm that has already been formed. After the formation of the tribofilm, the base oil containing ZDDP was replaced with oil without ZDDP and then the rubbing continued for extended periods of 6–24 h. The tribofilm showed a great thermal and mechanical stability even after rubbing for periods as long as 24 h. In addition, the results showed that rubbing the tribofilm in base oil has two main effects. The first one is that the tribofilm maintains a certain thickness above 30 nm without being completely removed by rubbing. The second effect is that the polyphosphate chains become shorter in the form of orthophosphate and pyrophosphate. Ancillary experiments of Suominen Fuller et al. [79] showed that when the oil of mature ZDDP tribofilm was replaced by base oil without ZDDP, no change occurred in the tribofilm thickness with further rubbing in the base oil. However, Parsaeian et al. [171] found that rubbing premature tribofilm (after 25 min rubbing: before reaching steady state thickness) in base oil without ZDDP causes an initial sharp decrease in the tribofilm thickness after the first few minutes of rubbing. The reduction was about 20–70 nm depending on load and temperature, i.e. the higher the load or temperature the higher the removal. The sharp decrease in the tribofilm thickness was followed by a steady state period during which the tribofilms maintained its thickness without any further removal. Adding fresh oil containing ZDDP again after 60 min of total rubbing time (35 min of rubbing in base oil) results in a fast growth of the tribofilm thickness similar to the initial growth rate before replacing the oil with base oil. On the other hand, rubbing mature tribofilm in base oil (after 180 min rubbing: after reaching steady state thickness) results in just a small decrease in the tribofilm thickness. As indicated before, increasing temperature or load after replacing oil with base oil resulted in more immediate removal of the tribofilm. However, temperature and load did not show any monotonic trend on the steady state thickness. The drop in the tribofilm thickness was investigated using XPS, which showed that before replacing the oil with base oil the top layer of the tribofilm contains longer phosphate chains. This layer seems to be the one removed after replacing the oil, which suggests that it is softer and less tenacious than the layers underneath it. It was also found using XPS analysis that if rubbing is continued again by adding fresh oil containing ZDDP, the removed long phosphate chains can be formed again.

6.6 Rheological Properties of Antiwear Films

The friction, lubrication and adhesion properties of any tribological surface are greatly affected by the rheological properties of the interface [172]. These properties can undergo changes when using additives such as ZDDP or DDP, which can form a protective tribofilm of transient thickness covering the interface. In order to quantify these changes, the rheological properties of the thin antiwear film should be measured. This can be mainly achieved using two methods. The first one is by generating a thick tribo- or thermal film that can be scratched and removed for ex situ analysis using the bulk or interfacial rheometry. The second possible method is by measuring these properties in situ. The in situ rheological measurements of the antiwear tribo- or thermal films or generally speaking any thin film on a substrate are performed, in essence, in the same way as the bulk rheological measurements, e.g. creep and shear, but using different experimental techniques such as SFA and AFM. For instance, Georges et al. [173] used SFA to measure the rheological and mechanical properties of the ZDDP physisorbed films by imposing an oscillatory motion in three different directions, i.e. x, y and z axes. This enabled them to measure the damping coefficient, normal stiffness and lateral stiffness at different separation distances between the ball and disc. They could then relate the measured damping coefficient ($A\omega$) at a certain oscillatory frequency (ω) and a separation distance (D) to viscosity (η) using the following relation:

$$A_{\omega} = \frac{6\pi\eta R^2\omega}{D - 2L_h} \quad (6.30)$$

where R is the ball radius and L_h is the total thickness of any adsorbed layers on each surface. Good agreement between the experimental data and the damping coefficient given by the previous equation was found as shown in Fig. 6.13. One interesting result is the approximately exponential increase in the viscosity at small separation distances, which has the effect of increasing the viscous resistance to sliding. Another important finding is that the elastic compressive and shear moduli appeared to increase with reducing the separation distance. The authors related this observation to a possible compaction of the heterogeneous physisorbed layers. In agreement with these results, Bec et al. [155] suggested that the contact pressure could compact the loose layers of the ZDDP tribofilm and transfer them into a solid polyphosphate.

Bec et al. [155] used SFA to measure the mechanical and rheological properties of the ZDDP tribofilm. The results showed that the sulphide and phosphate layers exhibit elastoplastic properties. This suggests that the ZDDP tribofilm can be polymeric in nature [51] and its rheological properties, similar to the mechanical properties, vary along the depth of the tribofilm [6]. This was confirmed by Pidduck and Smith [158] and Bec et al. [155] who reported that the base layer of sulphide resists flow significantly whereas the top layer of alkyl phosphate precipitates behaves as a viscous polymer, which does not resist flow and could be easily removed. Similarly, the bulk layers of polyphosphates forming the ridges of the tribofilm can also be removed

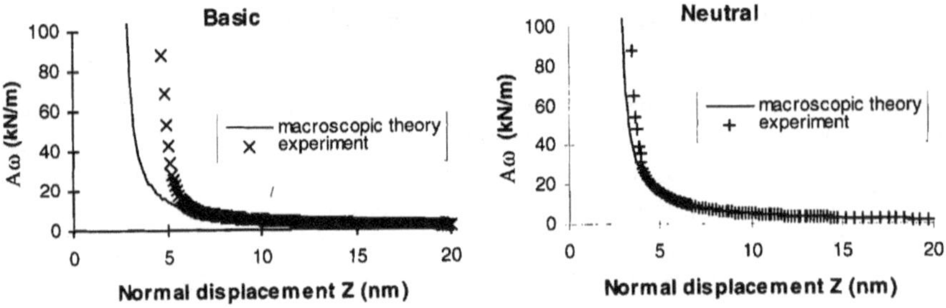

Fig. 6.13 Comparison between the measured and predicted damping coefficient of the basic and neutral ZDDP. Reprinted from Georges et al. [173]

easily leaving trough regions of bare sulphide/oxide layers. This suggested that the polyphosphate layers in the trough originated from the ridges material, i.e. alkyl phosphate precipitates and polyphosphates, that flowed into trough regions by shear flow. Bec et al. [155] reported a viscosity of 5×10^4 to 3×10^5 Pa s for the alkyl phosphate layer and 10^8 Pa s for the polyphosphate layer.

Based on the aforementioned discussion, it can be concluded that the P-based antiwear films have rich rheological properties. However, no elaborate study was conducted so far to unravel all these properties and to provide conclusive insights into the relation between the rheological, mechanical and tribological properties of the ZDDP or DDP tribo- and thermal films.

6.7 Tribological Properties of Antiwear Films

6.7.1 Friction Mechanism

Many studies aimed at understanding the genesis of friction of the P-based tribofilms using different techniques. For example, Ye et al. [163] used the nanoscratch method combined with AFM imaging and observed that the tribofilm exhibits different friction coefficients depending on the contact depth, which indicated that different levels of shear strength exist within the tribofilm. The lowest friction coefficient found was at few nanometres beneath the surface, which was attributed to the presence of an ultra-low friction inner skin layer that can act as a type of solid lubricant to reduce friction. In another study, Ye et al. [166] showed that the inner skin layer exhibits a low shear modulus and can yield easily, which can explain the capability of this layer to reduce friction. The heterogeneous friction behaviour along the depth is also accompanied by a heterogeneity along the surface as reported by Neitzel et al. [174] who observed that friction is different at different length scales. For instance, within a material exhibiting a high friction coefficient, areas of low friction at the nano- or

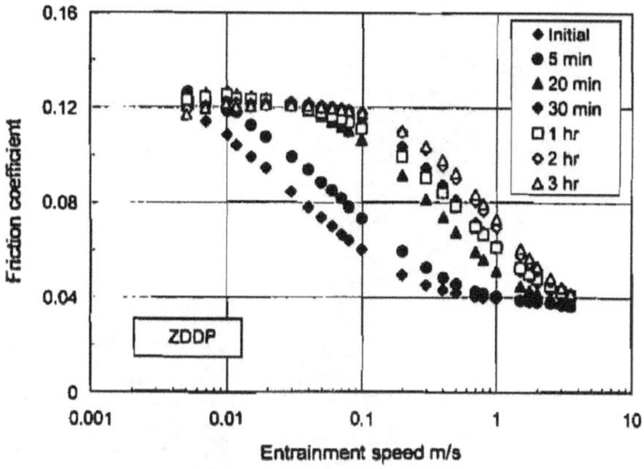

Fig. 6.14 Effect of the ZDDP tribofilm on the Stribeck curve after different rubbing periods. Reprinted from Taylor and Spikes [175]

microscopic scale might exist. These small domains of low friction force can affect the macroscopic friction or induce slip.

Taylor et al. [159] reported that ZDDP forms a thick solid-like film on the wear scar that produces an effective surface roughening. This increased roughness can inhibit the entrainment of the fluid film between the contacting surfaces and thus results in a higher friction than in the case of surfaces not covered with the ZDDP tribofilm, as shown in Fig. 6.14. However, the results of Taylor and Spikes [175] showed that even smooth ZDDP tribofilms can increase friction. This suggested that whether the ZDDP tribofilm is rough or smooth, it can inhibit the entrainment of the lubricating film between the rubbing contacts and hence increases friction. This apparent increase in friction was also suggested to represent a shift in the Stribeck curve to a higher speed, which means that the tribofilm is capable of maintaining the boundary lubrication condition up to a higher speed than in the case of bare contacts without a tribofilm.

A different explanation for the increase in friction force when the ZDDP tribofilm is present on the steel surface was provided by Suarez et al. [137]. Assuming that the friction force can be given by the following equation:

$$F_F = \tau_{\text{ZDDP}} A_{\text{ZDDP}} + \tau_{\text{steel}} A_{\text{steel}} \tag{6.31}$$

where τ_{ZDDP} and τ_{steel} are the mean shear strength of the ZDDP tribofilm and steel surface, respectively, and A_{ZDDP} and A_{steel} are the real contact area of the ZDDP tribofilm and steel surface, respectively. Therefore, increasing the tribofilm thickness increases the contact area (A_{ZDDP}) and thus increases the friction force.

The increased friction of the antiwear tribofilms over time can also be explained based on the observations of Mazuyer et al. [176] that after a certain threshold of contact time the layers covering the rubbing surfaces can interact and hence the mechanical properties, e.g. shear elastic modulus and interfacial shear strength, become a function of the contact time. In the case of the ZDDP tribofilm, this means that the longer the rubbing time, the thicker the tribofilm will be and thus the stronger the

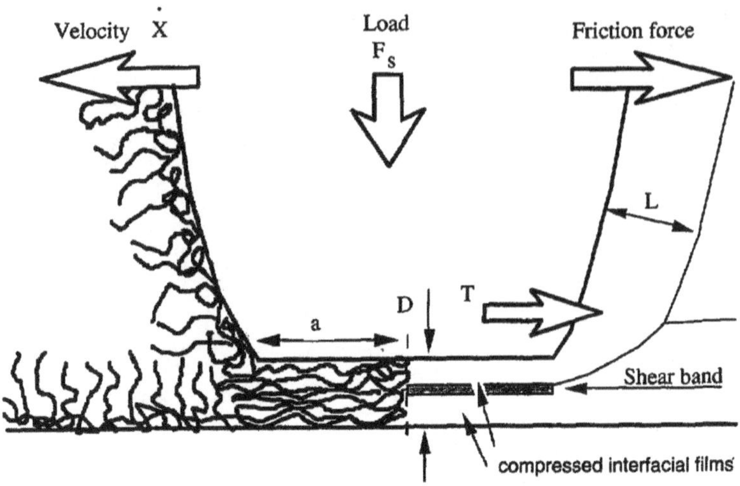

Fig. 6.15 Schematic of the ordered structure of the compressed hydrodynamic layer. Reprinted from Georges et al. [177]

interactions can occur between the tribofilms covering the contacting surfaces, which leads to higher friction.

Georges et al. [177] highlighted the steric action of the adsorbed layers on friction. When the separation distance is small, the pressure is high. In this case, the fluid starts to have an ordered structure in the direction of shear, as shown in Fig. 6.15. This was inferred from the observation that the stabilised friction coefficient was found to be inversely proportional to the sliding speed. In addition, Li et al. [178] showed that the strength of adhesion of the interfacial layers adsorbed on a substrate has great influence on friction. In the case of a single layer that adhered loosely, friction was maximum due to the puckering effect of the layer when probed by the AFM tip. On the other hand, in the case of a bulk material with strongly adhered layers, this puckering effect is suppressed and the thickness does not have any effect on friction after a certain threshold value. Ancillary results of Lee et al. [179, 180] suggested the same trend for different materials including MoS_2.

The fact that sulphur might be the first to reach the metal surface is very crucial in order to relate the mechanism of additive decomposition and tribofilm formation to friction and wear. In the case of sulphur containing deposits are formed initially at the metal surface, then we can infer two main points. Firstly, during the running-in period, the sulphur rich layer is the one that predetermines the observed friction and wear. Therefore, if the porous FeS_2 is the main component of this layer, which is suggested in the case of surfaces covered with Fe_2O_3 or Fe_3O_4 as shown in the potential diagram in Fig. 6.3, friction will be small. However, a larger friction is expected in the case of a base layer consisting of a dense and uniform FeS, which is likely to be formed on bare iron [106]. Furthermore, the existence of a sulphur rich layer as a base layer on the metal surface indicates that the subsequently formed layers should be deposited on this layer, whether fully or partially, rather than completely on the metal surface itself.

6.7.2 Antiwear Mechanism

The ZDDP additive when decomposed under high contact pressure or heat can form a superior antiwear tribofilm on contacting surfaces. The DDP additive can provide better [22, 23], comparable [24] or worse [25, 26] antiwear properties to the ones of ZDDP, which seem to depend greatly on the operating conditions, properties of contacting surfaces and chemistries of base oil and additives.

Since the first inception of the ZDDP in the late 1930s, extensive studies on the genesis of its antiwear mechanism have been carried out [21]. Table 6.4 summarises the several theories and mechanisms that were proposed to explain the capability of the formed antiwear tribofilms in protecting the contacting surfaces from severe wear. These different mechanisms can be categorised depending on whether the cause of protection is rheological, mechanical or chemical.

Initially, it was suggested that the antiwear mechanism of antiwear additives is based on forming a thicker hydrodynamic film. This thick film can reduce the stress at the contacting asperities and hence reduces wear [7]. The sulphides, phosphorus compounds and oxides were also found to have good lubricating properties that help mitigate wear [181]. Molina [45] suggested that the ratio between the crystalline and amorphous regions of the antiwear film might be important in evaluating the antiwear action of the tribofilm. Molina's results indicated that the amorphous pyrophosphate has better lubricating properties that the crystalline phosphate. In line with these results, it was suggested that the polyphosphate, which has a low melting temperature of 200–300 °C, melts and forms a viscous glass on the contacting surfaces that helps reduce wear [3]. In addition, the interfacial iron-oxide-sulphide eutectic system was also proposed to form a viscous film at the contacting surfaces under extreme conditions and hence separates the contacting surfaces and reduces wear [3, 5]. Same mechanism was also proposed for the FeS layer covering the asperity-asperity contacts [6]. Interestingly, the amount of sulphur in the tribofilm was found to increase with increasing load [7], which highlights the smart action of ZDDP to reduce wear. This smart action was explored by So and Lin [154] who indicated that the plastic deformation is responsible for increasing the temperature at the asperities and hence the decomposition of the additive. Additionally, this plastic deformation creates subsurface defects that enhances the mixing and reaction of the decomposition products, i.e. P, S, Zn and O, with the rubbing surface. Accordingly, So et al. [44] suggested that the antiwear mechanism depends closely on the ratio between the tribofilm formation and its removal. In addition, Habeeb and Stover [11] reported that the ZDDP tribofilm can also reduce wear by decomposing the peroxides. On the other hand, other studies [8–10] suggested that the antiwear action and the absence of severe abrasive wear is due to the digestion of the iron oxide generated during wear into the amorphous phosphate structure of the formed interfacial tribofilm.

Table 6.4 Summary of the proposed antiwear mechanisms of ZDDP

Authors	Mechanism	Details
Watkins [3]	Rheological and mechanical	Formation of a viscous glass on the contacting surfaces that flows and acts as a barrier
Gansheimer [181]	Rheological and mechanical	Good lubricating properties of sulphides, phosphorus compounds and oxides
Rounds [4]	Rheological and mechanical	Formation of oriented sacrificial layer that reduces shear stresses and acts as a barrier
Bec et al. [155]	Rheological and mechanical	Formation of a grease- or gel-like hydrocarbon-rich layer that lias low shear strength
Watkins [3], Glaeser et al. [5]	Mechanical	Formation of an interfacial iron-oxide-sulphide eutectic system as a barrier
Bell et al. [6]	Mechanical	Formation of FeS on the surface as a barrier
Bell et al. [6]	Mechanical	Formation of a low shear stress layers that reduce fatigue wear and delamination processes
So et al. [44]	Mechanical	Balance between the tribofilm formation and its removal
So and Liu [154]	Mechanical	Formation of a compliant protective tribofilm
Palacios [7]	Mechanical	Formation of a thick hydrodynamic film as a barrier
Habeeb and Stover [11]	Chemical	Decomposition of peroxides
Belin et al. [10], Martin et al. [8, 9]	Chemical	Digestion of sharp wear particles of iron oxide
Molina [45]	Chemical	Formation of an amorphous glass rather than a crystalline one

Bell et al. [6] suggested that a layer exists in the tribofilm with a low shear stress that helps reduce fatigue wear and delamination processes. In addition, they proposed that the adhesion between the different layers is expected to be crucial for the load carrying capability of the tribofilm and other wear and friction performance. On the other hand, Rounds [4] proposed that the P-based additives such as ZDDP can act by a mechanism similar to the one of the fatty acid by forming an oriented sacrificial layer covering the surface. This antiwear capability depends on the formed hydrocarbon chain length and chain branching. Accordingly, the hydrocarbon-rich layer at the outermost of the tribofilm seems to play a role in reducing wear [155]. This layer

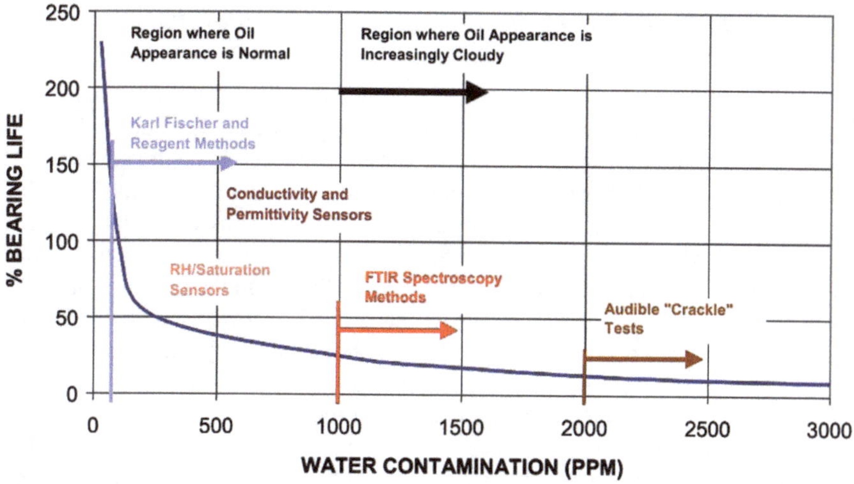

Fig. 6.16 Effect of water on the life of bearing based on 100% life at 0.01% water in addition to the available techniques to detect water and their range are indicated. Reprinted from Sheehan [185]

seemed to have a grease or gel-like nature that has a low shear strength, which helps in redistributing the concentrated loads at the asperities and therefore reduces the high shear stresses at the asperity-asperity contacts and thus mitigates wear.

In summary, the unique antiwear properties of the ZDDP or DDP tribofilms seem to be related not only to its adhesion, hardness and elasticity but also to the capability of the formed tribofilm to maintain local order on the molecular scale through the flow, rearrangement and change in composition of the interfacial layers.

6.8 Antiwear Films in Humid Environments

Corrosion phenomena typically involves water and oxygen [182], which are abundant at any interface. This makes water one the main elements in the corrosive environment. Water can affect the bearing performance in different ways [183]. On the one hand, it can shorten the bearing life, as shown in Fig. 6.16, due to rust, hydrogen embrittlement and oxidation. On the other hand, water can accelerate wear due to the oxidation of the lubricating film and the destabilisation of the protective tribofilm. Therefore, the relative degree of saturation of water in the oil is one of the important factors to consider. Water can enter the oil from the humid air or directly as a free water. The temperature of the oil determines whether this water will mix with the oil as a dissolved water, i.e. below the saturation temperature, or as a free water, i.e. above the saturation temperature [184].

In the following subsections, a review is provided on the effect of water on the composition and tribological properties of the P-based antiwear film.

6.8.1 Effect of Water on the Composition of Tribofilms

Rounds [72] showed that water accelerates the ZDDP decomposition and reduces S, P and Zn contents in the antiwear film. These changes are additive dependent and can be related to the formation of acidic hydrolysis products. In contrast, the results of Nedelcu et al. [144] showed that water inhibits the growth of the ZDDP tribofilm. This effect was manifested in the formation of shorter chains of polyphosphates. The authors attributed these effects to the depolymerisation reactions of the long polyphosphate chains and to the increased surface distress in the presence of water. Ancillary experiments carried out by Cen et al. [143] showed that water indeed inhibits the formation of the protective tribofilm. In addition, they noticed that shorter phosphate chains are formed with increasing the amount of water in the oil.

In line with these results, Parsaeian et al. [186] found that the smaller the relative humidity the thicker the tribofilm and the longer the formed phosphate chains. The trend holds true for the two tested temperatures, i.e. 80 and 98 °C. This was explained by the difficulty of ZDDP molecules to reach and react with the steel surface in an oil containing higher amount of free or dissolved water. In another study, Parsaeian et al. [148] studied the effect of water contamination in oil, i.e. 0, 0.5, 1.5 and 3 wt%, on wear and tribofilm thickness. The results showed that the larger the water content is present in oil the thinner the tribofilm thickness. Similar findings were also reported by Faut and Wheeler [187] for tricresylphosphate (TCP) additive.

All these results highlight the fact that the presence of water affects not only the rate of the additive decomposition but also the final composition of the protective film. Owing to the high affinity of phosphates to water [6], when water contamination occurs, some water molecules will react with the phosphate and replace ZnO, whereas some amount of water will stay unreacted. This will affect the ratio of ZnO/P_2O_5 and thus the phosphate chain length. Furthermore, the presence of unreacted water is expected to affect this ratio up or down depending on the localisation of water, i.e. at the metal surface or in the bulk. Therefore, to better understand the effect of water on the composition of ZDDP or DDP antiwear film, the focus should be given to trace the composition without water and study how water can change this morphology.

6.8.2 Effect of Water on the Tribological Properties of Tribofilms

There are only few studies [143, 144, 148, 149, 186] on the effect of water on the friction and wear performance of the P-based antiwear film, i.e. mainly of ZDDP. These studies mainly showed that water inhibits the formation of the protective tribofilm and increases wear. The authors attributed these effects to the depolymerisation reactions of the long polyphosphate chains and to the increased surface distress in the presence of water. Other reasons for the increased wear in the presence of water in the oil can also be related to corrosion, hydrogen embrittlement and accelerated fatigue due to the possible condensation of small amount of water in the microcracks.

Cen et al. [143] found that the effect of water on friction is less prominent than the case of wear. However, although their results showed that up to a moderate amount of water, friction is not affected, the experiments conducted in a humid atmosphere of a relative humidity 90%, showed that the friction coefficient drops by about 40%. It should be noted that this behaviour of friction can be related to the one of wear. Due to the larger wear in the case of large water contamination, surface smoothing can occur during the induction period that can lead to a reduction in friction coefficient. In addition, this reduction can also be related to the formation of thinner tribofilm of short polyphosphate chains when high water concentration is present in the oil.

These scarce results highlight the importance to study the effect of water on the decomposition of antiwear additive and the final composition. Understanding the effect of water on these processes helps control the wear and friction performance by just controlling the amount of water in the oil and its evaporation rate. In addition, understanding the effect of water provides insight into the reaction kinetics of the ZDDP or DDP tribofilm formation and its behaviour in different environments.

6.9 Concluding Remarks

Throughout this chapter an overview of the state-of-the-art knowledge of the different aspects concerning the P-based antiwear films of ZDDP and DDP additives were presented and discussed. At first, a review was provided on the chemical and physical nature of the antiwear film and the possible reactions during its formation. It was shown that the P-based additives can decompose in different mechanisms, i.e. thermally, hydrolytic or oxidative, to form a complex antiwear film consisting mainly of Zn/Fe, P, S and O. The formation kinetics of these elements are still not fully elucidated.

All the recent studies have confirmed that the formed P-based antiwear films have a patchy layered structure of large and small pads. The large pads exhibit a great elastic response. The elastic modulus, hardness and thickness of these pads are not uniform but vary along the surface and depth. As these properties change substantially from one position to another, this suggests that they should be considered as local properties rather than averaged values. Thus, one of the future works should focus on investigating the effect of the local parameters versus the average values on the estimation of the friction and wear using the different models available in the literature.

The P-based antiwear tribofilms help protect the contacting surfaces by different mechanisms including the digestion of sharp particles and formation of sacrificial rigid layers of good mechanical properties, e.g. elasticity and hardness. There is a lack of studies, however, focusing on the role of the rheological properties of the P-based tribofilms on their durability and antiwear capability.

A systematic study taking into account the aforementioned factors seems to be crucial, which should provide better insight into the origin of the superior antiwear properties of ZDDP and DDP tribofilms in order to be able to optimise their properties and eventually replace them with more environmentally friendly additives.

References

1. Z. Yin, M. Kasrai, M. Fuller, G.M. Bancroft, K. Fyfe, K.H. Tan, Application of soft x-ray absorption spectroscopy in chemical characterization of antiwear films generated by ZDDP part I: the effects of physical parameters. Wear **202**(2), 172–191 (1997)
2. G.M. Bancroft, M. Kasrai, M. Fuller, Z. Yin, K. Fyfe, K.H. Tan, Mechanisms of tribochemical film formation: stability of tribo- and thermally-generated ZDDP films. Tribol. Lett. **3**, 47–51 (1997)
3. R.C. Watkins, The antiwear mechanism of ZDDP's. Part II. Tribol. Int. **15**(1), 13–15 (1982)
4. F.G. Rounds, Effects of additives on the friction of steel on steel I. Surface topography and film composition studies. ASLE Trans. **7**(1), 11–23 (1964)
5. W.A. Glaeser, D. Baer, M. Engelhardt. In situ wear experiments in the scanning auger spectrometer. Wear 162–164. Part A(0), 132–138 (1993). *Wear of Materials: Proceedings of the 9th International Conference*
6. J.C. Bell, K.M. Delargy, A.M. Seeney, Paper IX (ii) the removal of substrate material through thick zinc dithiophosphate anti-wear films. Tribol. Ser. **21**, 387–396 (1992)
7. J.M. Palacios, Thickness and chemical composition of films formed by antimony dithiocarbamate and zinc dithiophosphate. Tribol. Int. **19**(1), 35–39 (1986)
8. J.M. Martin, Antiwear mechanisms of zinc dithiophosphate: a chemical hardness approach. Tribol. Lett. **6**(1), 1–8 (1999)
9. J.M. Martin, C. Grossiord, T. Le Mogne, S. Bec, A. Tonck, The two-layer structure of ZnDTP tribofilms: part I: AES, XPS and XANES analyses. Tribol. Int. **34**(8), 523–530 (2001)
10. M. Belin, J.M. Martin, J.L. Mansot, Role of iron in the amorphization process in friction-induced phosphate glasses. Tribol. Trans. **32**(3), 410–413 (1989)
11. J.J. Habeeb, W.H. Stover, The role of hydroperoxides in engine wear and the effect of zinc dialkyldithiophosphates. ASLE Trans. **30**(4), 419–426 (1986)
12. F. Rounds, Effects of hydroperoxides on wear as measured in four-ball wear tests. Tribol. Trans. **36**(2), 297–303 (1993)
13. P.A. Willermet, D.P. Dailey, R.O. Carter, P.J. Schmitz, W. Zhu, Mechanism of formation of antiwear films from zinc dialkyldithiophosphates. Tribol. Int. **28**(3), 177–187 (1995)
14. M.N. Webster, C.J.J. Norbart, An experimental investigation of micropitting using a roller disk machine. Tribol. Trans. **38**(4), 883–893 (1995)
15. C. Benyajati, A.V. Olver, C.J. Hamer, An experimental study of micropitting, using a new miniature test-rig. Tribol. Ser. **43**, 601–610 (2003)
16. V. Brizmer, H.R. Pasaribu, G.E. Morales-Espejel, Micropitting performance of oil additives in lubricated rolling contacts. Tribol. Trans. **56**(5), 739–748 (2013)
17. E. Lainé, A.V. Olver, T.A. Beveridge, Effect of lubricants on micropitting and wear. Tribol. Int. **41**(11), 1049–1055 (2008)
18. J. Andersson, M. Antonsson, L. Eurenius, E. Olsson, M. Skoglundh, Deactivation of diesel oxidation catalysts: vehicle and synthetic aging correlations. Appl. Catal. B Environ. **72**(1), 71–81 (2007)
19. C. Larese, F. Cabello Galisteo, M. López Granados, R. Mariscal, J.L.G. Fierro, M. Furió, R. Fernández Ruiz, Deactivation of real three way catalysts by $CePo_4$ formation. Appl. Catal. B Environ. **40**(4), 305–317 (2003)

20. M.A. Nicholls, P.R. Norton, G.M. Bancroft, M. Kasrai, T. Do, B.H. Frazer, G. De Stasio, Nanometer scale chemomechanical characterization of antiwear films. Tribol. Lett. **17**(2), 205–216 (2004)
21. H. Spikes, The history and mechanisms of ZDDP. Tribol. Lett. **17**(3), 469–489 (2004)
22. B.H. Kim, R. Mourhatch, P.B. Aswath, Properties of tribofilms formed with ashless dithiophosphate and zinc dialkyl dithiophosphate under extreme pressure conditions. Wear **268**(3), 579–591 (2010)
23. X. Fu, W. Liu, Q. Xue, The application research on series of ashless P-containing EP and AW additives. Ind. Lubr. Tribol. **57**(2), 80–83 (2005)
24. R. Sarin, A.K. Gupta, D.K. Tuli, A.S. Verma, M.M. Rai, A.K. Bhatnagar, Synthesis and performance evaluation of o, o-dialkylphosphorodithioic disulphides as potential antiwear, extreme-pressure and antioxidant additives. Tribol. Int. **26**(6), 389–394 (1993)
25. M.N. Najman, M. Kasrai, G.M. Bancroft, Chemistry of antiwear films from ashless thiophosphate oil additives. Tribol. Lett. **17**(2), 217–229 (2004)
26. Z. Zhang, E.S. Yamaguchi, M. Kasrai, G.M. Bancroft, Tribofilms generated from ZDDP and DDP on steel surfaces: part 1, growth, wear and morphology. Tribol. Lett. **19**(3), 211–220 (2005)
27. A.M. Barnes, K.D. Bartle, V.R.A. Thibon, A review of zinc dialkyldithiophosphates (ZDDPS): characterisation and role in the lubricating oil. Tribol. Int. **34**(6), 389–395 (2001)
28. M.A. Nicholls, T. Do, P.R. Norton, M. Kasrai, G. Michael Bancroft, Review of the lubrication of metallic surfaces by zinc dialkyl-dithiophosphates. Tribol. Int. **38**(1), 15–39 (2005)
29. D.R. Armstrong, E.S. Ferrari, K.J. Roberts, D. Adams, An examination of the reactivity of zinc di-alkyl-di-thiophosphate in relation to its use as an anti-wear and anti-corrosion additive in lubricating oils. Wear **217**(2), 276–287 (1998)
30. L.R. Rudnick, *Lubricant Additives: Chemistry and Applications* (CRC Press, Boca Raton, 2009)
31. P.G. Harrison, T. Kikabhai, Proton and phosphorus-31 nuclear magnetic resonance study of zinc (II) o, o′-dialkyl dithiophosphates in solution. J. Chem. Soc. Dalton Trans. (4), 807–814 (1987)
32. Z. Pawlak, *Tribochemistry of Lubricating Oils*, vol. 45 (Elsevier, Amsterdam, 2003)
33. D.R. Armstrong, E.S. Ferrari, K.J. Roberts, D. Adams, An investigation into the molecular stability of zinc di-alkyl-di-thiophosphates (ZDDPs) in relation to their use as anti-wear and anti-corrosion additives in lubricating oils. Wear **208**(1–2), 138–146 (1997)
34. B.H. Kim, V. Sharma, P.B. Aswath, Chemical and mechanistic interpretation of thermal films formed by dithiophosphates using XANES. Tribol. Int. **114**, 15–26 (2017)
35. H. Spedding, R.C. Watkins, The antiwear mechanism of ZDDP's. Part I. Tribol. Int. **15**(1), 9–12 (1982)
36. T.H. Handley, J.A. Dean, O, O′-dialkyl phosphorodithioic acids as extractants for metals. Anal. Chem. **34**(10), 1312–1315 (1962)
37. N.E. Gallopoulos, Thermal decomposition of metal dialkyldithiophosphate oil blends. ASLE Trans. **7**(1), 55–63 (1964)
38. P.A. Willermet, L.R. Mahoney, C.M. Bishop, Lubricant degradation and wear III. Antioxidant reactions and wear behavior of a zinc dialkyldithiophosphate in a fully formulated lubricant. ASLE Trans. **23**(3), 225–231 (1980)
39. A. Neville, A. Morina, T. Haque, M. Voong, Compatibility between tribological surfaces and lubricant additives—how friction and wear reduction can be controlled by surface/lube synergies. Tribol. Int. **40**(10), 1680–1695 (2007)
40. N.N. Gosvami, J.A. Bares, F. Mangolini, A.R. Konicek, D.G. Yablon, R.W. Carpick, Mechanisms of antiwear tribofilm growth revealed in situ by single-asperity sliding contacts. Science **348**(6230), 102–106 (2015)
41. J. Zhang, H. Spikes, On the mechanism of ZDDP antiwear film formation. Tribol. Lett. **63**(2), 1–15 (2016)
42. W. Tysoe, On stress-induced tribochemical reaction rates. Tribol. Lett. **65**(2), 48 (2017)

43. F.T. Barcroft, R.J. Bird, J.F. Hutton, D. Park, The mechanism of action of zinc thiophosphates as extreme pressure agents. Wear **77**(3), 355–384 (1982)
44. H. So, Y.C. Lin, G.G.S. Huang, T.S.T. Chang, Antiwear mechanism of zinc dialkyl dithiophosphates added to a paraffinic oil in the boundary lubrication condition. Wear **166**(1), 17–26 (1993)
45. A. Molina, Isolation and chemical characterization of a zinc dialkyldithiophosphate-derived antiwear agent. ASLE Trans. **30**(4), 479–485 (1986)
46. R.B. Jones, R.C. Coy, The chemistry of the thermal degradation of zinc dialkyldithiophosphate additives. ASLE Trans. **24**(1), 91–97 (1981)
47. D. Shakhvorostov, M.H. Müser, Y. Song, P.R. Norton, Smart materials behavior in phosphates: role of hydroxyl groups and relevance to antiwear films. J. Chem. Phys. **131**(4), 044704 (2009)
48. I.-M. Feng, Pyrolysis of zinc dialkyl phosphorodithioate and boundary lubrication. Wear **3**(4), 309–311 (1960)
49. I.-M. Feng, W.L. Perilstein, M.R. Adams, Solid film deposition and non-sacrificial boundary lubrication. ASLE Trans. **6**(1), 60–66 (1963)
50. R.C. Coy, R.B. Jones, The thermal degradation and EP performance of zinc dialkyldithiophosphate additives in white oil. ASLE Trans. **24**(1), 77–90 (1981)
51. R.J. Bird, G.D. Galvin, The application of photoelectron spectroscopy to the study of ep films on lubricated surfaces. Wear **37**(1), 143–167 (1976)
52. S.-H. Choa, K.C. Ludema, G.E. Potter, B.M. Dekoven, T.A. Morgan, K.K. Kar, A model of the dynamics of boundary film formation. Wear **177**(1), 33–45 (1994)
53. R.G. Pearson, Hard and soft acids and bases. J. Am. Chem. Soc. **85**(22), 3533–3539 (1963)
54. R.G. Pearson, J. Songstad, Application of the principle of hard and soft acids and bases to organic chemistry. J. Am. Chem. Soc. **89**(8), 1827–1836 (1967)
55. R.G. Pearson, Hard and soft acids and bases, HSAB, part 1: fundamental principles. J. Chem. Educ. **45**(9), 581 (1968)
56. R.G. Pearson, Hard and soft acids and bases, HSAB, part II: underlying theories. J. Chem. Educ. **45**(10), 643 (1968)
57. R.G. Pearson, Recent advances in the concept of hard and soft acids and bases. J. Chem. Educ. **64**(7), 561 (1987)
58. P.A. Willermet, R.O. Carter III, E.N. Boulos, Lubricant-derived tribochemical films—an infrared spectroscopic study. Tribol. Int. **25**(6), 371–380 (1992)
59. A.J. Burn, The mechanism of the antioxidant action of zinc dialkyl dithiophosphates. Tetrahedron **22**(7), 2153–2161 (1966)
60. J.A. Howard, Y. Ohkatsu, J.H.B. Chenier, K.U. Ingold, Metal complexes as antioxidants. I. The reaction of zinc dialkyldithiophosphates and related compounds with peroxy radicals. Can. J. Chem. **51**(10), 1543–1553 (1973)
61. J.-M. Martin, C. Grossiord, T. Le Mogne, J. Igarashi, Transfer films and friction under boundary lubrication. Wear **245**(1–2), 107–115 (2000)
62. M.L.S. Fuller, M. Kasrai, G. Michael Bancroft, K. Fyfe, K.H. Tan, Solution decomposition of zinc dialkyl dithiophosphate and its effect on antiwear and thermal film formation studied by X-ray absorption spectroscopy. Tribol. Int. **31**(10), 627–644 (1998)
63. Z. Yin, M. Kasrai, G.M. Bancroft, K. Fyfe, M.L. Colaianni, K.H. Tan, Application of soft X-ray absorption spectroscopy in chemical characterization of antiwear films generated by ZDDP part II: the effect of detergents and dispersants. Wear **202**(2), 192–201 (1997)
64. A. Morina, H. Zhao, J.F.W. Mosselmans, In-situ reflection-XANES study of ZDDP and MoDTC lubricant films formed on steel and diamond like carbon (DLC) surfaces. Appl. Surf. Sci. **297**, 167–175 (2014)
65. M. Nicholls, M.N. Najman, Z. Zhang, M. Kasrai, P.R. Norton, P.U.P.A. Gilbert, The contribution of XANES spectroscopy to tribology. Can. J. Chem. **85**(10), 816–830 (2007)
66. Z. Zhang, E.S. Yamaguchi, M. Kasrai, G.M. Bancroft, X. Liu, M.E. Fleet, Tribofilms generated from ZDDP and DDP on steel surfaces: part 2, chemistry. Tribol. Lett. **19**(3), 221–229 (2005)
67. J.S. Sheasby, T.A. Caughlin, A.G. Blahey, K.F. Laycock, A reciprocating wear test for evaluating boundary lubrication. Tribol. Int. **23**(5), 301–307 (1990)

68. P.A. Willermet, J.M. Pieprzak, D.P. Dailey, R.O. Carter, N.E. Lindsay, L.P. Haack et al., The composition of surface layers formed in a lubricated cam/tappet contact. J. Tribol. **113**(1), 38–47 (1991)
69. P.A. Willermet, D.P. Dailey, R.O. Carter III, P.J. Schmitz, W. Zhu, J.C. Bell, D. Park, The composition of lubricant-derived surface layers formed in a lubricated cam/tappet contact ii. effects of adding overbased detergent and dispersant to a simple ZDTP solution. Tribol. Int. **28**(3), 163–175 (1995)
70. M. Aktary, M.T. McDermott, G.A. McAlpine, Morphology and nanomechanical properties of ZDDP antiwear films as a function of tribological contact time. Tribol. Lett. **12**(3), 155–162 (2002)
71. N.E. Lindsay, R.O. Carter III, P.J. Schmitz, L.P. Haack, R.E. Chase, J.E. deVries, P.A. Willermet, Characterization of films formed at a lubricated cam/tappet contact. Spectrochim. Acta Part A Mol. Spectrosc. **49**(13–14), 2057–2070 (1993)
72. F.G. Rounds, Some factors affecting the decomposition of three commercial zinc organodithiophosphates. ASLE Trans. **18**(2), 79–89 (1975)
73. R. McClintock, Effect of lubricants on rear axle pinion bearing breakin. ASLE Trans. **6**(2), 154–160 (1963)
74. J.S. Sheasby, T.A. Caughlin, J.J. Habeeb, Observation of the antiwear activity of zinc dialkyldithiophosphate additives. Wear **150**(1–2), 247–257 (1991)
75. M. Kasrai, M. Puller, M. Scaini, Z. Yin, R.W. Brunner, G.M. Bancroft, M.E. Fleet, K. Fyfe, K.H. Tan, Study of tribochemical film formation using x-ray absorption and photoelectron spectroscopies. Tribol. Ser. **30**, 659–669 (1995)
76. E.H. Loeser, R.C. Wiquist, S.B. Twiss, Cam and tappet lubrication. IV—radioactive study of sulfur in the EP film. ASLE Trans. **2**(2), 199–207 (1959)
77. J. Ye, S. Araki, M. Kano, Y. Yasuda, Nanometerscale mechanical/structural properties of molybdenum dithiocarbamate and zinc dialkylsithiophosphate tribofilms and friction reduction mechanism. Jpn. J. Appl. Phys. **44**(7B), 5358–5361 (2005)
78. M. Fuller, Z. Yin, M. Kasrai, G. Michael Bancroft, E.S. Yamaguchi, P. Ray Ryason, P.A. Willermet, K.H. Tan, Chemical characterization of tribochemical and thermal films generated from neutral and basic ZDDPs using X-ray absorption spectroscopy. Tribol. Int. **30**(4), 305–315 (1997)
79. M.L.S. Fuller, L.R. Fernandez, The use of X-ray absorption spectroscopy for monitoring the thickness of antiwear films from ZDDP. Tribol. Lett. **8**, 187–192 (2000)
80. M.A. Nicholls, G. Michael Bancroft, P.R. Norton, M. Kasrai, G. De Stasio, B.H. Frazer, L.M. Wiese, Chemomechanical properties of antiwear films using X-ray absorption microscopy and nanoindentation techniques. Tribol. Lett. **17**(2), 245–259 (2004)
81. J.M. Martin, M. Belin, J.L. Mansot, H. Dexpert, P. Lagarde, Frictioninduced amorphization with ZDDP—an EXAFS study. ASLE Trans. **29**(4), 523–531 (1986)
82. M.N. Najman, M. Kasrai, G.M. Bancroft, A. Miller, Study of the chemistry of films generated from phosphate ester additives on 52100 steel using X-ray absorption spectroscopy. Tribol. Lett. **13**(3), 209–218 (2002)
83. M.N. Najman, M. Kasrai, G.M. Bancroft, B.H. Frazer, G. De Stasio, The correlation of microchemical properties to antiwear (AW) performance in ashless thiophosphate oil additives. Tribol. Lett. **17**(4), 811–822 (2004)
84. Z. Zhang, M. Najman, M. Kasrai, G.M. Bancroft, E.S. Yamaguchi, Study of interaction of EP and AW additives with dispersants using XANES. Tribol. Lett. **18**(1), 43–51 (2005)
85. M.N. Najman, M. Kasrai, G.M. Bancroft, Investigating binary oil additive systems containing P and S using X-ray absorption near-edge structure spectroscopy. Wear **257**(1), 32–40 (2004)
86. M. Crobu, A. Rossi, F. Mangolini, N.D. Spencer, Tribochemistry of bulk zinc metaphosphate glasses. Tribol. Lett. **39**(2), 121–134 (2010)
87. R. Heuberger, A. Rossi, N.D. Spencer, XPS study of the influence of temperature on ZnDTP tribofilm composition. Tribol. Lett. **25**(3), 185–196 (2006)
88. R. Heuberger, A. Rossi, N.D. Spencer, Reactivity of alkylated phosphorothionates with steel: a tribological and surface analytical study. Lubr. Sci. **20**, 79–102 (2008)

89. R. Heuberger, A. Rossi, N.D. Spencer, Pressure dependence of znDTP tribochemical film formation: a combinatorial approach. Tribol. Lett. **28**(2), 209–222 (2007)
90. J.R. Van Wazer, K.A. Holst, Structure and properties of the condensed phosphates. I. Some general considerations about phosphoric acids. J. Am. Chem. Soc. **72**(2), 639–644 (1950)
91. J.R. Van Wazer, Structure and properties of the condensed phosphates. III. Solubility fractionation and other solubility studies. J. Am. Chem. Soc. **72**(2), 647–655 (1950)
92. C. Minfray, J.M. Martin, C. Esnouf, T. Le Mogne, R. Kersting, B. Hagenhoff, A multi-technique approach of tribofilm characterisation. Thin Solid Films **447**, 272–277 (2004)
93. R. Heuberger, A. Rossi, N.D. Spencer, XPS study of the influence of temperature on ZnDTP tribofilm composition. Tribol. Lett. **25**(3), 185–196 (2007)
94. M. Crobu, A. Rossi, N.D. Spencer, Effect of chain-length and countersurface on the tribochemistry of bulk zinc polyphosphate glasses. Tribol. Lett. **48**(3), 393–406 (2012)
95. M. Crobu, A. Rossi, F. Mangolini, N.D. Spencer, Chain-length-identification strategy in zinc polyphosphate glasses by means of XPS and ToF-SIMS. Anal. Bioanal. Chem. **403**(5), 1415–1432 (2012)
96. E. Liu, S.D. Kouame, An XPS study on the composition of zinc dialkyl dithiophosphate tribofilms and their effect on camshaft lobe wear. Tribol. Trans. **57**(1), 18–27 (2014)
97. E.C. Onyiriuka, Zinc phosphate glass surfaces studied by xps. J. Non-Cryst. Solids **163**(3), 268–273 (1993)
98. J.W. Wiench, M. Pruski, B. Tischendorf, J.U. Otaigbe, B.C. Sales, Structural studies of zinc polyphosphate glasses by nuclear magnetic resonance. J. Non-Cryst. Solids **263**, 101–110 (2000)
99. B. Tischendorf, J.U. Otaigbe, J.W. Wiench, M. Pruski, B.C. Sales, A study of short and intermediate range order in zinc phosphate glasses. J. Non-Cryst. Solids **282**(2), 147–158 (2001)
100. G. Walter, U. Hoppe, J. Vogel, G. Carl, P. Hartmann, The structure of zinc polyphosphate glass studied by diffraction methods and 31P NMR. J. Non-Cryst. Solids **333**(3), 252–262 (2004)
101. M. Najman, M. Kasrai, G. Michael Bancroft, R. Davidson, Combination of ashless antiwear additives with metallic detergents: interactions with neutral and overbased calcium sulfonates. Tribol. Int. **39**(4), 342–355 (2006)
102. B. Dacre, C.H. Bovington, The effect of metal composition on the adsorption of zinc di-isopropyldithiophosphate. ASLE Trans. **26**(3), 333–343 (1983)
103. K. Ito, J.-M. Martin, C. Minfray, K. Kato, Lowfriction tribofilm formed by the reaction of ZDDP on iron oxide. Tribol. Int. **39**(12), 1538–1544 (2006)
104. K. Ito, J.M. Martin, C. Minfray, K. Kato, Formation mechanism of a low friction ZDDP tribofilm on iron oxide. Tribol. Trans. **50**(2), 211–216 (2007)
105. S. Plaza, The adsorption of zinc dibutyldithiophosphates on iron and iron oxide powders. ASLE Trans. **30**(2), 233–240 (1987)
106. M. Watanabe, M. Sakuma, T. Inaba, Y. Iguchi, Formation and oxidation of sulfides on pure iron and iron oxides. Mater. Trans. JIM **41**(7), 865–872 (2000)
107. Y.-R. Li, G. Pereira, M. Kasrai, P.R. Norton, The effect of steel hardness on the performance of ZDDP antiwear films: a multi-technique approach. Tribol. Lett. **29**(3), 201–211 (2008)
108. J.S. Sheasby, T.A. Caughlin, W.A. Mackwood, The effect of steel hardness on the performance of antiwear additives. Wear **201**(1–2), 209–216 (1996)
109. B. Vengudusamy, J.H. Green, G.D. Lamb, H.A. Spikes, Tribological properties of tribofilms formed from ZDDP in DLC/DLC and DLC/steel contacts. Tribol. Int. **44**(2), 165–174 (2011)
110. A. Erdemir, C. Donnet, Tribology of diamond-like carbon films: recent progress and future prospects. J. Phys. D Appl. Phys. **39**(18), R311 (2006)
111. I. Velkavrh, M. Kalin, J. Vizintin, The performance and mechanisms of DLC-coated surfaces in contact with steel in boundary-lubrication conditions: a review. Strojniški vestnik **54**(3), 189–206 (2008)
112. M. Kalin, I. Velkavrh, J. Vižintin, L. Ožbolt, Review of boundary lubrication mechanisms of DLC coatings used in mechanical applications. Meccanica **43**(6), 623–637 (2008)

113. D.M. Nuruzzaman, M.A. Chowdhury, A. Nakajima, M.L. Rahaman, S.M.I. Karim, Friction and wear of diamond like carbon (DLC) coatings—a review. Recent Pat. Mech. Eng. **4**(1), 55–78 (2011)
114. R. Zahid, M.B.H. Hassan, M. Varman, R.A. Mufti, M.A. Kalam, N.W.B.M. Zulkifli, M. Gulzar, A review on effects of lubricant formulations on tribological performance and boundary lubrication mechanisms of non-doped DLC/DLC contacts. Crit. Rev. Solid State Mater. Sci. **42**, 1–28 (2016)
115. S. Equey, S. Roos, U. Mueller, R. Hauert, N.D. Spencer, R. Crockett, Tribofilm formation from ZnDTP on diamond-like carbon. Wear **264**(3), 316–321 (2008)
116. S. Equey, S. Roos, U. Mueller, R. Hauert, N.D. Spencer, R. Crockett, Reactions of zinc-free anti-wear additives in DLC/DLC and steel/steel contacts. Tribol. Int. **41**(11), 1090–1096 (2008)
117. K. Topolovec-Miklozic, F. Lockwood, H. Spikes, Behaviour of boundary lubricating additives on dlc coatings. Wear **265**(11), 1893–1901 (2008)
118. M. Kalin, E. Roman, L. Ožbolt, J. Vižintin, Metal-doped (Ti, WC) diamond-like-carbon coatings: reactions with extreme-pressure oil additives under tribological and static conditions. Thin Solid Films **518**(15), 4336–4344 (2010)
119. M. Kalin, E. Roman, J. Vižintin, The effect of temperature on the tribological mechanisms and reactivity of hydrogenated, amorphous diamond-like carbon coatings under oil-lubricated conditions. Thin Solid Films **515**(7), 3644–3652 (2007)
120. S. Akbari, J. Kovač, M. Kalin, Effect of ZDDP concentration on the thermal film formation on steel, hydrogenated non-doped and Si-doped DLC. Appl. Surf. Sci. **383**, 191–199 (2016)
121. T. Haque, A. Morina, A. Neville, R. Kapadia, S. Arrowsmith, Nonferrous coating/lubricant interactions in tribological contacts: assessment of tribofilms. Tribol. Int. **40**(10), 1603–1612 (2007)
122. B. Podgornik, S. Jacobson, S. Hogmark, DLC coating of boundary lubricated components—advantages of coating one of the contact surfaces rather than both or none. Tribol. Int. **36**(11), 843–849 (2003)
123. B. Podgornik, J. Vižintin, S. Jacobson, S. Hogmark, Tribological behaviour of WC/C coatings operating under different lubrication regimes. Surf. Coat. Technol. **177**, 558–565 (2004)
124. B. Podgornik, J. Vižintin, Tribological reactions between oil additives and dlc coatings for automotive applications. Surf. Coat. Technol. **200**(5), 1982–1989 (2005)
125. M.I. de Barros' Bouchet, J.M. Martin, T. Le-Mogne, B. Vacher, Boundary lubrication mechanisms of carbon coatings by MoDTC and ZDDP additives. Tribol. Int. **38**(3), 257–264 (2005)
126. M. Ban, M. Ryoji, S. Fujii, J. Fujioka, Tribological characteristics of Si-containing diamond-like carbon films under oil-lubrication. Wear **253**(3), 331–338 (2002)
127. B. Podgornik, S. Jacobson, S. Hogmark, Influence of ep additive concentration on the tribological behaviour of dlc-coated steel surfaces. Surf. Coat. Technol. **191**(2), 357–366 (2005)
128. M. Kalin, J. Vižintin, J. Barriga, K. Vercammen, K. van Acker, A. Arnšek, The effect of doping elements and oil additives on the tribological performance of boundary-lubricated dlc/dlc contacts. Tribol. Lett. **17**(4), 679–688 (2004)
129. M. Kalin, J. Vižintin, Differences in the tribological mechanisms when using non-doped, metal-doped (Ti, WC), and non-metal-doped (Si) diamond-like carbon against steel under boundary lubrication, with and without oil additives. Thin Solid Films **515**(4), 2734–2747 (2006)
130. T. Haque, A. Morina, A. Neville, Tribological performance evaluation of a hydrogenated diamond-like carbon coating in sliding/rolling contact—effect of lubricant additives. Proc. Inst. Mech. Eng. Part J J. Eng. Tribol. **225**(6), 393–405 (2011)
131. L. Yang, A. Neville, A. Brown, P. Ransom, A. Morina, Effect of lubricant additives on the WDLC coating structure when tested in boundary lubrication regime. Tribol. Lett. **57**(2), 14 (2015)
132. C. Donnet, A. Grill, Friction control of diamond-like carbon coatings. Surf. Coat. Technol. **94**, 456–462 (1997)

133. M. Kalin, S. Jahanmir, G. Dražič, Wear mechanisms of glass-infiltrated alumina sliding against alumina in water. J. Am. Ceram. Soc. **88**(2), 346–352 (2005)
134. L. Lazzarotto, L. Dubar, A. Dubois, P. Ravassard, J. Oudin, Three selection criteria for the cold metal forming lubricating oils containing extreme pressure agents. J. Mater. Process. Technol. **80**, 245–250 (1998)
135. A. Naveira-Suarez, A. Tomala, M. Grahn, M. Zaccheddu, R. Pasaribu, R. Larsson, The influence of base oil polarity and slide–roll ratio on additive-derived reaction layer formation. Proc. Inst. Mech. Eng. Part J J. Eng. Tribol. **225**, 565–576 (2011)
136. A. Tomala, A. Naveira-Suarez, I.C. Gebeshuber, R. Pasaribu, Effect of base oil polarity on micro and nanofriction behaviour of base oil+ ZDDP solutions. Tribol. Mater. Surf. Interfaces **3**(4), 182–188 (2009)
137. A. Naveira Suarez, M. Grahn, R. Pasaribu, R. Larsson, The influence of base oil polarity on the tribological performance of zinc dialkyl dithiophospate additives. Tribol. Int. **43**(12), 2268–2278 (2010)
138. P. Kar, P. Asthana, H. Liang, Formation and characterization of tribofilms. J. Tribol. **130**(4), 042301 (2008)
139. I. Minami, K. Hirao, M. Memita, S. Mori, Investigation of anti-wear additives for low viscous synthetic esters: hydroxyalkyl phosphonates. Tribol. Int. **40**(4), 626–631 (2007)
140. I. Minami, S. Mori, Anti-wear additives for ester oils. J. Synth. Lubr. **22**(2), 105–121 (2005)
141. A. Dorgham, A. Neville, K. Ignatyev, F. Mosselmans, A. Morina, An in situ synchrotron XAS methodology for surface analysis under high temperature, pressure, and shear. Rev. Sci. Instrum. **88**(1), 015101 (2017)
142. E.S. Ferrari, K.J. Roberts, D. Adams, A multi-edge X-ray absorption spectroscopy study of the reactivity of zinc di-alkyl-di-thiophosphates (ZDDPS) anti-wear additives: 1. An examination of representative model compounds. Wear **236**(1), 246–258 (1999)
143. H. Cen, A. Morina, A. Neville, R. Pasaribu, I. Nedelcu, Effect of water on ZDDP anti-wear performance and related tribochemistry in lubricated steel/steel pure sliding contacts. Tribol. Int. **56**, 47–57 (2012)
144. I. Nedelcu, E. Piras, A. Rossi, H.R. Pasaribu, XPS analysis on the influence of water on the evolution of zinc dialkyldithiophosphate–derived reaction layer in lubricated rolling contacts. Surf. Interface Anal. **44**(8), 1219–1224 (2012)
145. E.S. Ferrari, K.J. Roberts, M. Sansone, D. Adams, A multi-edge xray absorption spectroscopy study of the reactivity of zinc di-alkyl-dithiophosphates anti-wear additives: 2. In situ studies of steel/oil interfaces. Wear **236**(1), 259–275 (1999)
146. C.H. Bovington, B. Dacre, The adsorption and reaction of decomposition products of zinc di-isopropyldiophosphate on steel. ASLE Trans. **27**(3), 252–258 (1984)
147. A. Ghanbarzadeh, P. Parsaeian, A. Morina, M.C.T. Wilson, M.C.P. van Eijk, I. Nedelcu, D. Dowson, A. Neville, A semi-deterministic wear model considering the effect of zinc dialkyl dithiophosphate tribofilm. Tribol. Lett. **61**(1), 12 (2016)
148. P. Parsaeian, A. Ghanbarzadeh, M. Wilson, M.C.P. Van Eijk, I. Nedelcu, D. Dowson, A. Neville, A. Morina, An experimental and analytical study of the effect of water and its tribochemistry on the tribocorrosive wear of boundary lubricated systems with ZDDP-containing oil. Wear **358**, 23–31 (2016)
149. P. Parsaeian, A. Ghanbarzadeh, M.C.P. Van Eijk, I. Nedelcu, A. Morina, A. Neville, Study of the interfacial mechanism of ZDDP tribofilm in humid environment and its effect on tribochemical wear; part II: numerical. Tribol. Int. **107**, 33–38 (2017)
150. G. Pereira, D. Munoz-Paniagua, A. Lachenwitzer, M. Kasrai, P.R. Norton, T. Weston Capehart, T.A. Perry, Y.-T. Cheng, A variable temperature mechanical analysis of ZDDP-derived antiwear films formed on 52100 steel. Wear **262**(3), 461–470 (2007)
151. H. Ji, M.A. Nicholls, P.R. Norton, M. Kasrai, T. Weston Capehart, T.A. Perry, Y.-T. Cheng, Zinc-dialkyldithiophosphate antiwear films: dependence on contact pressure and sliding speed. Wear **258**(5), 789–799 (2005)
152. Y. Shimizu, H.A. Spikes, The influence of slide–roll ratio on ZDDP tribofilm formation. Tribol. Lett. **64**(2), 19 (2016)

153. P.M.E. Cann, G.J. Johnston, H.A. Spikes, Formation of Thick Films by Phosphorus-Based Anti-wear Additives (Mechanical Engineering Publication Ltd, 1987), pp. 543–554
154. H. So, Y.C. Lin, The theory of antiwear for ZDDP at elevated temperature in boundary lubrication condition. Wear **177**(2), 105–115 (1994)
155. S.A.J.M.R.C.J.C. Bec, A. Tonck, J.-M. Georges, R.C. Coy, J.C. Bell, G.W. Roper, Relationship between mechanical properties and structures of zinc dithiophosphate anti–wear films, in *Proceedings of the Royal Society of London A: Mathematical, Physical and Engineering Sciences*, vol. 455 (The Royal Society, 1999), pp. 4181–4203
156. O.L. Warren, J.F. Graham, P.R. Norton, J.E. Houston, T.A. Michalske, Nanomechanical properties of films derived from zinc dialkyldithiophosphate. Tribol. Lett. **4**(2), 189–198 (1998)
157. J.F. Graham, C. McCague, P.R. Norton, Topography and nanomechanical properties of tribochemical films derived from zinc dialkyl and diaryl dithiophosphates. Tribol. Lett. **6**, 149–157 (1999)
158. A.J. Pidduck, G.C. Smith, Scanning probe microscopy of automotive anti-wear films. Wear **212**(2), 254–264 (1997)
159. L. Taylor, A. Dratva, H.A. Spikes, Friction and wear behavior of zinc dialkyldithiophosphate additive. Tribol. Trans. **43**(3), 469–479 (2000)
160. M. Aktary, M.T. McDermott, J. Torkelson, Morphological evolution of films formed from thermooxidative decomposition of ZDDP. Wear **247**(2), 172–179 (2001)
161. G.W. Canning, M.L. SuominenFuller, G.M. Bancroft, M. Kasrai, J.N. Cutler, G. De Stasio, B. Gilbert, Spectromicroscopy of tribological films from engine oil additives. Part I Films from ZDDP's. Tribol. Lett. **6**(3–4), 159–169 (1999)
162. M.A. Nicholls, P.R. Norton, G.M. Bancroft, M. Kasrai, G. De Stasio, L.M. Wiese, Spatially resolved nanoscale chemical and mechanical characterization of ZDDP antiwear films on aluminum-silicon alloys under cylinder/bore wear conditions. Tribol. Lett. **18**(3), 261–278 (2005)
163. J. Ye, M. Kano, Y. Yasuda, Evaluation of local mechanical properties in depth in MoDTC/ZDDP and ZDDP tribochemical reacted films using nanoindentation. Tribol. Lett. **13**(1), 41–47 (2002)
164. J. Ye, M. Kano, Y. Yasuda, Evaluation of nanoscale friction depth distribution in ZDDP and MoDTC tribochemical reacted films using a nanoscratch method. Tribol. Lett. **16**, 107–112 (2004)
165. J. Ye, M. Kano, Y. Yasuda, Friction property study of the surface of ZDDP and MoDTC antiwear additive films using AFM/LFM and force curve methods. Tribotest **9**(1), 13–21 (2002)
166. J. Ye, M. Kano, Y. Yasuda, Determination of nanostructures and mechanical properties on the surface of molybdenum dithiocarbamate and zinc dialkyl-dithiophosphate tribochemical reacted films using atomic force microscope phase imaging technique. J. Appl. Phys. **93**(9), 5113–5117 (2003)
167. K.T. Miklozic, J. Graham, H. Spikes, Chemical and physical analysis of reaction films formed by molybdenum dialkyl-dithiocarbamate friction modifier additive using raman and atomic force microscopy. Tribol. Lett. **11**(2), 71–81 (2001)
168. C.C. Chou, J.F. Lin, A new approach to the effect of EP additive and surface roughness on the pitting fatigue of a line-contact system. J. Tribol. **124**(2), 245–258 (2002)
169. M.A. Nicholls, T. Do, P.R. Norton, G. Michael Bancroft, M. Kasrai, T. Weston Capehart, Y.-T. Cheng, T. Perry, Chemical and mechanical properties of ZDDP antiwear films on steel and thermal spray coatings studied by XANES spectroscopy and nanoindentation techniques. Tribol. Lett. **15**(3), 241–248 (2003)
170. K.D. Costa, Single-cell elastography: probing for disease with the atomic force microscope. Dis. Markers **19**(2–3), 139–154 (2004)
171. P. Parsaeian, A. Ghanbarzadeh, M.C.P. Van Eijk, I. Nedelcu, A. Neville, A. Morina, A new insight into the interfacial mechanisms of the tribofilm formed by zinc dialkyl dithiophosphate. Appl. Surf. Sci. **403**, 472–486 (2017)

172. M.C. Friedenberg, C. Mathew Mate, Dynamic viscoelastic properties of liquid polymer films studied by atomic force microscopy. Langmuir **12**(25), 6138–6142 (1996)
173. J.-M. Georges, A. Tonck, S. Poletti, E.S. Yamaguchi, P.R. Ryason, Film thickness and mechanical properties of adsorbed neutral and basic zinc diisobutyl dithiophosphates. Tribol. Trans. **41**(4), 543–553 (1998)
174. I. Neitzel, V. Mochalin, J.A. Bares, R.W. Carpick, A. Erdemir, Y. Gogotsi, Tribological properties of nanodiamond-epoxy composites. Tribol. Lett. **47**(2), 195–202 (2012)
175. L.J. Taylor, H.A. Spikes, Friction-enhancing properties of ZDDP antiwear additive: part I—friction and morphology of ZDDP reaction films. Tribol. Trans. **46**(3), 303–309 (2003)
176. D. Mazuyer, A. Tonck, S. Bec, J.L. Loubet, J.M. Georges, Nanoscale surface rheology in tribology. Tribol. Ser. **39**, 273–282 (2001)
177. J.-M. Georges, A. Tonck, J.-L. Loubet, D. Mazuyer, E. Georges, F. Sidoroff, Rheology and friction of compressed polymer layers adsorbed on solid surfaces. J. Phys. II **6**(1), 57–76 (1996)
178. Q. Li, C. Lee, R.W. Carpick, J. Hone, Substrate effect on thickness-dependent friction on graphene. Physica Status Solidi (b) **247**(11–12), 2909–2914 (2010)
179. C. Lee, X. Wei, Q. Li, R. Carpick, J.W. Kysar, J. Hone, Elastic and frictional properties of graphene. Physica Status Solidi (b) **246**(11–12), 2562–2567 (2009)
180. C. Lee, Q. Li, W. Kalb, X.-Z. Liu, H. Berger, R.W. Carpick, J. Hone, Frictional characteristics of atomically thin sheets. Science **328**(5974), 76–80 (2010)
181. J. Gansheimer, On the lubricating properties of mixtures of mineral oil with certain inorganic phosphates, hydroxides, and sulfides. ASLE Trans. **15**(3), 201–206 (1972)
182. D. Landolt, S. Mischler, *Tribocorrosion of Passive Metals and Coatings* (Elsevier, Amsterdam, 2011)
183. M. Duncanson, Detecting and controlling water in oil, in *Proceedings of Noria Lubrication Excellence 2005 Conference*, San Antonio, Texas, 25–29 April 2005
184. A.C. Eachus, The trouble with water. Tribol. Lubr. Technol. **61**(10), 32–38 (2005)
185. P.E. Sheehan, The wear kinetics of NaCl under dry nitrogen and at low humidities. Chem. Phys. Lett. **410**(1), 151–155 (2005)
186. P. Parsaeian, M.C.P. Van Eijk, I. Nedelcu, A. Neville, A. Morina, Study of the interfacial mechanism of ZDDP tribofilm in humid environment and its effect on tribochemical wear; part I: experimental. Tribol. Int. **107**, 135–143 (2017)
187. O.D. Faut, D.R. Wheeler, On the mechanism of lubrication by tricresylphosphate (TCP)—the coefficient of friction as a function of temperature for TCP on M-50 steel. ASLE Trans. **26**(3), 344–350 (1983)

Chapter 7
In Situ Observation of Lubricating Films by Micro-FTIR

Shigeyuki Mori

Abstract Tribological properties closely depend on the structures of lubricating films between moving surfaces. Lubricating films are formed at lubricating contacts under dynamic conditions, including high temperature, high pressure and high shearing. Thus, structures of lubricating films are greatly affected by the conditions of the environment. In situ observation of lubricating films by micro-FTIR is described in order to characterize the film components and structure formed from lubricants under dynamic conditions.

7.1 Background

7.1.1 Tribological Contact

Tribological properties closely depend on the structures of lubricating films between moving surfaces. Lubricating films are formed at lubricating contacts under dynamic conditions, including high temperature, high pressure and high shearing. Thus, structures of lubricating films are greatly affected by the conditions of the environment (Fig. 7.1). A lubricant film under elasto-hydrodynamic lubrication (EHL) conditions can be formed using a lubricant having proper viscosity only during lubricating conditions. Under boundary lubrication conditions, lubricant films that are dynamically altered by friction and wear are formed via the balance of formation and removal of films during lubrication. Lubricating films are formed dynamically, and film structures should be altered during lubricating process. Therefore, in situ observation of a lubricating film is needed to fundamentally understand tribological phenomena. Because tribological phenomena occur under the effect of mechanical energy, in situ observation is one of the most important techniques to solve tribological problems [1].

S. Mori (✉)
Department of Chemical Engineering, Faculty of Engineering, Iwate University, 4-3-5 Ueda, Morioka, Iwate 020-8551, Japan
e-mail: mori@iwate-u.ac.jp

M. Dienwiebel and M.-I. De Barros Bouchet (eds.), *Advanced Analytical Methods in Tribology*, Microtechnology and MEMS,
https://doi.org/10.1007/978-3-319-99897-8_7

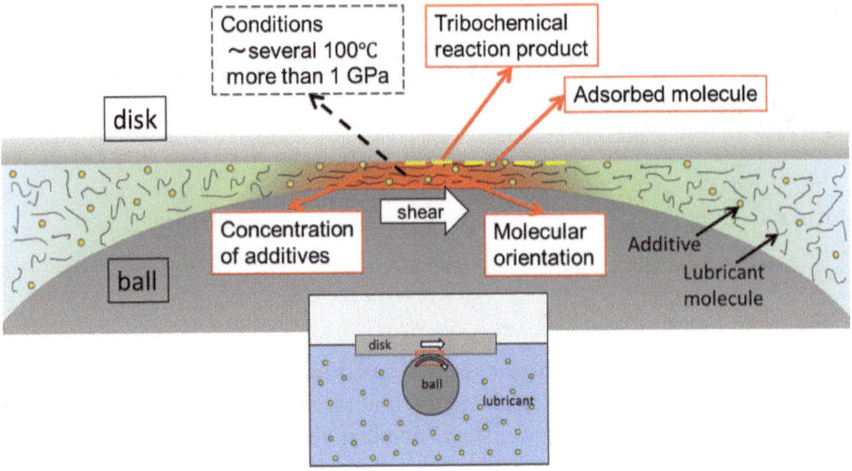

Fig. 7.1 Dynamic structure of lubricant film at sliding contact

7.1.2 In Situ Observation

There are many techniques for in situ observation of lubricating films. The methods for in situ observation using electromagnetic waves, such as infrared, visible, ultraviolet and X-ray light, are useful to investigate lubricating films in air, because air is transparent for electromagnetic waves (Fig. 7.2). At least one tribomaterial should be transparent for the electromagnetic wave to obtain the information of a lubricant film between two surfaces. The method using interferometry and a space layer disk makes it possible to determine the thickness of a lubricant film at nanoscale under EHL conditions [2]. However, the information on chemical structures of the film cannot be supplied, although tribological properties are closely dependent on the chemical structure of the film. Spectroscopic methods are better able to obtain chemical information from film spectra. For example, spectra of visible and ultraviolet regions provide chemical bonds, while infrared spectra contain vibration modes of chemical bonds. In this chapter, the application of infrared spectroscopy to chemical analysis of lubrication films is summarized.

In situ observation of lubricating films has been performed with infrared emission spectroscopy, and molecular orientation of a lubricant film has been observed [3]. However, the spectrum is not enough to analyze chemical structures of lubricant films. Fourier-transform infrared spectroscopy (FTIR) is developed to supply clear IR spectroscopic information. High spatial resolution of spectrum can be obtained with an infrared microscope. Then, micro-FTIR has been used for in situ observation of lubricating films under EHL conditions [4, 5].

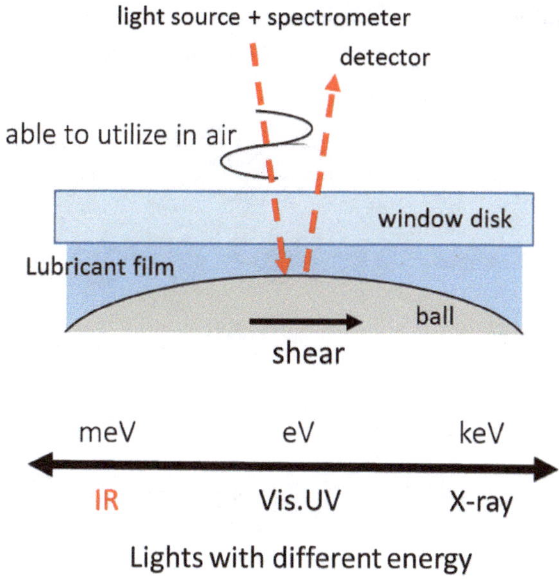

Fig. 7.2 In situ observation of lubricant film with light

7.2 Principle and Method

7.2.1 *Infrared Spectroscopy*

Molecular vibration absorbs infrared radiation. There are typical vibration modes such as C–H, C–O–C, C=O, O–H, and N–H at 3000–2840, 1260–1000, 1750–1735, 3650–3200, and 3400–3350 cm^{-1}, respectively. Their absorption peaks are observed at different positions, for example C–H stretching vibration at 2930 cm^{-1}. Organic liquid as a base oil can be detected from the peak of C–H stretching vibration. Additives such as oiliness agents can be analyzed by the vibrations of C=O, O–H, N–H and so on. Additives having organic and inorganic functional groups, such as phosphate, are also detected by IR absorption peak. Therefore, chemical components of a lubricating film can be assigned from IR absorption spectrum of the film (see Table 7.1) [6].

7.2.2 *Lubrication Tester*

Figure 7.3a, b shows a typical ball-on-disk tribotester for in situ IR analysis [5, 7]. Ball and disk are made of bearing steel AISI52100 and IR transparent material, respectively. Lubrication tests can be performed between ball and disk under EHL conditions. The IR beam comes from a Cassegrain mirror of a microscope through an IR transparent disk and is reflected on steel ball surface (Fig. 7.3b). The IR beam

Table 7.1 IR absorption peak positions of lubricant components

Lubricant component	Chemical structure	Peak position (cm^{-1})
Base oil hydrocarbon	C–H *st*	3000–2840
	C–H *bend*	1600–1400
Ether	C–O–C	1260–1000
Ester	COOR	1750–1735
		1300–1000
Oiliness agent	O–H	3650–3580
		3550–3200
	COOH	1720–1700
	N–H	3400–3350
		3330–3250
Viscosity index improver	COOR	1750–1735
		1300–1000
EP additive sulfide	S–S	500–400 *w*
	S–C	700–600 *w*
Phosphate	P=O	1350–1250
	O–P–O	970–940
Thickener soap	COO^-	1610–1550
		1420–1300
2nd-amide	N–H *assoc*	3100–3070
2nd-amide	C=O *assoc*	1650

st: stretching, *bend*: bending, *assoc*: associated, *w*: weak

crosses the lubricant film layer twice, and the IR spectrum of the oil film can be obtained during lubrication tests. The sampling position for IR spectrum is scanned as shown in Fig. 7.4a for mapping and (b) for rapid scan.

7.2.3 *Materials*

Disk window material can be selected from diamond, silicon, germanium, calcium fluoride (CaF_2), and sapphire (Al_2O_3), depending on the IR transparent region [5], strength, hardness, and cost. Because the mechanical property of silicon is similar to bearing steel, it is a convenient material to investigate an EHL lubrication film. Although CaF_2 has a wide range of IR transparency, it is weak against moisture and polar compounds such as soap. Silicon and germanium have a good transparency of IR, but they are not transparent for visible light.

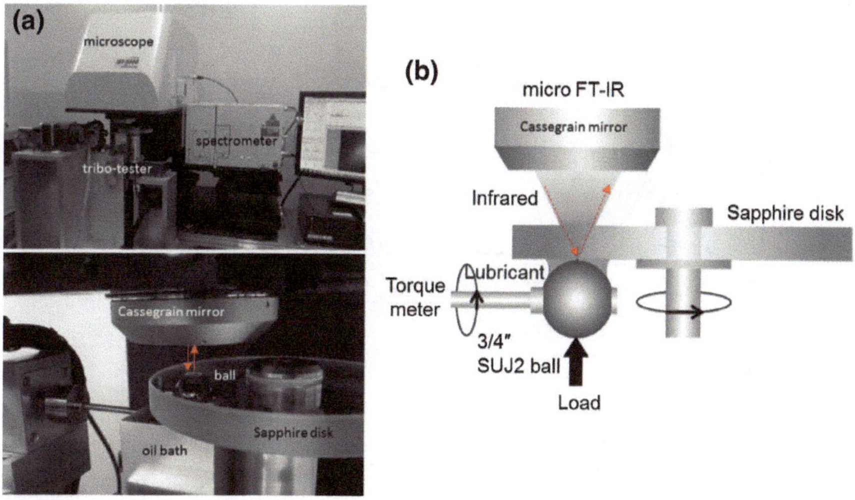

Fig. 7.3 A typical ball-on-disk tester with micro-FTIR; **a** Photograph of the tester **b** schematic diagram

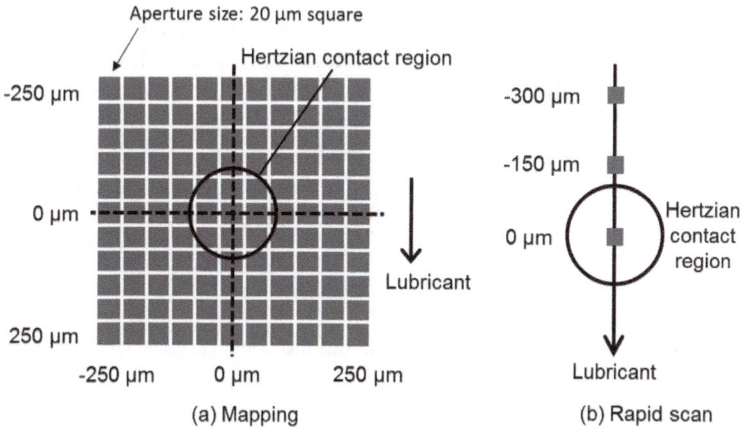

Fig. 7.4 Aperture size and data sampling position **a** Scanning mode, **b** rapid scan mode

7.3 Structure of EHL Film

7.3.1 Film Thickness and Pressure

Figure 7.5a shows the IR spectrum of a lubricant film formed with synthetic hydrocarbon oil (PAO:polyalphaolefin) under EHL conditions [8, 9]. The main peak at 2930 cm^{-1} is originated from C–H stretching vibration of PAO. According to the Beer-Lambert law (5.1), the thickness (t_F) of an oil film can be estimated from the

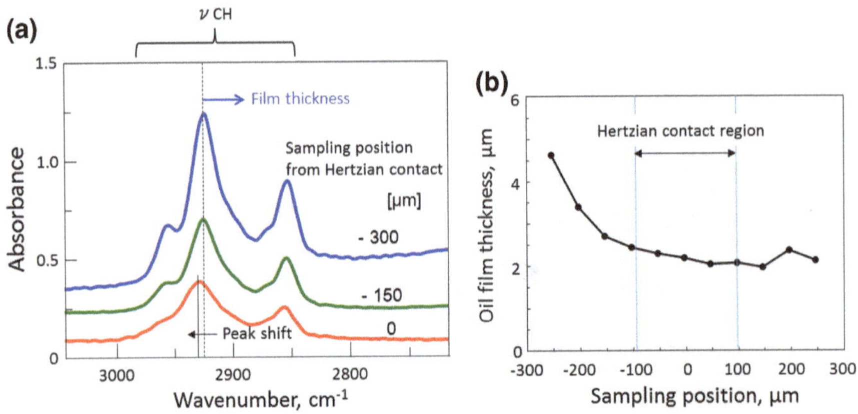

Fig. 7.5 **a** IR spectra of PAO at different sampling positions (cf. Fig 7.3d), **b** the profile of film thickness obtained from the absorbance. Conditions: 0.2 m/s, 10 N

absorbance (A) of the peak using a calibration curve obtained previously. Because an aperture size for an IR beam is 20–50 μm, the distribution of oil film thickness can be obtained by scanning the measuring position around the Hertzian contact, as shown in Fig. 7.5b. The sensitivity of IR measurement for thickness is less than 50 nm and depends on the absorption coefficient (ε) of the characteristic peak. The film thickness of hydrocarbon oil at even several microns, which is impossible to measure with interferometry, can be measured from the absorbance of C–H bending vibration peak. When the film is too thick, the IR peak intensity becomes saturated and then film thickness cannot be precisely obtained.

$$A = -\varepsilon c\, t_F \tag{5.1}$$

where ε and c are the absorption coefficient and the concentration of the component, respectively.

The peak position of C–H vibration in the IR spectrum at the center of Hertzian contact shifts slightly to a higher wavenumber as shown in Fig. 7.5a, which can be explained by the configuration of alkyl chains under high pressure. Because the shift rate has been reported as 15 cm^{-1}/GPa [8], the distribution of pressure around the contact area can be estimated from the peak shift as shown in Sect. 4.2.

7.3.2 *Component of Lubricant*

The lubricant is a mixture of a base oil and additives such as oiliness agents, friction modifiers, viscosity index improvers, extreme pressure additives, anti-wear additives and so on. The lubricating property is closely dependent on the lubricant component

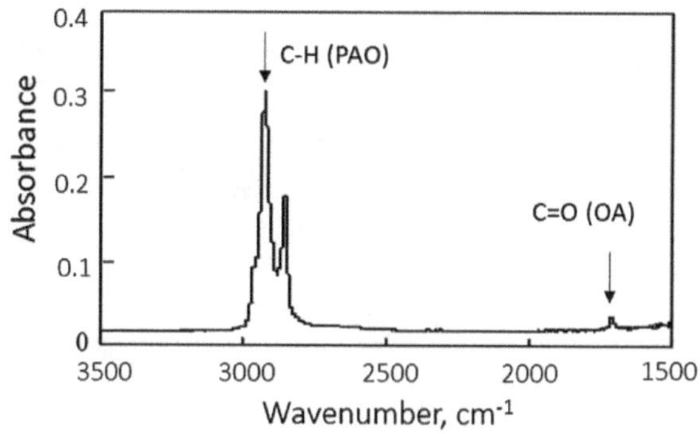

Fig. 7.6 IR spectrum of oleic acid containing lubricant obtained during lubrication under EHL condition

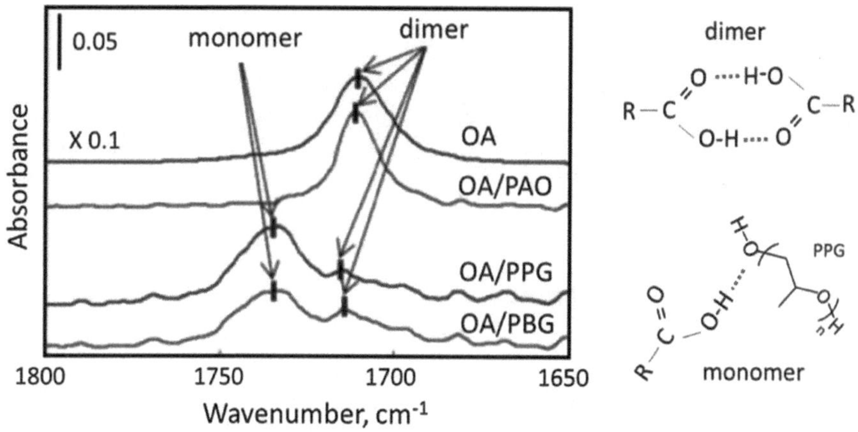

Fig. 7.7 IR spectra of a monomer and a dimer of oleic acid in polar and non-polar base oils obtained under static conditions of 3.4 N load

and the concentration of additives. IR spectrum provides chemical structures and concentrations of lubricant components. Figure 7.6 shows IR spectrum of a lubricant film containing oleic acid as an oiliness agent in PAO under a dynamic condition [9, 10]. In addition to the peak of C–H from PAO, there is a small peak of C=O from oleic acid. The concentration of oleic acid can be estimated from the intensity ratio of C=O and C–H, according to a calibration curve obtained previously. The peak position of C=O at 1720 cm^{-1} indicates that the carboxyl group of the oleic acid forms a dimer which can dissolve in non-polar base oil. When oleic acid is dissolved in a polar base oil such as polyalkylene glycol, the peak position of C=O is shifted to a higher wavenumber at 1730 cm^{-1}. This result indicates that the carboxyl group interacts with glycol as a polar molecule and then oleic acid exists as a monomer in a polar base oil as shown in Fig. 7.7 [7].

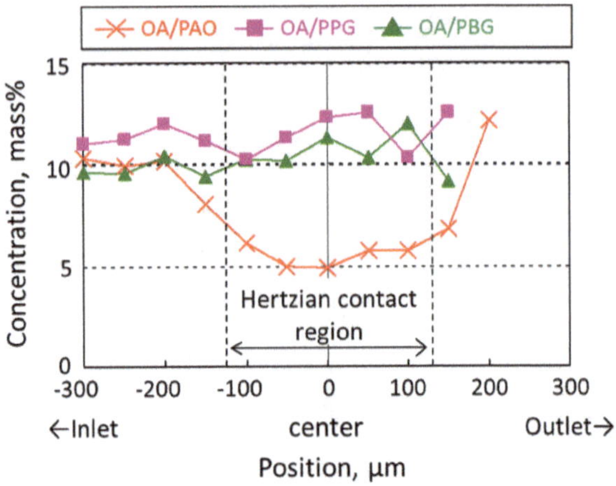

Fig. 7.8 Effect of base oil on concentration of oleic acid at EHL contact region. Conditions: load 10 N, entrainment speed 0.13 m/s, SRR 0%

A dynamic behavior of viscosity index improver (VII) can be observed by a micro-FTIR spectrometer [11]. The concentration of lubricant components such as alcohols, amines, amides, and water can also be estimated from their characteristic absorption peaks in IR spectra as shown in the following section.

7.3.3 *Chemical Information*

The chemical composition of a lubricant affects the tribological property through viscosity of the lubricant and the adsorption of additives on lubricating surfaces. The distribution of additive concentration can be obtained by scanning sampling position in IR spectra as shown in Fig. 7.4a, b. Figure 7.8 shows a concentration profile of oleic acid in PAO observed around the Hertzian contact during EHL lubricating tests [7]. The concentration change at the lubricating contact is dependent on the combination of the additive and base oil, as shown in Fig. 7.8. The concentration of oleic acid in PAO as a non-polar base oil is decreased to almost half of the original concentration. When oleic acid dissolves in polypropylene glycol (PPG) or polybutylene glycol (PBG) as a polar base oil, the concentration of oleic acid is not decreased in the contact zone. It can be explained that oleic acid comes into the contact zone with the base oil because the oleic acid molecule interacts with a polar base oil, as shown in Fig. 7.7. In the case of polybutene as a base oil, the concentration is not changed, although the base oil is non-polar. Polybutene has a large pressure-viscosity coefficient and is known to solidify through high pressure at the EHL contact. Then, the solidified lubricant with oleic acid comes into the contact zone without changing the concentration of oleic acid.

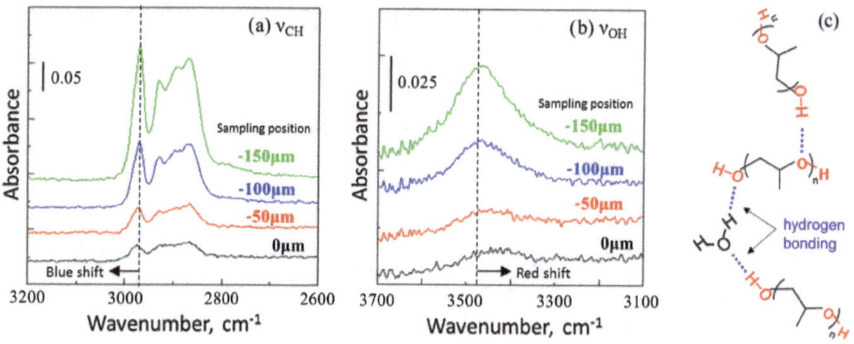

Fig. 7.9 Chemical shift of characteristic IR peak position of C–H (**a**) and O–H (**b**) vibrations. Molecular interaction of PPG through hydrogen bonding of the end-group OH (**c**)

7.4 Molecular Interaction

7.4.1 Interaction of Components

The molecular interaction between lubricant components affects the physical properties of lubricants. For example, the solubility of an additive to a base oil is closely dependent on the molecular interaction of the two components. The adsorption property of additives can be controlled by the balance of the molecular interaction of the additive with the surface and base oils. The viscosity of a lubricant is also affected by molecular interaction of the base oil itself. The molecular interaction is expected to be altered by high pressure at lubricating contacts.

Figure 7.10a shows IR spectra of polypropylene glycol inside (0 μm) and outside (−150 μm) of the EHL contact region. A blue shift of C–H vibration due to high pressure can be seen as described above (see Fig. 7.5a) [8]. On the other hand, the peak of O–H vibration originated from the end-group of polypropylene glycol shifts to a lower wavenumber as shown in Fig. 7.9b. It is opposite to the shift of C–H vibration. The result can be explained by the molecular interaction of O–H through hydrogen bonding at high pressure, as shown in Fig. 7.9c. The end-group O–H interacts with other O–H groups through hydrogen bonding. The interaction stabilizes at high pressure, resulting in a long bond distance of O–H bond. Because bond strength decreases by increasing bond distance between O and H, the peak position of O–H vibration shifts to a lower wavenumber. The molecular interaction of carboxylic acid, amine, and amide will be shown in the following section.

The molecular interaction of the base oil at high pressure affects the traction coefficient through the viscosity of a lubricant at high pressure. Two polypropylene glycols are selected to compare the lubrication property under EHL conditions. Polypropylene glycol has an O–H end-group, but the other has an O–CH_3 group instead. Their main chain structures are the same, and thus the viscosities of the two oils at atmospheric pressure are similar. The former molecules interact through

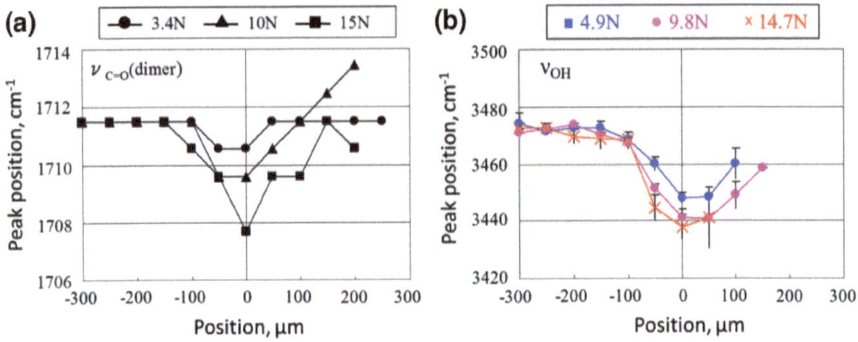

Fig. 7.10 Effect of load on peak position of PPG lubricant component around EHL contact

hydrogen bonding between end groups, but the latter molecules interact with each other without hydrogen bonding. The glycol with O–H exhibits a higher traction coefficient (0.06) than that without O–H (0.026) obtained by a ball on disk tester under the conditions of 16 N load, 0.27 m/s sliding speed and 40 °C [12]. One of the explanations for the high traction coefficient is the molecular interaction through hydrogen bonding by high pressure at contact area.

7.4.2 Effect of Load

Figure 7.10a shows the distribution of chemical shifts of C=O vibration of oleic acid in PAO around the Hertzian contact zone [7]. The shift increases at the center of the Hertzian contact under high load. The peak shift of O–H vibration of the end-group in polypropylene glycol is also observed, as shown in Fig. 7.10b [8].

These results clearly indicate that the hydrogen bonding of COOH and O–H groups can be stabilized by high pressure at the lubricating contact.

The peak shift at high pressure and temperature is confirmed through IR measurement using Diamond Anvil Cell (DAC) under the conditions of high pressure (up to 5 GPa) and high temperature (up to 100 °C) [12]. A red shift of peaks is observed at high pressure. On the other hand, a blue shift is observed at high temperature. The rate of peak shift of O–H group is $-25\ \text{cm}^{-1}$/GPa at 25 °C and $0.18\ \text{cm}^{-1}$/K under atmospheric pressure. The pressure has more effective on the peak shift than the temperature under these observation conditions.

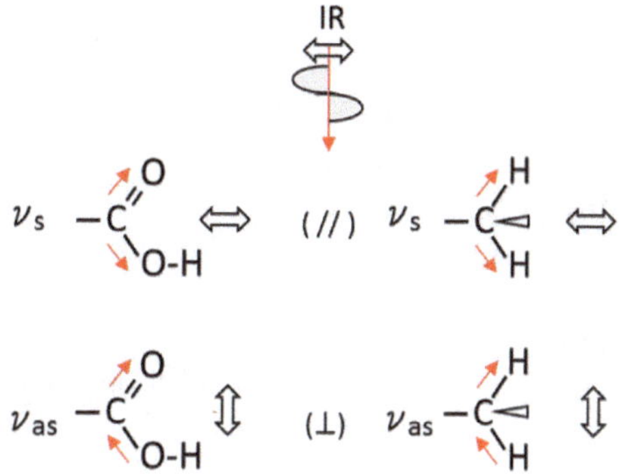

Fig. 7.11 Vibration modes and their vectors of electric vibration

7.5 Molecular Alignment

7.5.1 Effect of Shear

It is well-known that in non-Newtonian fluids, the viscosity of a lubricant decreases at high shear conditions, which is explained by the molecular alignment of lubricant molecules. The molecular alignment can be detected by IR or polarized IR spectroscopy as shown in Fig. 7.11. There are two stretching vibration modes in a carboxyl group, symmetric υ_s and anti-symmetric υ_{as}. Their electronic vectors are at right angles to each other. The stretching vibration of the methylene group CH_2 in the alkyl chain also has two vibrational modes as shown in Fig. 7.11. Polarized IR spectra reveal the molecular orientation under shearing condition from a linear dichroic ratio (*D*) of IR absorption as (5.2),

$$D = A(//)/A(\perp) \tag{5.2}$$

where *A(//)* and $A(\perp)$ are the absorbances for radiation polarized parallel and perpendicular, respectively (see Fig. 7.11).

7.5.2 Effect of Molecular Structure

Long alkyl chains of a lubricant can form a line along the sliding direction under shearing conditions as mentioned above. The molecular orientation of fatty acids can also be detected by a dichroic ratio of IR spectrum (5.2). Alkyl cyanobiphenyl (ACB) is one of the liquid crystals and exhibits a low traction coefficient (0.035) when a lubrication test with ACB is carried out under EHL conditions [13]. When alkyl cyanocyclohexylphenyl (ACC) is used as a lubricant, the traction coefficient (0.045) is slightly higher compared with ACB as shown in Fig. 7.12a. This result suggests that a rigid and flat structure of biphenyl group in ACB molecules plays an important

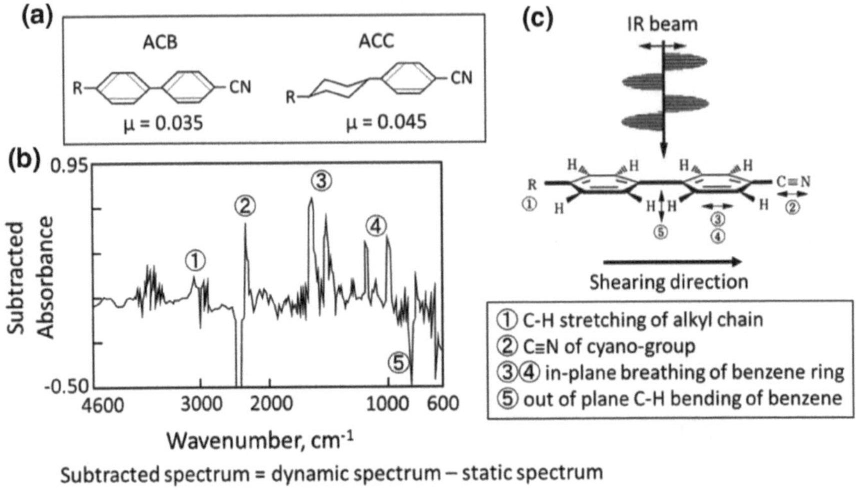

Fig. 7.12 In situ observation of molecular alignment of liquid crystal (5CB) under shearing condition by micro-FTIR **a** Molecular structure and traction coefficient, **b** subtracted IR spectrum, **c** vibration modes and IR beam

role for the low traction coefficient. Therefore, the molecular orientation of ACB is confirmed by in situ observation during the sliding condition. Figure 7.12b is the subtracted IR spectrum of ACB obtained by in situ observation during sliding. The IR spectrum obtained under a dynamic condition was subtracted by that obtained under a static condition, so that the intensity of C–H vibration that mainly originated from the alkyl chain disappears. There are three characteristic peaks at 1606 and 1494, 845 and 2227 cm^{-1} assigned as benzene ring breathing (in plane), benzene C–H bending (out of plane), and stretching vibration of cyano-group C≡N. It is noteworthy that the intensities of benzene (in plane) and cyano-group are increased, but that of benzene (out of plane) is decreased. This result clearly shows that the plane of diphenyl plane is oriented in the plane of shear during lubricating condition as shown in Fig. 7.12c. Therefore, one of the important factors for the low traction coefficient of ACB is the molecular orientation of a flat structure of the cyano-biphenyl group in the shearing plane. A similar behavior of molecular orientation can be seen with a mixture of benzoic acid derivatives and oleic acid which aggregate through hydrogen bonding with carboxyl groups [14].

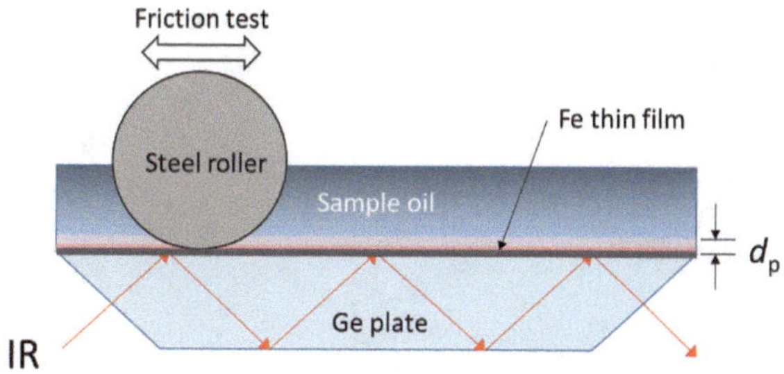

Fig. 7.13 In situ observation of lubricating film formed during friction with Fe film and steel rod using ATR method: penetration depth, d_p [see (5.3)]

7.6 Boundary Lubrication

7.6.1 Adsorbed Layer

The structure of the adsorbed layer plays an important role in the tribological performance in boundary lubrication. A lubricant film of 0.1–1 μm in thickness formed under EHL conditions is suitable to be analyzed using IR spectroscopy. This method, however, is not sensitive enough to detect nanoscale films, such as adsorption films of lubricant additives. In the case of films at the nanoscale, a highly surfaces-sensitive method, such as Sum Frequency Generation (SFG) spectroscopy, is needed, and the results obtained with SFG will be mentioned below. Attenuated Total Reflection (ATR) IR spectroscopy can be applied to in situ observation of the chemical structures of thin surface films. When the infrared beam reflects totally at the interface between IR transparent materials such as Ge and lubricant oil, the IR beam penetrates into the oil phase as shown in Fig. 7.13. As described in (5.3), the penetration depth (d_p) is determined by the incident angle (θ), the index of refraction of materials (n_1, n_2), and the wavelength (λ) of IR. In the case of Ge, lubricant and incident angle of 45°, the IR beam can penetrate into the lubricant phase with depth at submicron. When the Ge surface is covered with a thin layer (around 10 nm) of Fe prepared by sputter deposition, chemical interaction, including the chemical reaction of iron with the lubricant components, can be analyzed by ATR method [15].

$$d_p = \frac{\lambda}{2\pi n_1 \sqrt{\left[\sin^2\theta - (n_2/n_1)^2\right]}} \tag{5.3}$$

7.6.2 Reaction Layer

Boundary films formed through tribochemical reaction were also analyzed by ATR method. A reciprocating friction test of a steel roller and an Fe-coated ATR disk with ZnDTP-containing lubricant was carried out under ATR spectroscopic measurement. The tribochemical reaction product from ZnDTP was detected during a lubrication

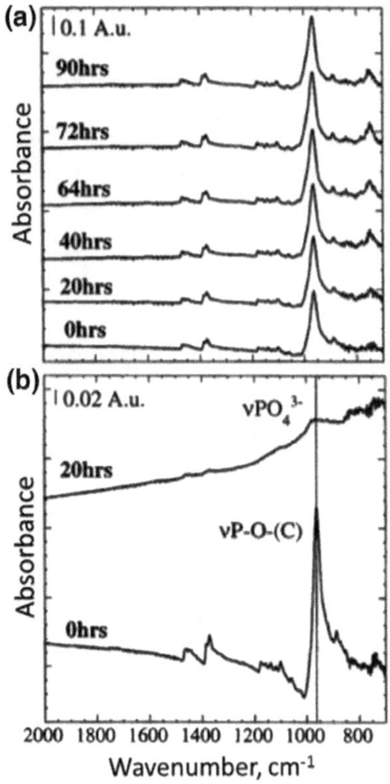

Fig. 7.14 In situ observation of lubricant film formed from ZnDTP at 150 °C under (**a**) static and (**b**) dynamic conditions at the sliding interface between Fe film and a steel rod

test by ATR method [16]. Although no spectral change is observed during heating at 150 °C for 90 h (Fig. 7.14a), phosphate as a tribochemical reaction product is detected by a lubrication test at 150 °C (Fig. 7.14b). A lubricating film was investigated under a boundary lubricating condition by ATR method [15].

7.7 Applications

IR spectroscopy has been applied to in situ observation of lubricant films mainly under EHL conditions. Other applications are described below.

7.7.1 Water-Based Lubricant

There are several lubrication fluids with water for practical use, such as O/W (oil in water) and W/O (water in oil) emulsions and water soluble lubricants. The behavior of oil and water at the sliding contacts is important to understand tribological properties

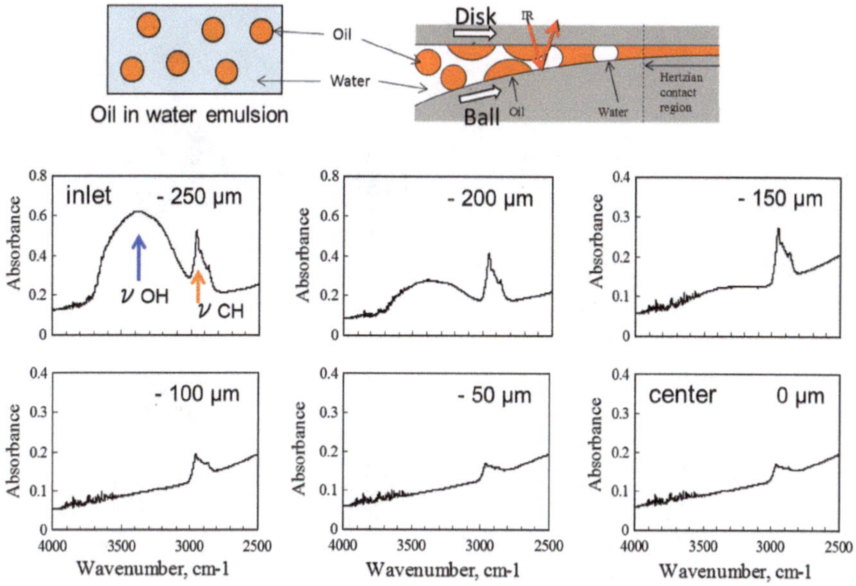

Fig. 7.15 Phase separation of o/w emulsion at lubricating contact under EHL condition

of water-based lubricants. For example, a base oil with the proper viscosity is needed to form lubricant film at EHL contact. Water, however, has too low viscosity to form water film. Therefore, phase separation of a water-based lubricant at the contact zone is necessary for lubrication with water-based lubricants.

Figure 7.15 shows IR spectra obtained by in situ observation of EHL film lubricating with O/W emulsion [17]. There are two main peaks of broad O–H peak from water and sharp C–H peak from the base oil at 3400 and 2960 cm^{-1}, respectively. It is clear that the absorbance of O–H decreases from the outside to the inside of the contact zone and no water is observed at the center of EHL contact. This suggests that phase separation occurs at the entrance of the contact zone, resulting in the formation of oil film at EHL contact. A similar oil separation is observed with w/o emulsion [18] and water soluble lubricants [19].

7.7.2 *Semi-Solid Lubricant*

Greases which are a mixture of thickener and base oil have been utilized as a semi-solid lubricant in a wide range of bearings and gears. Thickeners form fibrous micelles that can hold the base oil. The behavior of the thickener micelles around the contact zone, which affects the tribological property of grease, is not clear. In order to make clear the role of thickener, the lubricant film formed on a specimen is analyzed after lubrication test by micro-FTIR [20]. The concentration of thickener can be estimated

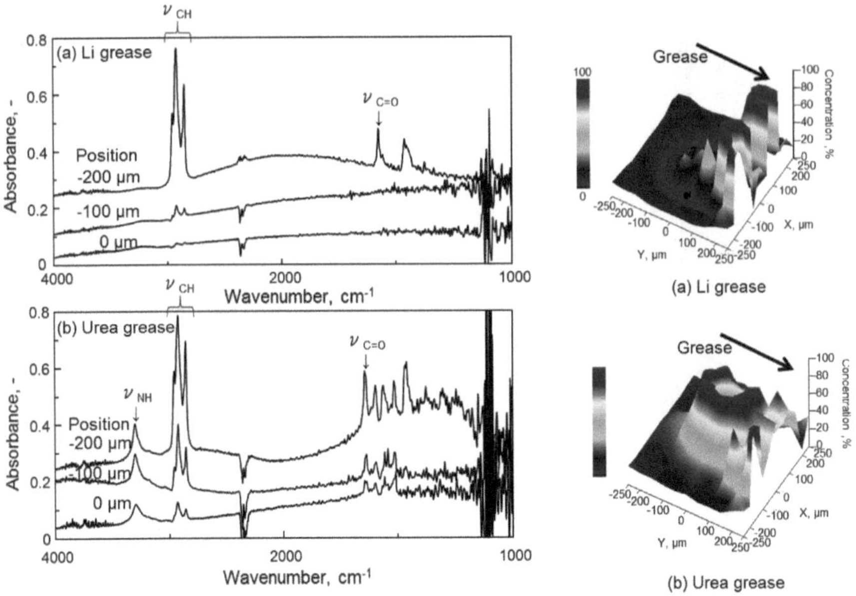

Fig. 7.16 In situ IR spectra and distribution of thickener concentration of Li-grease (**a**) and urea-grease (**b**)

from the absorbance ratio of N–H peak at 3300 cm^{-1} for urea-type thickener and C=O peak at 1580 cm^{-1} for soap-type thickener to the absorbance of C–H from the base oil, respectively. Figure 7.16 shows the increase in the concentration of the urea-type thickener at the contact zone during lubrication test [21]. The concentration behavior depends on the molecular structures of thickeners, i.e., soap-type and urea-type. Concentration of urea-type thickener increases more easily than that of soap-type thickener as shown in Fig. 7.16a, b, respectively.

In situ observation of thickener concentration during EHL test reveals that the concentration increases gradually with lubrication time under the lubricating condition, but the concentration decreases steeply by shearing (Slide Roll Ratio: SRR = 70%) under the sliding condition (Fig. 7.17) [22]. This result suggests that a boundary film of thickener is formed again at the sliding contact even when the film is removed with friction. The thickness of the boundary film is determined by the balance between formation and removal of the film. Increase in concentration of amide as an additive is also observed during lubrication of a gel-like lubricant with dialkyl amides [23, 24].

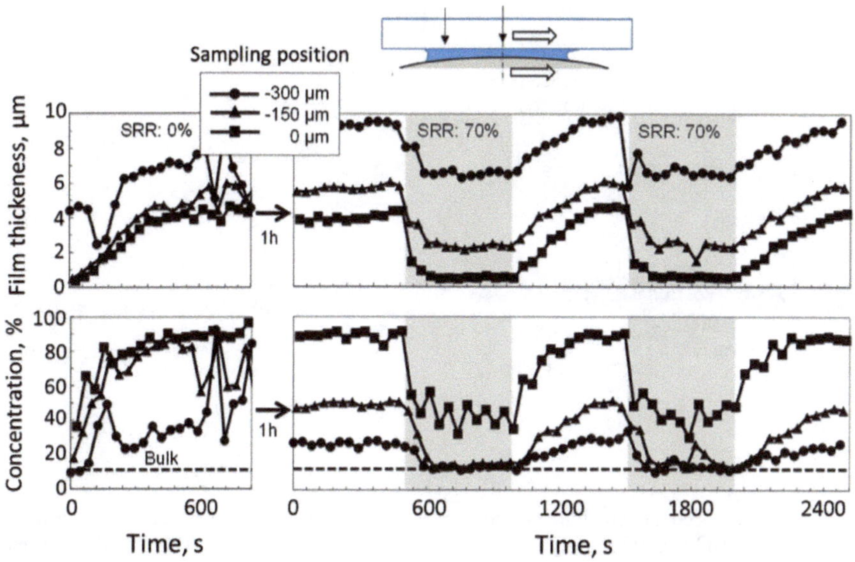

Fig. 7.17 Formation and removal of thickener film and effect of shearing on film formation of urea grease

7.7.3 *Paper Friction Material (Wet Clutch)*

One of the important factors to control the friction property of paper friction plates for wet clutches is the lubricant component between the paper material and the steel plates. The lubricant component cannot be analyzed by transmission or reflection IR spectroscopy, because the IR beam does not reflect on paper material. ATR was applied to analyze the additive concentration in PAO or paraffinic mineral oil (P) between the paper material and the diamond of the ATR unit as shown in Fig. 7.18a [25]. It is interesting that the concentrations of polybutenyl saccimide (Fig. 7.18b) and *N*-oleoylsulcosine (Fig. 7.18c) increase at 10 N of load under the effects of high and low sliding speeds, respectively. The results of additive concentrations affect the friction behavior of wet clutch with the additives. Water film between a hydrogel and a diamond of ATR unit can be observed by ATR method [26].

7.7.4 *Wire Drawing*

In metal working processes, such as cutting, drawing, ironing, and rolling, unique lubricants are demanded for high-energy efficiency, good finishing of metal surface, and higher reliability of the process. Because lubricity is controlled by the component and structure of the lubricant film between the metal and the working tool, in situ

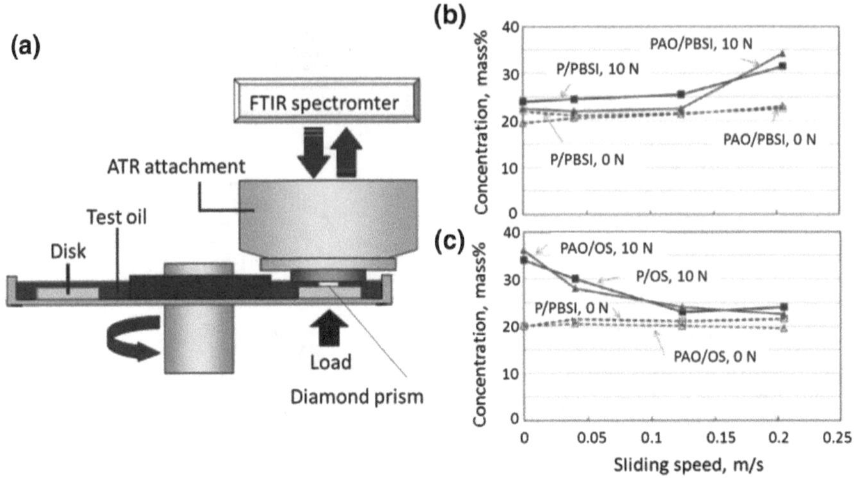

Fig. 7.18 Effect of contact conditions on concentration of additives at the interface of paper material for wet clutch. Additive; PBSI: polybutenyl succimide, OS: oleoylsarcosine, Base oil; P: paraffinic mineral oil, PAO: polyalphaolefin

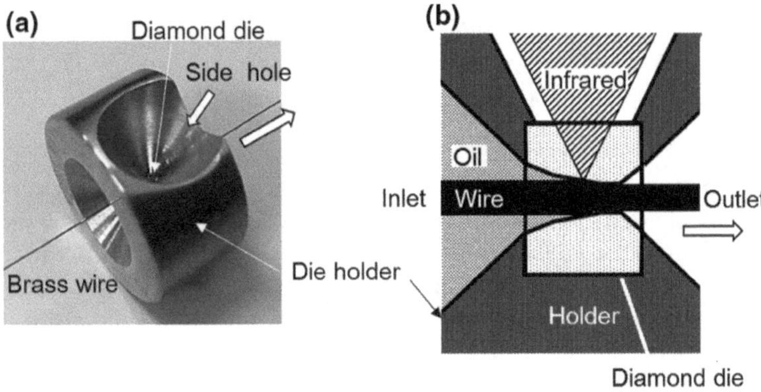

Fig. 7.19 Photo (**a**) and a schematic cross section (**b**) of diamond die in a holder

observation with IR has been applied to investigate the lubricant film between the metal and the working tool. Figure 7.19 shows a wire drawing die made of diamond [27]. The holder has a side window for an IR beam which reflects on a metal wire surface. The IR spectrum of the lubricant film can be taken during drawing with the die. Reduction of brass wire was carried out with oleic acid in PAO as a base oil. The oil film thickness, estimated from the absorbance of C–H vibration, linearly increases with the distance of the reduction region of the die. It is found that metal oleate, which is formed tribochemically from the oleic acid and brass during drawing, accumulated at the entrance of the bearing region of the die [28].

7.8 Future Work

There are many analytical tools to investigate lubricating films which affect tribological properties. Among them, IR spectroscopy is powerful for in situ analysis of lubricating films, especially for tests in air. However, IR spectroscopy has several disadvantages, such as limited sensitivity, spatial resolution, and chemical information. IR spectroscopy is more suitable to analyze organic compounds, such as lubricant components, rather than inorganic compounds, such as tribofilm formed from additives. Surface Enhanced Infrared Reflection Absorption Spectroscopy (SEIRAS) has been applied to chemical analyses of nano-film for electrochemistry and catalysis [29], and can be applied to the chemical analysis of lubricant films.

Because Raman spectroscopy can analyze many inorganic compounds, such as MoS_2, at high spatial resolution of 1 μm, it has been applied for in situ observation of lubricant films [30–32]. The advantage of Raman spectroscopy is the use of a glass plate as a window instead of IR transparent materials such as diamond and calcium fluoride for tribology test equipment. A combination of IR and Raman spectroscopies can be applied as powerful tools to investigate a wide range of lubricating films through in situ observation of lubricating films during EHL and boundary lubrication conditions.

A new optical technique for in situ observation is SFG spectroscopy, which is so sensitive enough to analyze the chemical structure of monolayer films with high spatial and time resolutions [33]. SFG measurement of polyethylene reveals surface segregation of organic additives in polymers, which affects surface mechanical properties such as friction [34]. SFG was applied to investigate lubricating interface of n-dodecane, n-$C_{12}H_{26}$, with stearic acid under shearing condition. A strong peak of CH_2-ss (symmetric stretching vibration) is observed at 2840 cm^{-1}, only with stearic acid containing oil under dynamic conditions (Fig. 7.20). When deuterated n-dodecane (n-$C_{12}D_{26}$) was used as a base oil, the peak at 2840 cm^{-1} disappears. SFG spectra reveal that the interfacial structure of stearic acid changes negligibly, but n-dodecane is aligned on the adsorbed film of stearic acid under shearing condition [35]. To understand tribological phenomena fundamentally, a suitable combined system of SFG spectrometer and a proper tribotester will provide fruitful information on the chemical structures of lubricant films under dynamic conditions through in situ observation of tribo-interface at high spatial and time resolutions.

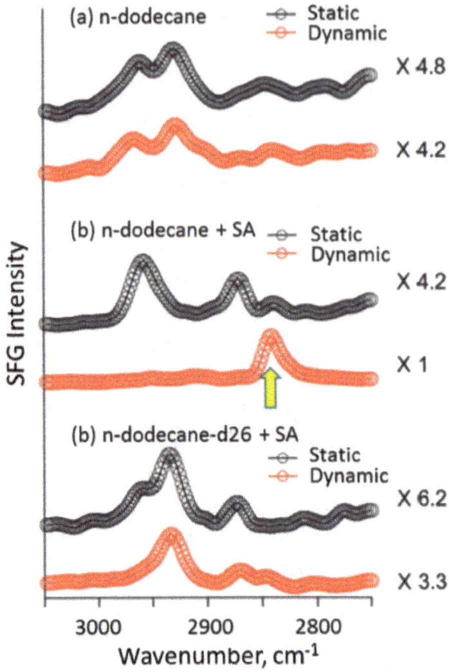

Fig. 7.20 Molecular orientation of base oil, n-dodecane at the interface between oleic acid adsorbed layer and lubricant observed by SFG spectroscopy

References

1. H.A. Spikes, In situ methods for tribology research. Tribol. Lett. **14**(1), 1 (2003)
2. G.J. Johnston, R. Wayte, H.A. Spikes, The measurement and study of very thin lubricant films in concentrated contacts. STLE Tribol. Trans. **34**(2), 187 (1991)
3. J. Lauer, K. Vincent, Fourier emission infrared microspectrophotometer for surface analysis. I. Application to lubrication problems. Infrared Phys. **19**(3–4), 395 (1979)
4. P.M. Cann, H.A. Spikes, In lubro studies of lubricants in EHD contacts using FTIR absorption spectroscopy. Tribol. Trans. **34**(2), 248 (1991)
5. P.M. Cann, H.A. Spikes, In-contact IR spectroscopy of hydrocarbon lubricants. Tribol. Lett. **19**(4), 289 (2005)
6. G. Socrates, *Infrared and Raman Characteristic Group Frequencies, Table and Charts*, 3rd edn. (Wiley, New York, 2004)
7. K. Takiwatari, H. Nanao, S. Mori, Effect of high pressure on molecular interaction between oleic acid and base oils at elastohydrodynamic lubrication contact. Lubr. Sci. **22**, 89 (2010)
8. K. Takiwatari, H. Nanao, E. Suzuki, S. Mori, Stabilization of hydrogen bonding in polypropylene glycol at EHL contact region. Lubr. Sci. **22**(9), 367 (2010)
9. Y. Hoshi, N. Shimotomai, M. Sato, S. Mori, Change of concentration of additives under EHL condition. J. Japanese Soc. Tribologists **44**(9), 736 (1999)
10. K. Takiwatari, H. Nanao, I. Minami, S. Mori, The interaction between oleic acid and base oils at elastohydrodynamic lubrication. J. Japanese Soc. Tribologists **54**(1), 48 (2009)
11. R.W.M. Wardle, R.C. Coy, P.M. Cann, H.A. Spikes, An 'in lubro' study of viscosity index improvers in end contacts. Lubr. Sci. **31**(1), 45 (1990)
12. K. Takiwatari, H. Nanao, Y. Hoshi, S. Mori, Molecular interaction originating from polar functional group in lubricants and its relationship with their traction property under elasthydrodynamic lubrication. Lubr. Sci. **27**, 265 (2015)

13. S. Mori, H. Iwata, Relationship between tribological performance of liquid crystals and their molecular structure. Tribol. Int. **29**(1), 35 (1996)
14. R. Lu, S. Mori, H. Tani, N. Tagawa, S. Koganezawa, Tribol. Int. **113**, 36 (2017)
15. F. Mangolini, A. Rossi, N.D. Spencer, Tribol. Lett. **45**, 207 (2012)
16. F.M. Piras, A. Rossi, N.D. Spencer, Growth of tribological films: In situ characterization based on attenuated total reflection infrared spectroscopy. Langmuir **18**, 6606 (2002)
17. N. Shimotomai, S. Mori, In situ observation of the lubrication film I O/W emulsion by mocro FT-IR. J. Japanese Soc. Tribologists **56**(1), 47 (2011)
18. Y. Shitara, S. Yasutomi, S. Mori, Direct observation of W/O emulsion in concentrated contact by FT-IR microspectroscopy. J. Japanese Soc. Tribologists **55**(10), 736 (2010)
19. Y. Hoshi, I. Minami, S. Mori, Change in concentration of water-glycol hydraulic fluid near the EHL contact region–Observation by micro FT-IR-. J. Japanese Soc. Tribologists **49**(11), 878 (2004)
20. P.M. Cann, Grease lubrication of rolling element bearings—Role of the grease thickener. Lubr. Sci. **19**(3), 183 (2007)
21. Y. Hoshi, K. Takiwatari, H. Nanao, H. Yashiro, S. Mori, In situ observation of EHL films of greases a micro infrared spectroscopy. J. Japanese Soc. Tribologists **60**(2), 153 (2015)
22. Y. Hoshi, K. Takiwatari, H. Nanao, S. Mori, In situ observation of transient responses in grease lubrication by a micro infrared spectroscopy. J. Japanese Soc. Tribologists **61**(11), 784 (2016)
23. K. Takahashi, Y. Shitara, S. Mori, Direct observation of thermo-reversible gel-lubricants in EHL by FT-IR micro-spectroscopy. Tribol. Online **3**(2), 131 (2008)
24. K. Takiwatari, Y. Hoshi, H. Nanao, T. Hojo, S. Mori, In situ observation of lubricant film of semi-solid lubricants at EHL contact using micro-FTIR. Tribol. Online **11**(2), 346 (2016)
25. T. Ichihashi, M. Kudo, S. Mori, Relation between the friction characteristics of wet clutches and the concentration of additives obtained by in-situ observation of oil film. J. Japanese Soc. Tribologists **58**(8), 581 (2013)
26. Y. Hoshi, K. Takiwatari, H. Nanao, H. Yashiro, M. Wada, H. Furukawa, S. Mori, In situ observation of hydrogel frictional interface by a micro FTIR-ATR spectroscopy. J. Japanese Soc. Tribologists **60**(1), 68 (2015)
27. H. Nanao, Y. Hoshi, T. Shizuku, K. Takiwatari, S. Mori, Direct observation of lubricant components between wire and diamond die for wire drawing with a micro-FTIR. Tribol. Lett. **60**(1), 12 (2015)
28. Y. Hoshi, H. Nanao, K. Takiwatari, S. Mori, T. Shizuku, In situ observation of lubricant film in a diamond die for wire drawing by micro-FTIR. Tribol. Online **11**(2), 88 (2016)
29. M. Osawa, In-situ surface-enhanced infrared spectroscopy of the electrode/solution interface, in *Electrochemical Science and Engineering*, ed. by R.C. Alkire, D.M. Kolb, J. Lipkowski, P.N. Ross, p. 269 (2006)
30. C.U. Amanda Cheong, P.C. Stair, In situ measurements of lubricant temperature and pressure at a sliding contact. J. Phys. Chem. **111**, 11314 (2007)
31. S. Zhang, Y. Liu, J. Luo, In situ observation of the molecular ordering in the lubricating contact area. J. Appl. Phys. **165**, 014302 (2014)
32. H. Okubo, S. Sasaki, In situ Raman observation of structural transformation of diamond-like carbon films lubricated with MoDTC solution: Mechanism of wear acceleration of DLC films lubricated with MoDTC solution. Tribol. Int. **113**, 399 (2017)
33. G.A. Somorjai, K.C. Chou, M. Yang, Sum frequency generation vibrational spectroscopy characterization of surface monolayers: Catalytic reaction intermediates and polymer surfaces. J. Surf. Sci. Nanotech. **2**, 106 (2004)
34. D.H. Gracias, D. Zhang, Y.R. Chen, G.A. Somorjai, Surface chemistry-mechanical property relationship of low density polyethylene: IR+visible sum frequency generation spectroscopy and atomic force microscopy study. Tribol. Lett. **4**, 231 (1998)
35. S. Watanabe, M. Nakano, K. Miyake, S. Sasaki, Analysis of the interfacial molecular behavior of n-dodecane containing stearic acid under lubricating conditions by sum frequency generation spectroscopy. Langmuir **32**, 13649 (2016)

Chapter 8
Micro-scale Real-Time Wear Dynamics Investigated by Synchroton Radiation

M. Belin, Jean-Michel Martin, J. Schou, I. L. Rasmussen, R. Feidenhans'l, T. Straasø and N. J. Mikkelsen

Abstract In situ wear measurements on a hard coating of TiAlN and CrN layers deposited on vitreous carbon have been carried out with synchroton radiation. The results show that wear dynamics can be successfully monitored on a lateral micrometer scale and with a submicrometer depth resolution. The wear process is highly irregular and the local wear rate may vary strongly from one position to another in the same wear track. Most of the ridges and grooves are generated within the first 500 nm and exist over several micrometers.

Friction and wear play an overwhelming role in our everyday life from motion of human bodies, car driving, tool handling and manufacturing [1]. Nevertheless, since the contact between two macroscopic bodies is of extremely complex nature, the basic mechanisms at macroscopic level are generally described on empirical basis

M. Belin (✉) · J.-M. Martin
Laboratoire de Tribologie et Dynamique des Systèmes, CNRS UMR 5512, Ecole Centrale de Lyon, 69134 Ecully, France
e-mail: michel.belin@ec-lyon.fr

J.-M. Martin
e-mail: jean-michel.martin@ec-lyon.fr

J. Schou · I. L. Rasmussen · T. Straasø
DTU Fotonik, Technical University of Denmark, Risø Campus, 4000 Roskilde, Denmark
e-mail: josc@fotonik.dtu.dk

I. L. Rasmussen
e-mail: inge.lise.rasmussen@regionh.dk

T. Straasø
e-mail: straasoe@gmail.com

R. Feidenhans'l · T. Straasø
Niels Bohr Institute, University of Copenhagen, Universitetsparken 5, 2100 Copenhagen, Denmark
e-mail: robert@nbi.ku.dk

N. J. Mikkelsen
CemeCon AS, Lokesvej 5, DK–8230 Åbyhøj, Denmark
e-mail: njm@lindberg.com

M. Dienwiebel and M.-I. De Barros Bouchet (eds.), *Advanced Analytical Methods in Tribology*, Microtechnology and MEMS,
https://doi.org/10.1007/978-3-319-99897-8_8

[2–4]. On the nanoscale, the atomic force microscope (AFM) has greatly advanced the investigation of wear. AFM is ideally suited to gain insight in nanoscale wear as the tip can act as a counterpart and simultaneously yield detailed information of adhesion of atoms to the tip and single asperity contact in vacuum experiments [5–7]. In situ fast scans over micrometer areas of the size up to 180 μm^2 have been demonstrated as well [3]. In contrast, macroscopic studies of the surface topography during wear have been carried out primarily with ex situ images from scanning electron microscopy (SEM) and energy dispersive X-ray (EDX) measurements [8, 9]. These techniques hardly allow us to scan a wear track continuously during the material removal or modification by the counterpart, and profilometry at the mm-scale has been conducted [10]. However, the gap between nanoscale AFM-based wear and real-time wear dynamics on a micrometer scale is practically unexplored. In the present work, for the first time, we have measured the wear thanks to a focused X-ray beam of high brilliance, that enables us *instantaneously* to observe the diversity of wear features through different layers of nitrides with micrometer lateral resolution. The extraordinary combination of synchrotron radiation and controlled wear gives us a non-invasive depth resolution of ~100 nm, which is unique compared with macroscopic wear experiments.

Removal of material during abrasive wear by a repeated sliding motion of a counterpart on a solid generates a track with grooves and ridges on the micrometer level. These features are not stationary, but change dynamically under the wear process [10]. In addition, debris from one microstructure can be transferred to other pits or grooves [11, 12]. Systematic macroscopic investigations of wear are complicated because of the statistical fluctuations and because of the comparatively small size (~μm) of the features. Our investigations tell us that once a ridge or a groove of width (>10 μm) larger than the grain size in the counterpart are created, their chance of surviving through several micrometers through layers of different nitrides is large. The interfaces do not significantly disturb the propagation of any of the features. However, a significant fraction of features also disappears within the wear of 1 μm depth.

We have designed a coating with a marker layer of CrN which gives us a significant change in X-ray transmission during thickness loss, and has a significant peak in the XANES (X-ray absorption near-edge structure) spectrum, which disappears when the chromium is removed from the sample. The wear of the samples was carried out in atmospheric air by a nomad instrument, a portable linear reciprocating tribometer constructed for in situ experiments [13], which keeps a constant normal force on the counterpart of 3 N during reciprocating motion. The counterpart itself is a hemispherical molded resin pad with embedded diamond grains (Fig. 8.1). The grains are of size of 2–4 μm (from Diprotex diamond powder) dispersed in a two-component resin (Mecaprex KM-EM and KMR). This mixture produces a wear rate of the hard coating on a substrate within a reasonable period of time. A typical wear time for removing all nitride layers is 6–10 h corresponding to 60,000–80,000 cycles including intermissions for measuring. The counterpart was moving back and forth over the 1.5 mm long track with an average velocity of 15 mm/s and a repetition rate of 5 Hz.

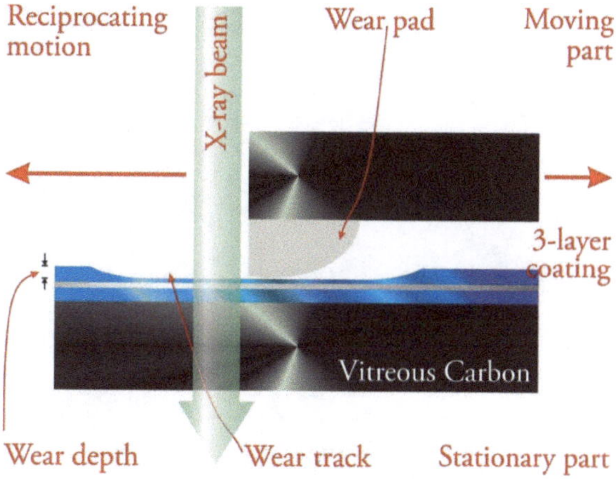

Fig. 8.1 Experimental setup for sample with a 3-layer coating. The thickness of the coatings is exaggerated for clarity (TiAlN, blue, and CrN, grey). The counterpart with the wear pad (white) was mounted on vitreous carbon as well. The absorption measurements including the XANES spectra were performed, when the counterpart was kept outside the beam line as shown in Fig. 8.1

The measurements were carried out at beam line ID 24 at the European Synchrotron Radiation Facility (ESRF) [14] with a beam of a lateral spatial extension of $5 \times 8\ \mu m^2$ FWHM (Full Width at Half Maximum) to pursue a detailed record of wear features at a half-cylinder-like track of length 1.5 mm and width 0.8 mm. The worn sample is kept at exactly the same position, while the counterpart with was moving back and forth in a reciprocating motion (Fig. 8.1). The counterpart was moved to the end of the track every time X-ray transmission and XANES spectra should be recorded.

The substrates were made in vitreous carbon disks (thickness 1.0 ± 0.1 mm, disk-shape of diameter 18 ± 0.1 mm) from SIGRADUR G. Subsequently, a hard coating of nitride layers of thickness from 0.5 to 1.5 μm were deposited on the carbon discs with PVD at CemeCon Scandinavia [15, 16]. The hard coating consisted of a top-layer of $t_1 = 1.51 \pm 0.07$ μm TiAlN (titanium aluminum nitride) on a layer $t_2 = 0.57 \pm 0.03$ μm of CrN (chromium nitride), which in turn covers a bottom layer of $t_3 = 1.27 \pm 0.03$ μm TiAlN (Fig. 8.1). The precise thickness of the layers indicated above was determined from a cross section in a Scanning Electron Microscope.

The wear depth was monitored by the transmission in the region from 5.996 to 6.037 keV (see the indication in Fig. 8.2) in 80 track positions, of which those from 30 to 50 were in the bottom of the track. The instantaneous coating thickness was determined from the transmitted intensity I from the standard expression of absorption:

$$I = I_0 \exp(-\mu_1 t_1) \exp(-\mu_2 t_2) \exp(-\mu_1 t_3) \exp(-\mu_4 t_4) \tag{8.1}$$

where I_0 is the initial intensity, μ_1 the absorption coefficient of TiAlN, μ_2 the coefficient of CrN and μ_4 the coefficient of vitreous carbon, available in [17]. The dominant part of the absorption takes place in the relatively thick carbon, such that only 0.218 of the intensity is transmitted through the carbon disc. The transmission through the coated sample rises from 0.142 at the very beginning until all three coatings are

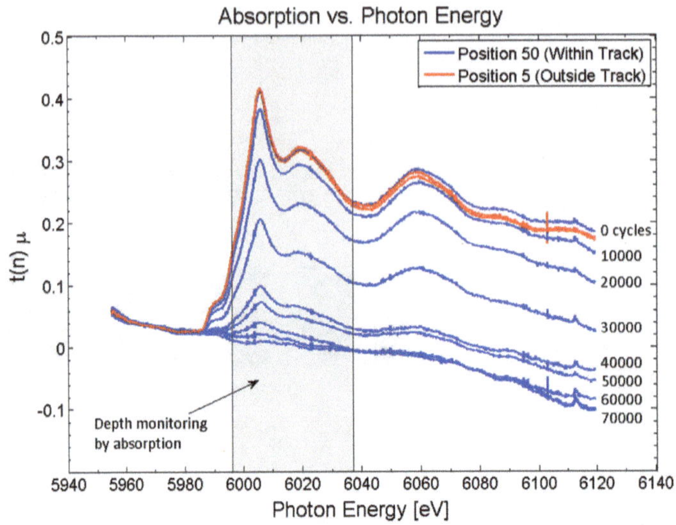

Fig. 8.2 XANES (X-ray absorption near edge spectrum) around the Cr K-edge. The red spectrum with the peak for the edge at 6.007 keV is taken for a position outside the wear track, while the blue one in the middle of the track is recorded for each 10,000 cycles. The hatched area indicates the region in which the absorption measurements were taken

removed during the wear, which is seen when the transmission reaches the value of 0.218. The monitoring of the wear thus relies on a sufficiently distinct transmission change (in our case 0.076), which is limited by the fact that the carbon disc has to be so thick that it can sustain the wear from the counterpart without breaking or bending.

An independent control of the wear depth was acquired with XANES (X-ray absorption near edge spectra) around the chromium *K*-edge (Fig. 8.2) between 5.955 and 5.119 keV. The K-edge of chromium in the nitride binding (CrN) is located about 10 eV higher than the edge of metal chromium—as also reported in the literature [18]). As seen from Fig. 8.2, once the first interface with the CrN layer is reached, the spectrum starts to decrease, and the second interface is reached when the absorption has vanished. The final depth profile of the tracks was controlled by ex situ profilometry with a Dektak instrument.

The cross section of the wear track with increasing depth during wear is shown in Fig. 8.3 as a function of the number of wear cycles. The wear track is about 800 μm wide and monitored with 80 positions across the track ranging from the unworn part on one side to the unworn part on the other side with about 25 positions in the track. Since the distance between each position is less than 20 μm, it means that we identify solely the features on a microscale level. On the other hand, these are the most severe ones for wear control, while asperities of the order of few μm usually are of minor importance.

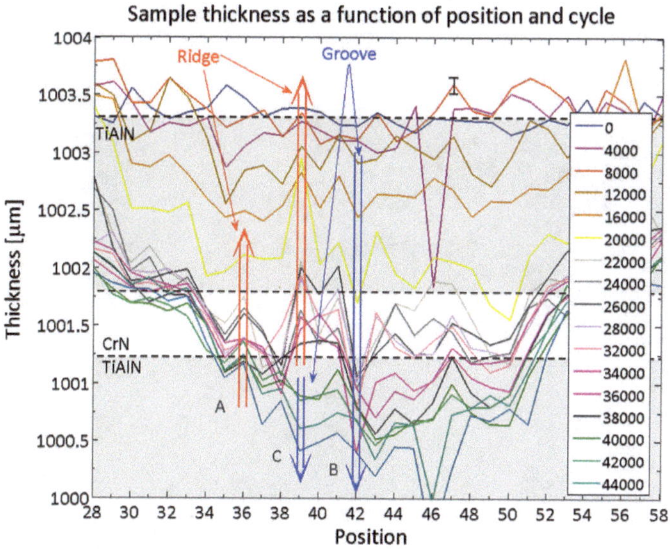

Fig. 8.3 Instantaneous wear cross section of the track. The lines indicate the instantaneous depth after the number of wear cycles for 25 positions across the track. The depth is calculated from (8.1) starting from the surface at 1003.4 μm. The TiAlN-carbon interface is at 1000 μm. "A" is an example of a ridge, "B" a groove and "C" a ridge which changes to a groove during the wear. The distance between each position is 13.5 μm

Within the track there are twice as many ridges as grooves that persist after wear of more than 2 μm. Most of them are more than 0.5 μm higher or deeper than the surrounding level. For each of these major features there is approximately one short-living feature that persists for less than 1 μm depth. In all cases, the interfaces do not play a significant role, but there is a tendency to extend wear in the second layer.

The stochastic nature of the wear process is demonstrated in Fig. 8.4, which shows the instantaneous depth as a function of cycles for three different positions (36, 37 and 42) in the middle of the track (Fig. 8.3). The behavior in each individual position is highly irregular and far from any linear dependence. Initially nothing happens, but after 10,000 cycles the wear starts with a more and less regular depth rate with pronounced deviations from a regular behavior. In some cases the thickness increases locally—probably because of transport of debris [11, 12], which is carried into the position by the counterpart. Generally, in the upper layer of TiAlN there is no substantial difference between the maximum value (0.074 ± 0.01 nm/cycle) and the minimum value (0.065 ± 0.02 nm/cycle) of the wear depth rate, defined as the first time and last time the interface between the upper layer and the CrN layer is reached. This value agrees actually very well with the control value (0.063 ± 0.01 nm/cycle) obtained from the K-edge of the XANES spectrum. The irregular behavior of the wear is even more pronounced in the CrN layer, and each position behaves differently. For the blue squares (position 42) the CrN is worn out in a few thousands cycles, while for green triangles (position 37) the wear rate is more and less constant. However, for the

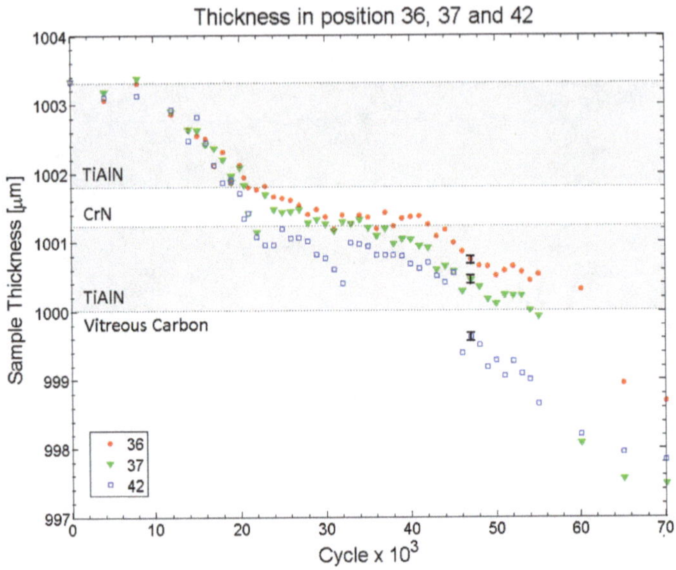

Fig. 8.4 Instantaneous depth for three positions in the track. Red position (36), green position (37) and blue position (42). The wear starts on the surface at 1003.4 μm and with the carbon-TiAlN interface at 1000 μm. The surface and the two other interfaces are shown in the figure. The depth in the vitreous carbon is indicated in an equivalent TiAlN absorption depth, since the exact composition of bottom of the grooves, of debris of TiAlN and carbon, is not known

red circles (position 36) the wear seems to stop around the interface for almost 15,000 cycles, which presumably indicates slow wear and continuous supply of debris. Once the wear has reached the lower TiAlN layer, the wear process continues more or less regularly with similar slope. At the interface between the TiAlN layer and the carbon substrate there is a large amount of material removed within 1000 cycles except for the green position that exhibits large jumps presumably resulting from mass transfer out of and into the position. The depth below the initial interface between the bottom layer of TiAlN and vitreous carbon is indicated in equivalent TiAlN depth, since the groove contains an unknown mixture of carbon and debris of TiAlN. The latter component is clearly observed by SEM as well.

We have shown that wear dynamics can be monitored on a lateral micrometer scale using an X-ray beam in the synchrotron radiation. The wear process is highly irregular and the local wear rate varies strongly from one position to another in the wear track. The ridges, grooves as well as the transport of debris from the track were for the first time followed on a micrometer level, confirming the stochastic nature of this type of wear. Most of the features are generated within the first 500 nm and exist over several micrometers. However, also a significant fraction of grooves and ridges disappear by wear over less than one micrometer. The results from transmission were supported by results from XANES measurements of the *K*-edge of Cr from the-CrN marker layer. The XANES spectra also possess a great potential for identifying oxidation during wear, since the position of the K-edge depends on the chemical environment.

Acknowledgements The authors thank Cornelius Strohm and other staff members at beam line ID 24 at ESRF for competent assistance. This work has been supported by the Danish Strategic Research Council with the NABIIT grant 2106-05-0035.

References

1. B.N.J. Persson, U. Tartaglino, O. Albohr, E. Tosatti, Nat. Mater. **3**, 882–885 (2004)
2. K. Holmberg, A. Matthews, *Coatings Tribology. Properties, Mechanisms, Techniques and Applications in Surface Engineering* (Elsevier, Amsterdam, The Netherlands, 2009)
3. B. Gotsmann, M.A. Lantz, Phys. Rev. Lett. **101**, 125501 (2008)
4. A. Erdemir, C. Donnet, J. Phys. D: Appl. Phys. **39**(18), (2006)
5. H. Bhaskaran et al., Nat. Nanotechnol. **5**, 181 (2010)
6. A. Filippov, V.L. Popov, M. Urbakh, Phys. Rev. Lett. **106**, 025502 (2011)
7. N.N. Gosvami et al., Phys. Rev. Lett. **107**, 144303 (2011)
8. Q. Luo, PEh Hovsepian, Thin Solid Films **497**, 203–209 (2006)
9. X. Li, C. Li, Y. Zhang, H. Tang, G. Li, C. Mo, Appl. Surf. Sci. **256**, 4272 (2010)
10. Q. Luo, Tribol. Lett. **37**, 529 (2010)
11. N. Hiraoka, E. Yamane, Tribol. Lett. **41**, 479 (2011)
12. M. Duarte, I. Vragovic, J.M. Molina, R. Prieto, J. Narciso, E. Louis, Phys. Rev. Lett. **102**, 045501 (2009)
13. L. Joly-Pottuz, J.-M. Martin, M. Belin, F. Dassenoy, G. Montagnac, B. Reynard, Appl. Phys. Lett. **91**, 153107 (2007)
14. S. Pascarelli, O. Mathon, M. Munoz, T. Mairs, J. Susini, J. Synchroton Rad. **13**, 351–358 (2006)
15. I.L. Rasmussen, N. Pryds, H.C. Pedersen, J. Schou, N.J. Mikkelsen, J. Phys. D Appl. Phys. **41**, 135307 (2008)
16. I.L. Rasmussen, M. Guibert, J.-M. Martin, M. Belin, N.J. Mikkelsen, H.C. Pedersen, J. Schou, Tribol. Lett. **37**, 15 (2010)
17. www.cxro.lbl.gov
18. S. Kimura, S. Emura, H. Ofuchi, Y.K. Zhou, S. Hasegawa, H. Asahiet, J. Crystal Growth **301–302**, 651 (2007)

Part III
In-Situ Microscopy and Topography Measurements

Chapter 9
Electrochemical Friction Force Microscopy

Florian Hausen

Abstract Tribology is a very interdisciplinary area of research where physicists, engineers, materials scientists and chemists all contribute to obtain a better understanding and control of friction and wear of surfaces in sliding contacts. As friction and wear is strongly affected by the chemical nature of the interacting surfaces electrochemistry plays an important role in tribology. Within this chapter the use of established electrochemical methods to control surface chemistry and how this affects tribological properties will be discussed.

9.1 Introduction

Tribology is a very interdisciplinary area of research where physicists, engineers, materials scientists and chemists all contribute to obtain a better understanding and control of friction and wear of surfaces in sliding contacts. As friction and wear is strongly affected by the chemical nature of the interacting surfaces electrochemistry plays an important role in tribology. Within this chapter the use of established electrochemical methods to control surface chemistry and how this affects tribological properties will be discussed.

Let us begin with the following question: When is the study of friction forces in an electrochemical environment of relevance? One might divide this question into two parts: On one hand the solid-liquid interface which has been shown to significantly influence friction and sliding and on the other hand the controlled charging of the

F. Hausen (✉)
Forschungszentrum Jülich, Institute of Energy and Climate Research, IEK-9,
52425 Jülich, Germany
e-mail: f.hausen@fz-juelich.de

F. Hausen
Institute of Physical Chemistry, RWTH Aachen University, Landoltweg 2,
52074 Aachen, Germany

F. Hausen
Jülich-Aachen Research Alliance, Section JARA-Energy, Jülich, Germany

M. Dienwiebel and M.-I. De Barros Bouchet (eds.), *Advanced Analytical Methods in Tribology*, Microtechnology and MEMS,
https://doi.org/10.1007/978-3-319-99897-8_9

interface. Regarding the first part, most sliding and hence, frictional processes occur at solid-liquid interfaces, especially when liquid lubricated contacts are involved. Such lubrication was fundamental for the industrialization as it secured long-life of machinery parts by reducing friction and wear. It is still of utmost importance in the world we are living in as demonstrated by increasing sales of lubricants. However, when thinking about smaller scales such as micrometer contacts which are relevant for microelectromechanical systems (MEMS), different strategies must be considered as many surfaces exhibit a thin layer of adsorbed water under ambient conditions. Although this water layer is typically extremely thin, it can significantly alter the friction and wear properties of the materials present in the sliding contact. Coming to the question of charges at the interface: When two non-identical surfaces are sheared against each other, charge separation might occur. This leads to building up of opposite charges on the surfaces of the two materials depending on their electron affinity. The same effect is known from rubbing hair with a balloon. The separation of charges is still present when removing the balloon from the hair causing an attractive force and spiky hair as long as there is no conductive medium in between enabling discharge. It can be easily seen that this effect diminishes as soon as electrically conductive materials or liquids are present. However, most of the conventional lubricants in the market today are mineral or synthetic oils and thus electrically insulating. Buildup of charges in machinery parts can cause severe damages when the potential becomes high enough that a spontaneous and unforeseeable discharge occurs. In such a lightning event extreme temperatures and currents can be reached, destroying the material. Thus, understanding and controlling the effect of surface charges are of significance for tribological applications. Making use of established electrochemical methods enables the experimentalist to easily realize various surface morphologies and structures in a controlled environment. Further, it has been demonstrated in the history of electrochemistry that electrochemical experiments possess comparable control of surface properties with respect to cleanliness as in ultra-high vacuum conditions.

The aim of this chapter is to provide the reader with an overview of current research in electrochemical friction studies. Various effects on friction that have been studied electrochemically are reviewed and recent research as well as open questions are discussed. The overall discussion is focused on experiments performed by means of atomic force microscopy (AFM). Additionally, electrochemical pin-on-disk experiments are included where appropriate. Questions that are addressed include: Does an applied potential to a surface change the overall friction of the tribopair directly? Can electrochemical methods be used to control friction and switch between certain friction regimes? Is the substrate or is the electrolyte more important for controlling friction and what are the underlying mechanisms in each case?

The chapter is organized as follows: First basic electrochemical principles that are crucial for understanding the later described tribological experiments are introduced. Second, a general discussion about electrochemical cell design and the adoption of electrochemical techniques into standard tribological tests is outlined. Subsequently, the main part of the chapter is separated into two parts: Following a discussion of tribological experiments performed in aqueous electrolytes, focusing on ion adsorption and oxidation, ionic liquid electrolytes and their special behavior in electrochemically controlled friction studies are presented.

Despite the fact that corrosion is undoubtedly an electrochemical process and similar techniques are used to study the effect, the subject of simultaneous occurrence of wear and corrosion, known as *tribocorrosion* [1], will not be addressed in this chapter.

9.2 Electrochemical Foundations

When an interface between a solid and a liquid is formed the chemical potentials μ_i of both phases are typically not equal. The system intends to achieve chemical equilibrium and as a result an electrochemical double layer (EDL) is formed at the interface, which consists of charges of opposite sign. In a classical view the interface is characterized by electrostatic interactions between the ions in the electrolyte and the electrode. Hence, the ions are attracted towards the electrode surface as close as possible. According to Helmholtz this picture leads to the formation of two parallel rigid layers of opposing charges, as depicted in Fig. 9.1. However, the formation of a hydration shell can hinder ions in the electrolyte from reaching the electrode surface and force them to remain at a distinct distance from the electrode. The plane through the center of these molecules is called the outer Helmholtz plane (OHP) and is defined by the radius of the non-specifically adsorbed hydrated ions. In the case of a specific adsorption the hydration shell is partially removed and the center of the ions lies in the inner Helmholtz plane (IHP) defined by the radius of the ions. This differentiation becomes important in order to understand the influence of different anions on friction forces.

More developed models of the electrochemical double layer take the thermal motion of the ions into account. Gouy and Chapman concluded that the double layer is completely diffuse with no immobilized counter ions on the electrolyte side, i.e. no IHP or OHP is formed. According to Stern the electrostatic interactions in the first layer are that strong that this layer is still rigid. Hence, this model describes a combination of the Helmholtz and Gouy-Chapman model. All of the presented models of the electrochemical double layer have been developed in order to describe the behavior of ions in diluted aqueous electrolytes. Therefore, these models only consider interactions between ions and the electrode but neglect the interactions between ions themselves. Theoretical work regarding classical models has been reviewed by Henderson and Boda [3].

In addition to the electrochemical double layer, ions, solvent dipoles or molecules can be adsorbed at the electrode's surface by van-der-Waals interactions. Depending on the potential of the electrode such adsorptions can be strengthened, weakened or completely abolished due to Coulomb interactions. The point of zero charge (PZC) of an electrode is defined as the potential where no excess charge is accumulated at the surface or in the double layer. Hence, the exact value of the PZC depends strongly on the nature of each electrochemical system [4]. Generally, it is found that the characteristics of the surface have a larger influence than the electrolyte. As the surface charge is sensitive to the pH of the electrolyte, the PZC can be probed by approaching two surfaces towards each other in solutions of varying pH-values. It was shown experimentally that the forces between the two surfaces approaching each other change from repulsive to attractive at the PZC [5]. The ability to control friction by adjusting surface charges via the pH is illustrated in Fig. 9.5b.

Despite the successful description of EDL for dilute aqueous electrolytes, no classical model of the electrochemical double layer is applicable in the case of ionic liquids [6]. In this special class of electrolytes either a distinct layering of alternating anions and cations is observed or a checkerboard arrangement of anions and cations is formed at charged and solid interfaces, exhibiting an exponentially decaying profile into the bulk liquid. The concept of layering as a generic feature of ionic liquids at solid and charged interfaces, as shown in Fig. 9.2, was firstly introduced by Mezger et al. by using X-ray reflectivity measurements [7]. Since its initial observation it has been regularly observed by Atomic Force Microscopy [8–10] and Surface Force Apparatus [11, 12]. However, the exact nature of the EDL in ionic liquids is still controversial [13].

The potential difference ΔE across the EDL, that is between the bulk potentials of the solution and the electrode is called *galvani potential*. It cannot be measured

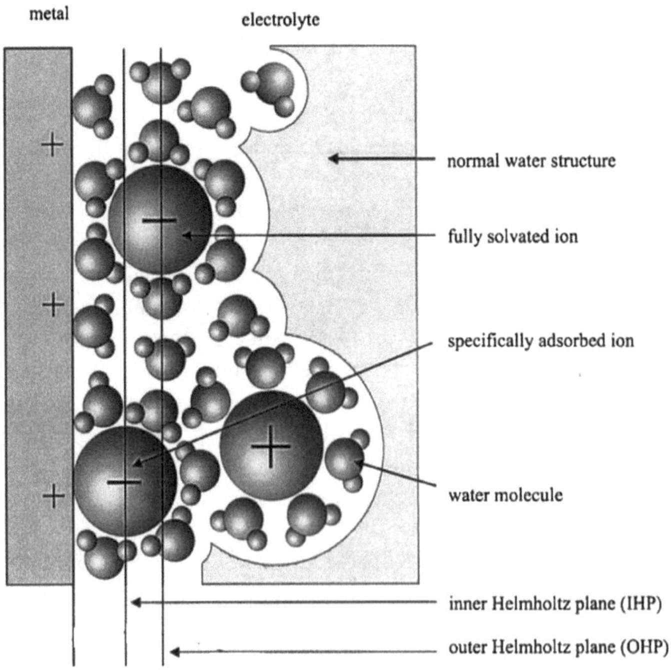

Fig. 9.1 Classical view of the electrochemical double layer (EDL). For details see text. Taken from [2]

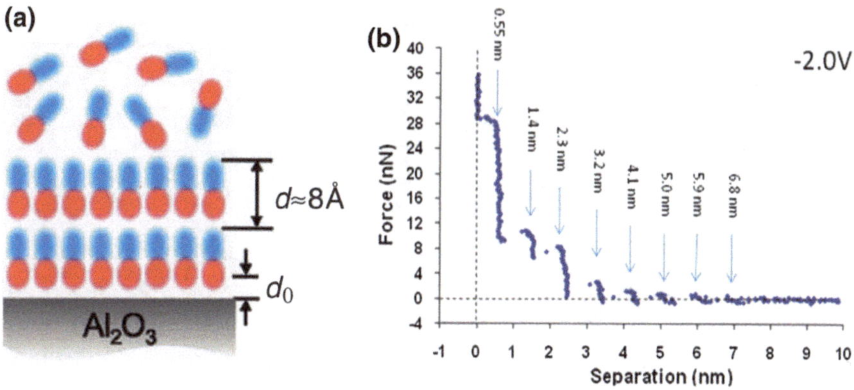

Fig. 9.2 **a** Scheme of the double layer arrangement of ionic liquids, from [7] and **b** as typically observed by AFM, from [9]

directly and a reference electrode with defined potential must be added to the system. As the electrode with a half-cell potential of 0.00 V at standard conditions (activity of protons are 1 mol/L) the standard hydrogen electrode (SHE) was chosen. For a wide variety of electrodes the standard potentials E^0 have been measured against SHE and are tabulated [14]. However, the SHE is not practicable in all experiments and thus a variety of other reference electrodes are commonly used. In aqueous systems reliable reference electrodes are the Saturated Calomel Electrode (SCE) exhibiting a standard potential of +0.24 V and the widely used Ag/AgCl—Electrode in 3 M KCl with E^0 = 0.21 V. In electrolytes containing Cu ions, a Cu wire is often used as reference electrode, resulting in a Cu/Cu2+ reference electrode (E^0 = 0.34 V). In non-aqueous systems potentials are frequently quoted with respect to the ferrocene redox couple Fc^+/Fc with a nominal shift of +400 mV versus NHE. It is important to note that reference potentials depend critically on the experimental conditions [15]. A platinum wire is often introduced as a quasi-reference electrode (QRE) when extreme clean conditions are required. It is frequently used when the work involves ionic liquids where aqueous reference electrodes would cause impurities. Such a QRE should be calibrated against a reference electrode with known potential. However, often only qualitative experiments are performed and the calibration procedure is omitted so that the exact reference potential remains unclear. A discussion about reference electrodes to use with ionic liquids has been published recently by Bonnaud et al. [16].

Cyclic voltammetry (CV) is one of the most prominent techniques to follow electrochemical processes [17]. In a CV experiment the potential applied to the electrode surface, typically being the working electrode (WE), is linearly cycled between an upper and a lower potential value and then reversed with a certain scan rate. In order to measure the potential at the WE accurately, the second electrode is split into the reference electrode (RE) and the counter electrode (CE). This layout is known as three—electrode—arrangement and has the advantage that no current

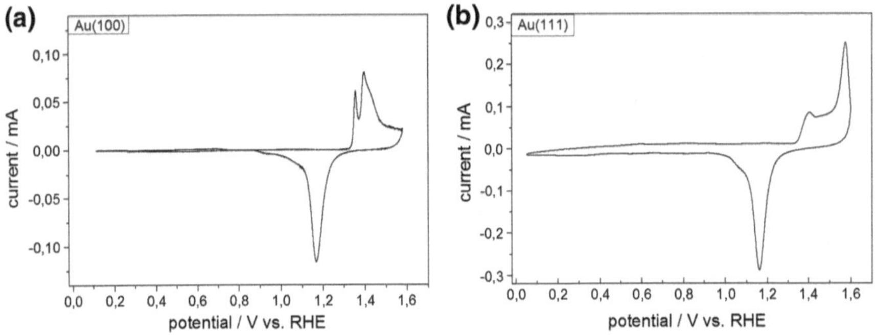

Fig. 9.3 Cyclic Voltammograms of **a** Au(100) and **b** Au(111) single crystal electrodes in 0.05 M H_2SO_4. Scan rate: 50 mV s^{-1}. The difference in the signature of the two crystallographic orientations is remarkable. For discussions of the corresponding peaks see text

flows through the reference electrode. In a two electrode setup the reference electrode is simultaneously acting as CE and RE. All electrodes are connected to a potentiostat which adjusts the measured potential to a predefined desired potential by sending a current through the counter electrode. In a cyclic voltammogram this current is plotted as a function of the applied potential as demonstrated in Fig. 9.3 and is characteristic for each electrochemical system. Each electrochemical process like the formation of adlayers, oxidation steps, or redox reactions causes a distinct signal in the CV. It is worth mentioning that the exchange of electrons between electrode and electrolyte and thus a faradaic current is not necessary for the appearance of peaks in cyclic voltammetry. Capacitive currents originating from a re-orientation of the EDL, e.g. due to the lifting of a reconstruction or phase changes of adsorbates, also generate signals in the respective CV.

As previously mentioned, the peaks in a CV can be related to the processes occurring at the electrode, e.g. the peaks at 1.35 and 1.39 V as well as at 1.40 and 1.57 V for Au(100) and Au(111) respectively are due to an electrochemical oxidation of the gold surfaces. Only the first atomic layers are electrochemically oxidized enabling the possibility to reduce the surface subsequently. Such a reduction corresponds to the peak at 1.16 V and is very similar for both surfaces, as expected from the amorphous character of the gold oxide. Individual aspects of CVs for different electrochemical systems will be discussed where appropriate. A thorough analysis of cyclic voltammograms and the respective underlying reactions in various electrolytes for gold surfaces can be found in [18]. For further insights into electrochemistry in closer detail the interested reader is referred to well-written textbooks on this topic [19, 20].

9.2.1 Electrochemical Cell Design

Nowadays, most AFM manufacturers offer liquid cells and electrochemical standard cells as an optional add-on to their commercial systems. Such cells are often designed to match the instrumental requirements very well but do not allow for more sophisticated electrochemical conditions like exchange of electrolyte during experimentation or special electrode geometries, which might be required. This has led to a variety of home-build electrochemical cells for electrochemical AFM (EC-AFM) measurements. Generally, an electrochemical cell is a container in which electrodes can be fixed and immersed in the electrolyte. The geometry of the electrodes should ensure a homogeneous electric field between WE and CE and specified distance between WE and RE. Thus, the perfect electrochemical cell must be adapted for individual needs and instrumentation. As a general guideline the counter electrode should be placed parallel or in form of a ring surrounding the working electrode. Furthermore, it is advantageous if the CE is larger in size compared to WE and from a material already used in the setup as contamination might occur by dissolving ions from the counter electrode. As the counter electrode must only supply currents, the actual material it is composed of is not critical. Much more care with respect to impurities must be taken with the reference electrode as it might contain unwanted ions. The electrode of choice must ensure a stable potential throughout the experiment which in some cases requires putting the reference electrode in a separate compartment of the electrochemical cell. Even very small amounts of impurities can contribute significantly to the overall shape of a cyclic voltammogram. A good example for this effect is the appearance of Fe^{2+}/Fe^{3+} redox-peaks at around 0.5 V in CVs of Au(111) single crystals caused by the release of iron from the cantilever holder of the AFM into the solution as recognizable in Fig. 9.7b and c. As cleanliness is crucial in electrochemical experiments any contamination must be avoided. Thus, the electrochemical cell itself should also be easily cleanable.

A good summary of design principles and how to incorporate them in electrochemical cells for scanning probe microscopy is given by Valtiner et al. who introduced a very versatile still widely applicable electrochemical cell [21]. In principle, the same considerations hold for implementing an electrochemical cell in a pin-on-disk configuration or for implementing EC capabilities into the Surface Force Apparatus. Pin-on-disk tribometers with included electrochemical capabilities have been realized by Argibay et al. [22] and Ismail et al. [23]. Also examples of electrochemically modified Surface Force Apparatus for friction studies are realized [24].

A good starting point for friction force microscopy experiments using a commercial AFM under controlled potentials would be to use the manufacturer's electrochemical cell and adapt it subsequently to individual needs. Labuda et al. went one step further and introduced an atomic force microscope which was especially designed for friction force microscopy under potential control with extremely high resolution down to the atomic scale as depicted in Fig. 9.4 [25].

Even though it might be convenient using a built-in potentiostat provided by the AFM manufacturer in order to control the WE potential, it is not desired. As elec-

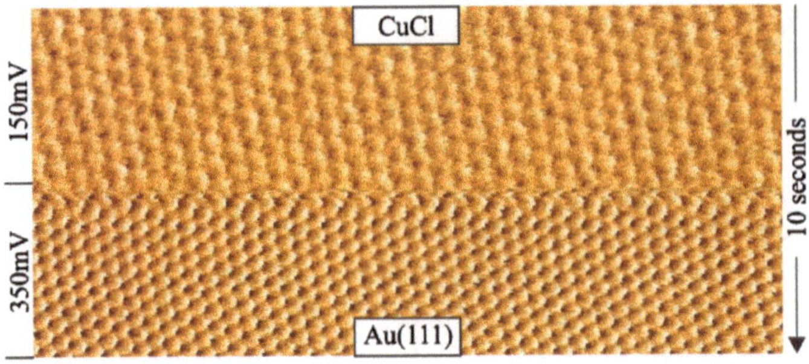

Fig. 9.4 High resolution lateral force map of the desorption of a copper adlayer on Au(111) by switching the applied electrode potential from 150 to 350 mV at 2.0 nN normal load (Electrolyte: 0.1 M $HClO_4$ + 1.5 mM $Cu(ClO_4)_2$ + 10^{-5} M HCl, Potentials are given with respect to the Ag/AgCl reference electrode). The respective structure and difference in lattice dimensions is clearly visible. From [25]

trochemical control of the sample is in principal separated from the operation of the AFM, all commercially available potentiostats can be used. The electrochemical signals can be recorded separately from the AFM environment but being displayed by using auxiliary input channels as supported by most AFMs. The latter allows the experimentalist to combine all signals spatially and temporally to precisely reveal the potential/current—structure/friction relationships for a given system. It is of importance to realize that the electrochemical signals are typically resulting from the entire sample (or even include artifacts from the cell design) rather than a local variation as probed by the AFM tip. Depending on the exact conditions the surface area exposed to the applied pressure of the tip might exhibit variances in electrochemical behavior compared to the overall sample. As is the case for most friction force microscopy experiments, contact mode cantilevers with rather small force constants should be employed. While it is common to use coated cantilevers in order to increase the reflectivity and gain a more intense signal on the photodiode, particularly if the cantilever is immersed in liquids, such metallic adlayers (usually aluminum or gold) might be attacked by the electrolyte resulting in dissolution or delamination of the coating. In both cases the properties of the cantilever will likely vary over the course of an experiment, making the interpretation of the results complex. Hence, special care must be taken when using such cantilevers or only uncoated cantilevers should be used in EC-FFM experiments. Secondly, it is important to think about the electrical conductivity of the cantilever. Electrostatic interactions may contribute significantly to the recorded forces. Such electrostatic interaction influences the overall normal load which affects the measured friction and can vary as a function of pH for identical surfaces. In order to diminish electrostatic interaction the ionic strength of the electrolyte should be rather high (>50 mmol) and the cantilever should be made of highly doped material enabling dissipation of charges.

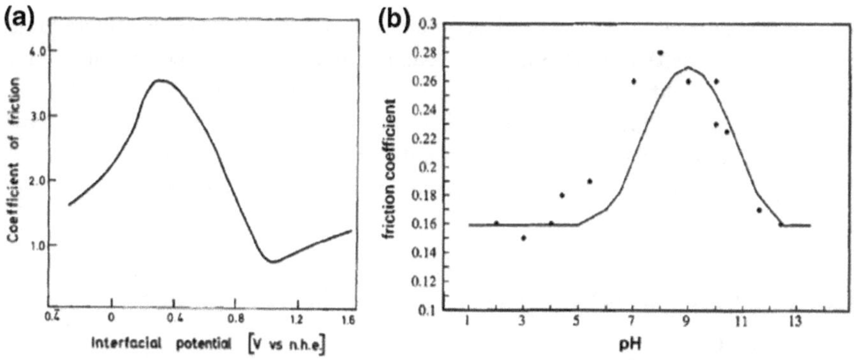

Fig. 9.5 **a** COF between two Pt sheets as a function of applied potential in diluted sulfuric acid. From [29] **b** COF between Al_2O_3 and Fe_2O_3 as a function of pH. The solid line represents the calculated COF while the points are measured values. For discussion see text. From [30]

9.3 Historical Development

The interest in the role of an applied potential to an electrode on sliding friction, nowadays consolidated in the term "triboelectrochemistry" [1, 22, 26] dates back to the pioneering work of Thomas Edison in 1875 [27]. In his patent for an improved telegraph he described that a potential-induced decomposition of a surface oxide resulted in a change of friction and thus, the mechanical forces of moving surfaces. One of the first scientifically motivated studies was conducted in the mid-20th century when the coefficient of friction (COF) measured between two platinum surfaces immersed in 0.1 N (or 0.05 M) sulfuric acid was observed to depend significantly on the applied potential [28]. As depicted in Fig. 9.5a the COF is maximized at potentials where no ions are adsorbed on the surface (at about 0.35 V vs. NHE). For both, larger cathodic or anodic potentials, i.e. with adsorption of layers of hydrogen or oxygen, respectively, the coefficient of friction was observed to decrease.

As early as in 1970 R.B. Waterhouse stated in his review entitled "Tribology and Electrochemistry" that chemisorbed adlayers on electrodes are expected to influence the COF more strongly than physisorbed molecules [29]. This statement will be reviewed in the discussion of anion adsorption onto electrodes. Furthermore, Waterhouse concluded, based on the result of a potential-independent COF in the presence of adsorbed impurities that the nature of the adsorbate is of major importance and not the applied potential itself.

Kelsall et al. studied the influence of potentials between an Al_2O_3 pin on a plate of various metal oxides as a function of pH [30]. The authors developed a mathematical equation to relate the electrostatic force between two surfaces to the coefficient of friction. In Fig. 9.5b the calculated coefficient of friction is shown next to the measured data for the combination of Al_2O_3 on Fe_2O_3 as a function of pH. The PZC of Al_2O_3 is at pH = 9 and the PZC of Fe_2O_3 is at pH = 6.3. Hence, for pH values between the PZCs, a strong attractive force is acting resulting in an "additional" normal load and thus higher friction. The work clearly demonstrates that electrostatic interactions are an important aspect in tribology. Xie et al. reviewed the influences on lubrication under charged conditions by separating them according to different lubrication states [31]. Controlling surface charges by pH is one way to influence friction, however, the use of classical electrochemical methods allow for a direct control of surface charges precisely without changing the composition of the electrolyte.

9.4 Friction Force Microscopy Under Electrochemical Conditions

9.4.1 Aqueous Electrolytes

Soon after the invention of the atomic force microscope in 1986 by Binnig, Quate and Gerber [32] the potential of such apparatus for studying friction at the nanoscale was first demonstrated by Mate et al. [33]. In their ground-breaking work the friction between a tungsten tip and a graphite surface was measured under ambient conditions, allowing the authors to visualize the atomic structure of the graphite lattice the friction signal of the AFM. The same surface was examined for electrochemical friction force microscopy in 0.1 M $NaClO_4$ solution as pioneered by Binggeli et al. [34, 35]. In these early electrochemical friction force microscopy (EC-FFM) experiments the friction at monoatomic high step edges was analyzed and separated into a geometrical and a chemical contribution. The authors found that the overall friction is predominate by the chemical component and attributed this to a ploughing effect of adsorbed water layers under the assumption that the resistance of such near-surface water layers to ploughing is potential dependent. A similar argument of increased electro-viscosity has been adopted by Valtiner et al. who found increased friction for mica—gold contact in 10^{-3} M nitric acid upon charging the gold surface to 400 mV [24]. The authors could exclude changes in the separation between the sliding surfaces by following the absolute distance between mica and gold in situ by means of surface force apparatus.

Weiland et al. extended the previously described experiments on HOPG in 0.1 M $NaClO_4$ and analyzed friction at atomic steps under applied potentials of 1.0 and −1.0 V [36]. While the topography remains unaffected by a change of the potential the friction force at the step edge is significantly altered. This result was later attributed to the intercalation of perchlorate ions into the steps of HOPG as a function of potential [37]. In agreement with this conclusion no change of friction was observed on the basal plane. Thus, the applied potential itself does not change the sliding between the AFM tip and the substrate in any form as it was already concluded by Waterhouse [29]. Weiland et al. also found that friction increased independent whether the tip slips up or down the step [36]. Similar observations have been made for monoatomic high steps on Au(111) in sulfuric acid [38] and for steps up to 5 layers in height on graphite, MoS_2 and NaCl under ambient conditions [39, 40].

The effect is attributed to the Schwoebel-Ehrlich barrier which is known to hinder diffusion across step edges [41, 42]. This impressively illustrates that friction is very sensitive to surface effects. However, if the step height becomes considerably larger than atomic dimensions, topographic influences dominate as suggested by Sundarajan et al. [43]. HOPG is among the best examined system in electrochemical friction studies beside gold surfaces. The reason for the popularity of gold lies in its well-known electrochemical behavior which offers numerous possibilities to alter the surface properties in situ [44–46].

9.4.1.1 Deposition of Ionic Adsorbates on Surfaces

One advantage of electrochemistry is the possibility to adsorb (sub)monolayers of cationic or anionic adsorbates onto electrodes in a very controlled way by the applied potential [44, 47]. For studying fundamental friction processes, this effect can be used to deposit and alter surface films in a distinct way.

As discussed in the electrochemical foundation section anions can adsorb onto an electrode at various potentials in the inner (IHP) or outer Helmholtz plane (OHP). Even at a negative electrode anions can adsorb due to attractive van der Waals interactions. The effect of anions being adsorbed in the IHP or OHP on friction has first been described by Kautek et al. who studied the adsorption of halogenides on silver by means of friction force microscopy [48]. Only specifically adsorbing anions in the IHP, such as bromide and iodide, showed an influence on friction while fluoride as weakly bound species in the OHP cause no significant change in friction. The authors assumed an easier displacement of the weakly bound anion by the AFM tip compared with the greater required force to displace specifically adsorbed ions, which hinder the movement of the tip, enhancing dissipation.

Another example of a strongly binding anion in the inner Helmholtz plane is the adsorption of sulfate onto gold electrodes which is a well-studied system in electrochemistry [45, 49]. In order to yield a more comprehensive view of the situation the sulfate ion SO_4^{2-} and the isoelectric perchlorate ion ClO_4^- is compared. Both ions are used in standard electrochemical electrolytes. The adsorption of sulfate onto Au(111) can be monitored by cyclic voltammetry as depicted in Fig. 9.6. The black curve shows the same CV as in Fig. 9.3a while the red curve is a magnification of the potential region between 0.0 and 0.9 V versus Ag/AgCl. At 0.3 V a first peak in the anodic scan occurs that can be attributed to the lifting of the reconstruction of the Au(111) surface. This peak is followed by a very broad peak between 0.4 and 0.7 V corresponding to the adsorption of sulfate onto the electrode in a random fashion. At 0.85 V a small spike is detected in the CV which is assigned to a phase transition of the adsorbed sulfate forming a regular $\left(\sqrt{3}x\sqrt{7}\right)R19.1°$-adlayer [50] which can be probed by EC-STM [51] and EC-AFM [52]. The cathodic scan demonstrates the reversibility of the transitions.

With the adsorption of sulfate an increase of friction was observed at rather high normal loads of 65 nN [38]. By means of high resolution EC-FFM the same system

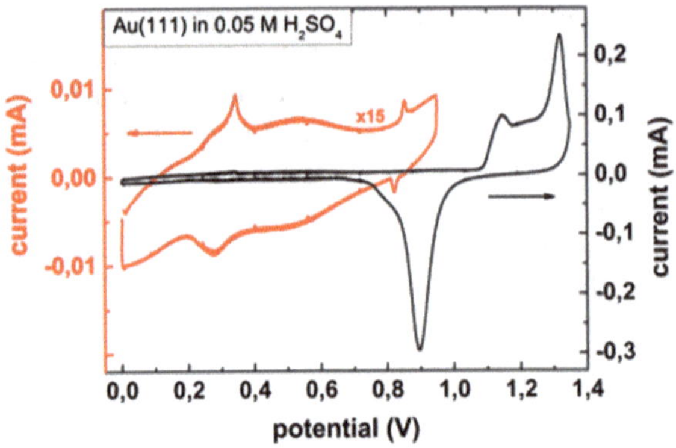

Fig. 9.6 Cyclic voltammograms of Au(111) in 0.05 M H_2SO_4. Potentials are given with respect to Ag/AgCl. The red curve represents a magnification of the CV between 0.0 and 0.9 V in order to visualize the signatures of sulfate adsorption in CV. For discussion see text after [52]

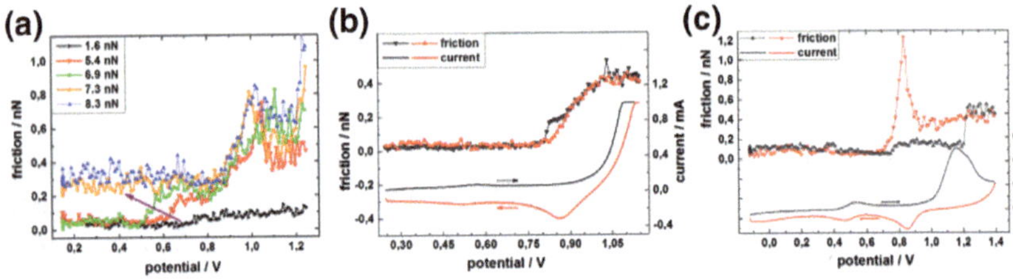

Fig. 9.7 **a** Dependency of friction on applied potential and normal load for Au(111) in 0.05 M H_2SO_4, potentials are given versus Ag/AgCl. After [52] The arrow indicates the shift of the onset of friction towards more cathodic potentials with increasing normal load. **b** Frictogram of Au(111) in 0.1 M $HClO_4$ at a normal load of 6.5 nN. Friction remains unchanged in a potential regime between 0.3 and 0.8 V. After [53] **c** Frictogram of Au(100) in 0.05 M H_2SO_4 at a normal load of nN. Sulfate is less strongly adsorbed compared to the Au(111) surface and no clear effect on friction is observed

has further been investigated in greater detail [52] and it was revealed that the detection limit of friction is a complex interplay between applied potential and normal load as illustrate in Fig. 9.7a. While for low normal loads of 1.0 nN no modulation of friction resulting from adsorption of sulfate on the gold surface was detected in the potential regime between 0.2 and 0.6 V, the onset of friction for higher normal loads is shifted to lower potentials. The effect was attributed to forcing the sulfate adlayer into the regular $\left(\sqrt{3}x\sqrt{7}\right)R19.1°$-adlayer under the pressure of the tip. This conclusion was supported by atomically resolved lateral force maps of the sulfate super lattice at rather negative potentials at which this structure cannot be resolved by EC-STM methods [52]. As it can also be seen from Fig. 9.7a the increase of friction due to the electrochemical oxidation of the gold surface is not influenced by the normal load and occurs for all loads at about 0.8 V.

How can it be proven that the observed potential dependent characteristics on friction are indeed a signature of the adsorption state of the anion? The perchlorate ion is isoelectrical to the sulfate ion but undergoes only a weak adsorption on Au(111). Labuda et al. present frictograms for Au(111) in 0.1 M $HClO_4$ as shown in Fig. 9.7b

[53]. Throughout this chapter the term frictogram will be used for simultaneously recorded CVs and friction data if presented in an overlaid diagram such as shown in Fig. 9.7b and c. The authors cannot find any detectable influence on friction by the weakly adsorbed perchlorate ion.

Another way of verifying the importance of specific binding of ions to influence friction is to utilize a Au(100) surface on which the sulfate anion binds less strongly because of geometrical constraints. When examined on this surface, no influence on friction was found for sulfate adsorption on Au(100) as revealed by frictograms of the system as depicted in Fig. 9.7c.

For the case of cationic adsorbates the deposition of copper on gold surfaces is among the best studied system and act as a model system as it is electrochemically well understood. Figure 9.8b shows a cyclic voltammogram of the deposition of copper on Au(111) and the corresponding friction values for various normal loads. We begin the discussion of the CV at a potential of 0.4 V versus Cu/Cu^{2+} where a bare gold surface is present as also indicated by the high-resolution friction force map shown in Fig. 9.8a. When lowering the potential from 0.4 V a first broad peak is visible which is assigned to the formation of a submonolayer Cu on Au(111) [54]. It was found that sulfate is co-adsorbed onto this submonolayer [54]. In the friction force map a change of the structure of the glide plain is visible with a clearly larger lattice constant (potential regime 0.17–0.03 V). Shortly before 0.0 V versus Cu/Cu^{2+} a full monolayer of Cu is deposited on the gold electrode.

The frictograms depicted in Fig. 9.8b reveal various interesting characteristics of the Cu/Au system when sliding against a silicon AFM tip. Comparing the frictograms at various normal loads, the transitions between distinct friction regimes can be correlated with the peaks in the cyclic voltammogram. This correlation highlights that the applied potential to the electrode does not change the overall friction of the tribo-pair. Modulations of friction are resulted by electrochemically induced changes of the surface structure or composition. More specifically, friction depends critically on the adsorbed species (bare gold—submonolayer Cu with coadsorbed sulfate—monolayer Cu). Friction is found to be low on the submonolayer coverage until the monolayer is formed shortly before the applied potential reaches 0.0 V. The formation of the monolayer is accompanied by an increase in friction. The higher level of friction is maintained when reversing the potential sweep until the monolayer is dissolved at about 0.07 V. Interestingly, during the transition from submonolayer coverage to bare gold a very strong increase in friction is observed before the same friction level as before the electrochemical cycle is reached on bare Au. A clear explanation for this behavior is still lacking but similar observations of an increase in friction during electrochemically induced reformation of a bare gold surface can be found in [38] as well as for the reduction of gold-oxide on Au(100) as shown in Fig. 9.7c.

When comparing the friction versus normal load behavior of the submonolayer and the monolayer of Cu on Au, a clear deviation is observed as illustrated in Fig. 9.8c and d. While friction remains low and almost independent of normal load for the submonolayer a linear increase of friction with normal load is observed for the full monolayer coverage. This result indicates that the coadsorbed sulfate plays a vital

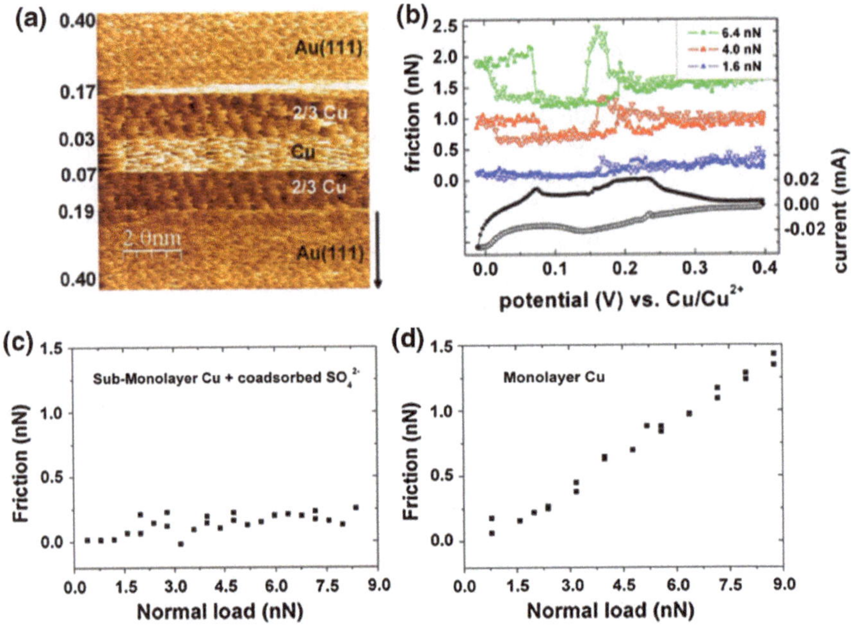

Fig. 9.8 **a** High resolution friction force map of the deposition of Cu on Au(111). The numbers on the left indicate the applied potential. **b** Corresponding frictograms recorded at various normal loads as indicated. Open symbols correspond to the cathodic shift of the potential, closed symbols correspond to the anodic shift of the potentials. **a** and **b** are taken from [55]. **c** The friction versus normal load plots of the submonolayer coverage and the monolayer coverage of copper on gold reveal distinct differences and indicate a strong lubricating effect of the coadsorbed sulfate layer on the submonolayer of copper on gold

role, acting as an effective lubricant for low loads. Nielinger et al. reported friction results for Cu deposition on Au(111) for considerably higher normal loads of 30 and 120 nN [56]. A strong increase in friction for all examined copper coverages on Au(111) was observed. However, the distinction between submonolayer and monolayer copper coverage was only possible at 120 nN. This observation was rationalized under the hypothesis that variations in the measured coefficient of friction as a function of normal load was likely the result of the AFM tip penetrating the lubricating sulfate adlayer.

9.4.1.2 Oxidized Surfaces

When studying tribological interactions oxidized surfaces, either formed naturally or intentionally, are often of particular interest because of their typically high hardness and wear resistance. Oxides are also easily accessible under electrochemical conditions. Friction studies using electrochemical oxidation has been reported for various

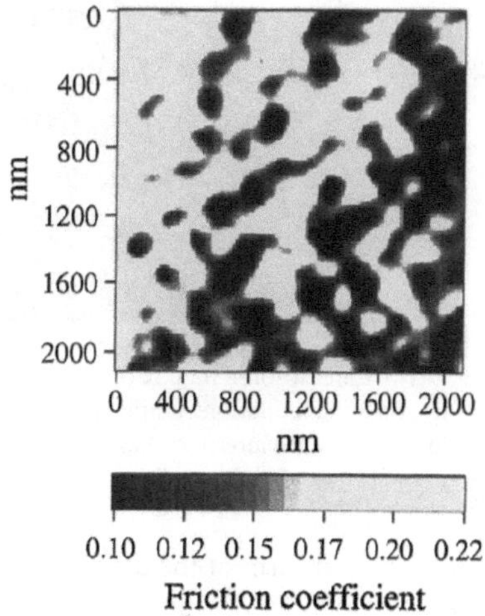

Fig. 9.9 AFM friction image of an iron surface in 0.02 M $NaSO_4$ (pH = 10) after three potential cycles between −1.2 and 0.4 V versus SCE. As clearly visible a very heterogeneous surface composition with respect to the COF arose after electrochemical treatment. From [30]

metals like Fe [26], Ta [57], Cu [22] or Au [53]. When iron is immersed in 0.1 M NaOH electrolyte it can be oxidized in a two-step process to FeOOH which can be easily followed by CV [26]. While the pure Fe surface exhibits relatively high COF of 0.15 Zhu et al. find a strong decrease to $\mu < 0.1$ accompanied with a smoothening of the surface [26]. As shown in Fig. 9.9 the authors find patches of low and high friction values next to each other across the surface after three voltammetric cycles. When forming a Fe-octanate rather than the peroxide the authors found only a less pronounced decrease of the COF to about 0.05 but could follow the growth of a surface film modifying friction.

This work impressively shows that friction is affected in several directions under electrochemical control. In addition to the change of the chemical nature of the surface film, that is being the bare metal, an oxide film or an adsorbant, also the roughness of the surface can influence the overall friction. The spatial distribution of a film generated as a result of the chemical reactions mediated by the shearing action of the two surfaces, a *tribofilm*, must also be taken into consideration. Labuda et al. studied the electrochemical oxidation of gold and observed roughening of the surface upon oxidation as well as a strong increase in friction as compared to the bare gold surface [53]. As the electrogenerated gold oxide is only in the range of a few atomic layers the authors could demonstrate an impressive reversibility between low and high friction states corresponding to bare or oxidized gold surface. High resolution friction force maps are shown in Fig. 9.10 and reveal the lattice parameters for the Au(111) surface and a roughened structure of the gold oxide.

The authors attributed the higher friction observed on the gold oxide to its increased roughness. This experiment illustrates that the structure of the surface is as important as its chemical nature, which is further underlined by friction experiments by Hausen et al. on Au(100) in sulfuric acid electrolyte [58]. The structure of the Au(100) surface was electrochemically controlled between a so-called pseudo-hexagonal phase at low potentials (<0.26 V vs. Ag/AgCl) which corresponds to the

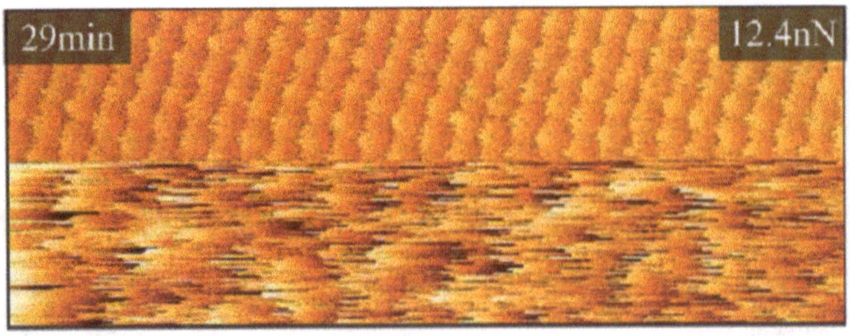

Fig. 9.10 Lateral force map revealing the atomic structure of bare Au(111) (upper part) and after electrochemical oxidation. While the Au(111) shows a regular lattice, the oxidized gold film exhibits irregular stick slip pattern, revealing its amorphous character. From [53]

reconstructed form of the electrode and the normally expected quadratic 1×1 structure at potentials positive of 0.26 V versus Ag/AgCl. Upon the transition, which can be followed by CV, the chemistry of the surface remained unchanged while the atomic arrangement was alternating between a more open 1×1 structure and a densely packed hexagonal form.

In summary, it has been identified that friction can be controlled electrochemically by intercalation and adsorption of ions to electrode surfaces while the adsorption state determines the effect on friction. The chemical nature of surface adlayers contributes to the overall friction and leads to spatially different patterns when adsorbed heterogeneously. At the same time, such inhomogeneities can introduce roughness to the sliding surfaces typically correlating with enhanced friction. Roughness plays a crucial role in controlling friction electrochemically on different length scales from subtle effects of reconstructed surfaces to significant changes of the roughness when electrochemically roughening a surface by oxidation. It is striking that friction was always observed to increase except in the case of electrochemically induced oxidation of iron. The difference in that one case is that the surface was smoothened upon oxidation forming a more "perfect" surface afterwards, while in all other cases the surface was roughened. Given that in real applications the expectation of a perfect surface is rather low indicates that electrochemical methods might open paths to not only modify but also to reduce friction during the application by targeted surface finishing.

All the processes mentioned so far clearly demonstrate that friction is extremely sensitive to subtle changes of the sliding contact which sets the requirements for a controlled environment. However, for most realistic applications it is typically unwanted to alter the glide plane dramatically, or it is not feasible to change the chemistry or structure of the glide plane due to boundary conditions given by the surrounding. This brings us to the idea of not changing the surfaces in contact through the electrochemical potential but rather finding a way of controlling the properties of the electrolyte, or more specifically in tribology the lubricant, by electrochemical means.

9.4.2 Ionic Liquids as Electrolytes

Ionic liquids (ILs), despite having been known for more than hundred years [59], are recently gaining a lot of interest in tribology [60–63], in energy storage applications [64], as solvents in synthesis [65], in catalysis or for cellulose [66], as well as in electrochemistry [67]. ILs show remarkable properties such as non-flammability, high resistivity against oxidation and reduction, as well as negligible vapor pressure and high thermal stability. ILs solely contain ions and hence, they are conductive and offer a broad electrochemical window. When working with ionic liquids the quantity of available compounds and the endless possibilities in tailoring task-specific new ionic liquids make a proper selection of the IL ideal for the system challenging. In the last 15 years the interest of using ILs in tribological applications has increased due to their special properties and strong film formation capabilities. The anti-wear and lubrication properties have been attributed to the charged induced adsorption of anions and cations onto the sliding surfaces, building up a tribofilm. Most often hydrophobic anions such as Bis(trifluoromethylsulfonyl)imide (Tf_2N) or Tris(pentafluoroethyl)trifluorophosphate (FAP) are used for EC-FFM experiments in ILs. Historically the first air-and humidity stable ILs, so-called second generation of ILs, consist of Tetrafluorobarate (BF_4) and Phosphorhexafluorate (PF_6) anions and imidazolium-based cations [68], and are therefore widely distributed. A drawback of these ions for tribological applications is their tendency of (partial) hydration when in contact with water, forming corrosive HF. Table 9.1 illustrates the chemical structure of common anions and cations for ionic liquids.

Figure 9.11 shows friction force maps of Au(111) in the ionic liquid 1-Butyl-1-methylpyrrolidiniumtris(pentafluoroethyl)trifluorophosphat ($[Py_{1,4}]$FAP) at −2 V and 0.0 V versus Pt-QRE. Monoatomic high steps are present at the electrode surface as well as some islands, most probably generated by lifting the herringbone reconstruction of Au(111), are visible in Fig. 9.11a. The color represents the overall friction with brighter colors corresponding to higher friction. Upon changing the electrode potential by 2.0–0.0 V the overall friction becomes larger and the image gets blurred (Fig. 9.11b). However, the monoatomic high steps on the gold surface are still visible, clearly indicating that the observed change in friction as represented by the change in color is caused by changes of the ionic liquid electrolyte. It is worth mentioning that such a potential step in aqueous electrolytes would cause the formation of gases due to the limited electrochemical window of water.

As discussed previously ionic liquids form alternating layers of anions and cations on charged surfaces [69]. Without controlling surface charges Smith et al. showed that friction between two atomically smooth mica surfaces depended strongly on the number of layers between the interacting surfaces [70]. As a consequence, various friction regimes exist for a given load. The ability of ionic liquids to form mechanically stable layers adjacent to charged surfaces and there conductivity, resulting from the fact that they solely contain ions, makes it straightforward to control the surface layers by an applied electrode potential. It has been shown by Hayes et al. that indeed the interfacial structure of ionic liquids is sensitive to the potential of

Table 9.1 Common cations and anions for ionic liquids

Cations	
Immidazolium – based $[R_1R_2IM]$	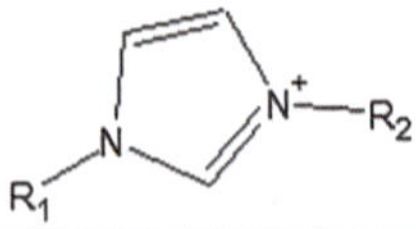
Pyrrolidinium – based $[Py_{R1R2}]$	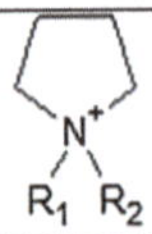
Anions	
Tetrafluoroborate (BF_4)	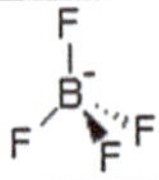
Hexafluorophosphate (PF_6)	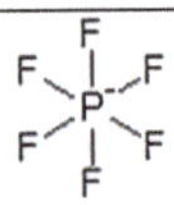
Bis(trifluoromethylsulfonyl)imide $(NTf_2$; TFSI)	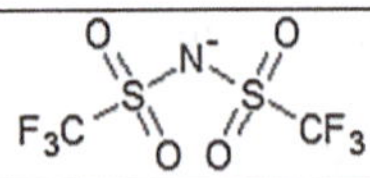
Tris(pentafluoroethyl)trifluorophosphate (FAP)	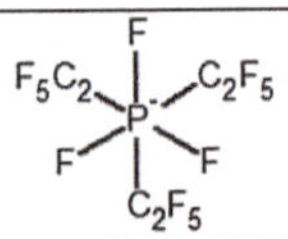

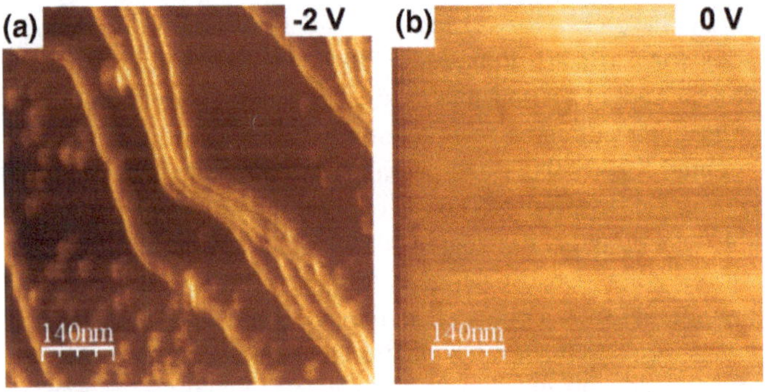

Fig. 9.11 Friction force maps of Au(111) in $[Py_{1,4}]$FAP at various applied potentials. Monoatomic high steps are clearly visible at −2 V versus Pt-QRE. The color represents the friction with brighter colors higher friction. For 0 V versus Pt-QRE the overall friction is clearly increased compared to −2 V as indicated by the color. However, the surface steps are still visible indicating that the change in friction is not resulted by a modification of the substrate but rather originated from changes in the ionic liquid interfacial layer

the electrode [9]. Numerous theoretical works have verified an oscillating nature of anion and cation layers at charged surfaces, either electrochemically controlled or naturally charged [69, 71–73].

The first tribological study of ionic liquids by means of EC-FFM was presented by Sweeney et al. in 2012 where the authors analyzed the friction between a gold surface and silicon AFM-tips in [$Py_{1,4}$]FAP [74]. A clear relation between the applied potential to the gold surface and friction was observed. As shown in Fig. 9.12a low friction was measured for low potentials (i.e. < 1.5 V vs. Pt-QRE) whereas friction was considerably higher for positive applied potentials. As no other substances were present in the electrolyte, the authors attributed the effect to changes in the interfacial structure of the ionic liquid and the intrinsically different lubrication properties of the respective anions or cations. It is worth mentioning that the cyclic voltammogram of [$Py_{1,4}$]FAP exhibits various peaks in the cathodic as well as in the anodic scan. As it can be clearly seen in Fig. 9.12a friction does not change linearly as a function of applied potential but corresponds to the peaks observed in the electrochemical signal. Such peaks in the CV are likely caused by re-orientations of ions of the ionic liquid at the surface of the electrode. Sum frequency generation (SFG) measurements have shown that the tilt angle between the ionic liquid cation and the surface normal direction depends critically on the applied potential at the electrode [75, 76]. A detailed study of the influence of the tilt angle on friction has recently been provided by Watanabe et al. in a pin-on-plate configuration using various imidazolium-based ionic liquids, exhibiting different tilt angles due to attractive electrostatic forces [77]. The authors found a strong correlation between tilt angle and friction, where larger tilt angles resulted in lower coefficients of friction. However, various other factors of the individual interfacial properties of the tribopair, such as the roughness of the shear plane or molecular densities of the adsorbed layers and thus variations in ion mobility, were also discussed and may contributed to the observed friction. This difficulty is further underlined by experiments conducted by Li and coworkers [78, 79].

In a first set of experiments the authors investigated the influence of the cation alkyl chain length by studying friction as function of normal load on Au(111) in 1-ethyl-3-methyl-imidazolium ([EMIM]), 1-butyl-3-methyl-imidazolium ([BMIM]) and 1-hexyl-3-methyl-imidazolium ([HMIM]) in combination with the tris(pentafluoroethyl)trifluorophosphate (FAP) anion. The results are depicted in Fig. 9.12b. In all cases, the highest friction was found at positive applied potentials, corresponding to an anion enriched interface. The authors attributed this to a relatively weak interaction of the FAP anion with the gold surface and disordered coil morphology. Such disorder results in a rougher shear plane of the sliding contact and thus, high friction. For the stronger adsorbed imidazolium-based cations, the trend is less clear, which was explained by the influence of the cation chain length with respect to the tilt angle and the formation of more resistant shear planes. In Fig. 9.12b also the opposite behavior, that is lower friction at positive potentials, is shown if the anion is exchanged with iodide. Although the change in shape of the anion (iodine is more spherical than FAP) seems to be a reasonable explanation for the observed difference in friction, a clear understanding that this is indeed the phys-

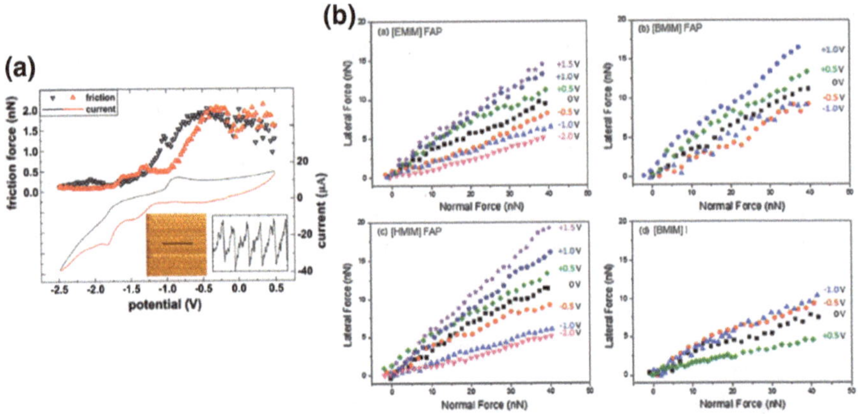

Fig. 9.12 **a** Frictogram of an Au(111) surface in [$Py_{1,4}$]FAP. Variations of friction are related to the electrochemical behavior of the IL. The inset shows a lateral force map and the corresponding line section as indicated at an applied electrode potential of −1.3 V versus Pt-QRE, revealing the lattice parameters of the underlying gold surface. From [74] **b** Lateral force as a function of normal load for various ionic liquid compositions as indicated in the graphs and electrode potentials on Au(111) surfaces. While for all FAP anion containing ILs the highest friction values are obtained for positive applied electrode potentials (**a–c**), this effect is reversed when iodide is used as anion (**d**). From [79]

ical mechanism has not yet established. Results from the same research group were published on the lubricating properties for a sliding an AFM silicon tip on a HOPG surface immersed in [HMIM]FAP [78]. In stark contrast to the results of [HMIM]FAP on Au(111), as discussed previously, a superlubricating regime, or a COF of <0.01, is found for positive applied potentials, where a FAP anion enriched interface would be expected. Negative applied potentials correspond to higher friction. In this case, the authors assigned their observation of reduced friction at positive potentials to reorientations of ions in the shear plane resulting in a smoother interface in the case of adsorbed anions. The huge variation in the COF for the same ionic liquid on different surfaces demonstrate that next to the overall electrochemical control of the layered structure of the ionic liquid at the electrode interface, direct substrate—ionic liquid interactions must be taken into account in order to reach a comprehensive view of ionic liquid lubrication. However, as the electrode potentials are quoted with respect to Pt-QRE, a direct comparison is limited.

As discussed previously, electrochemically controlled friction in ionic liquids has not yet been fully understood and it has been identified that numerous effects play a role. Further experiments are required before general guidelines can be expressed. It seems undoubtedly that it is very important to understand the various contributions in ionic liquid nanotribology such as the influence of chain length, cation and anion type, interaction with the surface, tilt angle, as well as shape of the ions. As it becomes very ambiguous to investigate the various contributions in experiments separately, theoretical work are an essential part in contributing to further understanding of the lubricating properties of ionic liquids. Canova et al. showed that subtle variations of IL composition lead to different lubrication mechanisms. Comparing the shape of NTf_2 and BF_4 anions for 1-butyl-3-methyl-imidazolium ([BMIM]) cations, the authors found larger fluctuations in the case of NTf_2 and consequently irregular shearing dynamics [80]. Fajardo et al. extended the theoretical work towards electrochemical friction studies and predicted a shift of the shear plane from the solid-liquid interface into the ionic liquid layers for larger surface charge densities, leading to overall lower friction [81].

When considering ionic liquid lubrication it is of utmost importance to control the environmental conditions very precisely. It has been shown that even very small traces of water changes the interfacial properties of ionic liquids significantly as well as their frictional behavior [82, 83]. The determination of the water content by Karl-Fischer titration can be misleading as the amount of water in the near-surface region and thus, in the region responsible for the tribological behavior, can be strongly deviating from the measured bulk concentrations. As many experiments studying the lubrication properties of ionic liquids are performed under ambient conditions, the water content can vary dramatically between different experiments, even for the same ionic liquid. Also, due to the very deviating hydrophilicity of ionic liquids composed of different ions direct comparisons between experiments are challenging.

Thinking about applications, ionic liquid lubrication becomes interesting only in niche markets because of the currently very high costs of ionic liquids. The interest is based on the strong surface films formed by ILs which are not repelled from the contact even under very high loads as demonstrated by atomic force microscopy, where a last strongly bound layer withstands high pressures of a sharp AFM tip [9]. Secondly, it seems possible to use only atomically thin layers of ionic liquids for lubrication, making them interesting as additives in conventional lubricants. Further experimental research is necessary to elucidate the individual contributions to the overall friction force by help of theoretical studies.

9.5 Summary

Combined electrochemical methods with tribological instruments are a powerful method to elucidate underlying principles of friction in situ, and support efforts to gain a more physical and scientific understanding of the origins of friction. Electrochemical friction force microscopy has been shown to significantly contribute to the understanding of tribological processes and lubrication at solid-liquid interfaces. Within a single experiment, various factors influencing friction can be investigated and isolated, including surface morphology, chemical identity as well as interfacial properties. Furthermore, these properties can be probed in a controlled and reversible manner. The control of surface charges by electrochemical means on metal surfaces has been discussed in this chapter. However, electrochemical effects on friction are not limited to metallic surfaces. While they have been excluded from the current discussion, the impacts of electrochemistry on friction on other surfaces, such as ceramic materials, are of great interest [84].

In summary, the applied potential to an electrode contributes considerably to the nanoscopically and macroscopically observed friction. Various aspects, including roughness, chemistry, adsorption state, electrolyte composition all influence friction and make the interpretation of the data challenging. However, given that friction is extremely sensitive to even subtle changes of the sliding surface, tribological examination of materials under electrochemical control cannot only be used in order to understand frictional processes better, but also to study such electrochemical phenomena by following the frictional response upon changing the electrochemical conditions. With the merging interest in precise control of interfaces, particularly at very small scales EC-FFM will shed light onto various aspects of nanotechnology, e.g. analyzing processes in energy storage systems or in the design of novel materials.

References

1. S.Ã. Mischler, Triboelectrochemical techniques and interpretation methods in tribocorrosion: a comparative evaluation. Tribol. Int. **41**, 573 (2008)
2. D.M. Kolb, An atomistic view of electrochemistry. Surf. Sci. **500**, 722 (2002)
3. D. Henderson, D. Boda, Insights from theory and simulation on the electrical double layer. Phys. Chem. Chem. Phys. **11**, 3822 (2009)
4. D.M. Kolb, Reconstruction phenomena at metal-electrolyte interfaces. Prog. Surf. Sci. **51**, 109 (1996)
5. A.C. Hillier, S. Kim, A.J. Bard, Measurement of double-layer forces at the electrode/electrolyte interface using the atomic force microscope: potential and anion dependent interactions. J. Phys. Chem. **100**, 18808 (1996)
6. A.A. Kornyshev, Double-layer in ionic liquids: paradigm change? J. Phys. Chem. B **111**, 5545 (2007)
7. M. Mezger et al., Molecular layering of fluorinated ionic liquids at a charged sapphire (0001) surface. Science **322**, 424 (2008)
8. R. Hayes, G.G. Warr, R. Atkin, At the interface: solvation and designing ionic liquids. Phys. Chem. Chem. Phys. **12**, 1709 (2010)
9. R. Hayes et al., Double layer structure of ionic liquids at the Au(111) electrode interface: an atomic force microscopy investigation. J. Phys. Chem. C **115**, 6855 (2011)
10. X. Zhang et al., Probing double layer structures of Au (111)-BMIPF6 ionic liquid interfaces from potential-dependent AFM force curves. Chem. Commun. **48**, 582 (2012)
11. S. Perkin, Ionic liquids in confined geometries. Phys. Chem. Chem. Phys. **14**, 5052 (2012)
12. R.M. Espinosa-Marzal, A. Arcifa, A. Rossi, N.D. Spencer, Microslips to 'Avalanches' in confined, molecular layers of ionic liquids. J. Phys. Chem. Lett. **5**, 179 (2014)
13. D. Henderson, J. Wu, Electrochemical properties of the double layer of an ionic liquid using a dimer model electrolyte and density functional theory. J. Phys. Chem. B **116**, 2520 (2012)
14. *CRC Handbook of Chemistry & Physics*. (CRC Press; Taylor & Francis group, 2017)
15. V.V. Pavlishchuk, A.W. Addison, Conversion constants for redox potentials measured versus different reference electrodes in acetonitrile solutions at 25 °C. Inorg. Chim. Acta **298**, 97 (2000)
16. C. Bonnaud, I. Billard, N. Papaiconomou, E. Chainet, J.C. Lepre, Rationale for the implementation of reference electrodes in ionic liquids. Phys. Chem. Chem. Phys. **18**, 8148 (2016)
17. R.G. Compton, C.E. Banks, *Understanding Voltammetry*. (World Scientific, 2007)

18. B.E. Conway, Electrochemical oxide film formation at noble-metals as a surface-chemical process. Prog. Surf. Sci. **49**, 331 (1995)
19. C.H. Hamann, W. Vielstich, *Elektrochemie, 4. Auflage*. (Wiley-VCH, 2005)
20. A.J. Bard, L.R. Faulkner, *Electrochemical Methods—Fundamentals and Applications*. (Wiley)
21. M. Valtiner, G.N. Ankah, A. Bashir, F.U. Renner, Atomic force microscope imaging and force measurements at electrified and actively corroding interfaces: challenges and novel cell design. Rev. Sci. Instrum. **82**, 23703 (2011)
22. N. Argibay, W.G. Sawyer, Frictional voltammetry with copper. Tribol. Lett. **46**, 337 (2012)
23. M.N.F. Ismail, T.J. Harvey, J.A. Wharton, R.J.K. Wood, A. Humphreys, Surface potential effects on friction and abrasion of sliding contacts lubricated by aqueous solutions. Wear **267**, 1978 (2009)
24. M. Valtiner, X. Banquy, K. Kristiansen, G.W. Greene, J.N. Israelachvili, The electrochemical surface forces apparatus: the effect of surface roughness, electrostatic surface potentials, and anodic oxide growth on interaction forces, and friction between dissimilar surfaces in aqueous solutions. Langmuir **28**, 13080 (2012)
25. A. Labuda et al., High-resolution friction force microscopy under electrochemical control. Rev. Sci. Instrum. **81**, 83701 (2010)
26. Y. Zhu, G.H. Kelsall, H.A. Spikes, Triboelectrochemistry on a nanometre scale. Tribol. Lett. **2**, 287 (1996)
27. T. Edison, US Patent 158787 (1875)
28. F.P. Bowden, D. Tabor, *The Friction and Lubrication of Solids*. (Clarendon Press, 2008)
29. R.B. Waterhouse, Tribology and electrochemistry. Tribology **3**, 158 (1970)
30. G.H. Kelsall, Y. Zhu, H.A. Spikes, Electrochemical effects on friction between metal oxide surfaces in aqueous solutions. J. Chem. Soc., Faraday Trans. **89**, 267 (1993)
31. G. Xie, D. Guo, J. Luo, Lubrication under charged conditions. Tribology Int. **84**, 22 (2015)
32. G. Binnig, C.F. Quate, C. Gerber, Atomic force microscope. Phys. Rev. Lett. **56**, 930 (1986)
33. C.M. Mate, G.M. McClelland, R. Erlandsson, S. Chiang, Atomic-scale friction of a tungsten tip on a graphite surface. Phys. Rev. Lett. **59**, 1942 (1987)
34. M. Binggeli, R. Christoph, H.-E. Hintermann, J. Colchero, O. Marti, Friction force measurements on potential controlled graphite in an electrolytic environment. Nanotechnology **4**, 59 (1993)
35. M. Binggeli, R. Christoph, H.-E. Hintermann, Observation of controlled, electrochemically induced friction force modulations in the nano-Newton range. Tribol. Lett. **1**, 13 (1995)
36. E. Weilandt, A. Menck, O. Marti, Friction studies at steps with friction force microscopy. Surf. Interface Anal. **23**, 428 (1995)
37. B. Schnyder, D. Alliata, R. Kötz, H. Siegenthaler, Electrochemical intercalation of perchlorate ions in HOPG: an SFM/LFM and XPS study. Appl. Surf. Sci. **173**, 221 (2001)
38. F. Hausen, M. Nielinger, S. Ernst, H. Baltruschat, Nanotribology at single crystal electrodes: influence of ionic adsorbates on friction forces studied with AFM. Electrochim. Acta **53**, 6058 (2008)
39. H. Hölscher, D. Ebeling, U. Schwarz, Friction at atomic-scale surface steps: experiment and theory. *Phys. Rev. Lett.* **101** (2008)
40. P. Egberts, et al., Environmental dependence of atomic-scale friction at graphite surface steps. *Phys. Rev. B—Condens. Matter Mater. Phys.* **88**, 1 (2013)
41. R.L. Schwoebel, E.J. Shipsey, Step motion on crystal surfaces. J. Appl. Phys. **37**, 3682 (1966)
42. G. Ehrlich, Atomic displacements in one- and two-dimensional diffusion. J. Chem. Phys. **44**, 1050 (1966)
43. S. Sundararajan, B. Bhushan, Topography-induced contributions to friction forces measured using an atomic force/friction force microscope. J. Appl. Phys. **88**, 4825 (2000)
44. E. Herrero, L.J. Buller, H.D. Abruna, Underpotential deposition at single crystal surfaces of Au, Pt, Ag and other materials. Chem. Rev. **1001**, 1897 (2001)
45. J. Lipkowski, Z. Shi, A. Chen, B. Pettinger, C. Bilger, Ionic adsorption at the Au(111) electrode. Electrochim. Acta **43**, 2875 (1998)

46. H. Angerstein-Kozlowska, B.E. Conway, A. Hamelin, L. Stoicoviciu, Elementary steps of electrochemical oxidation of single-crystal planes of Au Part II. A chemical and structural basis of oxidation of the (111) plane. J. Electroanal. Chem. Interfacial Electrochem. **228**, 429 (1987)
47. O.M. Magnussen, Ordered anion adlayers on metal electrode surfaces. Chem. Rev. **102**, 679 (2002)
48. W. Kautek, S. Dieluweit, M. Sahre, Combined scanning force microscopy and electrochemical quartz microbalance in-situ investigation of specific adsorption and phase change processes at the silver/halogenide interface. J. Phys. Chem. B **101**, 2709 (1997)
49. Z. Shi, J. Lipkowski, Investigations of So42- Adsorption at the Au(111) electrode in the presence of underpotentially deposited copper adatoms. J. Electroanal. Chem. **364**, 289 (1994)
50. F.C. Simeone, D.M. Kolb, S. Venkatachalam, T. Jacob, Die Au(111)-Elektrolyt-Grenzschicht: eine Tunnelspektroskopie- und DFT-Untersuchung. Angew. Chemie **119**, 9061 (2007)
51. D.M. Kolb, Elektrochemische oberflächenphysik. Angew. Chem. **113**, 1198 (2001)
52. F. Hausen, N.N. Gosvami, R. Bennewitz, Anion adsorption and atomic friction on Au(111). Electrochim. Acta **56**, 10694 (2011)
53. A. Labuda et al., Switching atomic friction by electrochemical oxidation. Langmuir **27**, 2561 (2011)
54. Z. Shi, S. Wu, J. Lipkowski, Coadsorption of metal atoms and anions: Cu upd in the presence of SO42–, Cl– and Br–. Electrochim. Acta Surf. Struct. Electrochem. React. **40**, 9 (1995)
55. Bennewitz, R., Hausen, F., N.N. Gosvami, Nanotribology of clean and modi fi ed gold surfaces. *J. Mater. Res.* **28**, 1279 (2013)
56. M. Nielinger, H. Baltruschat, Nanotribology under Electrochemical Conditions: influence of a copper (sub)monolayer deposited on single crystal electrodes on friction forces studied with atomic force microscopy. PhysChemChemPhys **9**, 3965 (2007)
57. D. Huitink, F. Gao, H. Liang, Tribo-electrochemical surface modification of tantalum using in situ AFM techniques. Scanning **32**, 336 (2010)
58. F. Hausen, J.A. Zimmet, R. Bennewitz, Surface structures and frictional properties of Au(100) in an electrochemical environment. Surf. Sci. **607**, 20 (2013)
59. P. Walden, Ueber die Molekulargrösse und elektrische Leitfähigkeit einiger geschmolzenen Salze. Bull. Acad. Sci. St. Petersbg. **8**, 405 (1914)
60. M.-D. Bermudez et al., Ionic liquids as advanced lubricant fluids. Molecules **14**, 2888 (2009)
61. A. Somers, P. Howlett, D. MacFarlane, M. Forsyth, A review of ionic liquid lubricants. Lubricants **1**, 3 (2013)
62. F. Zhou, Y. Liang, W. Liu, Ionic liquid lubricants: designed chemistry for engineering applications. Chem. Soc. Rev. **38**, 2590 (2009)
63. I. Minami, Ionic liquids in tribology. Molecules **14**, 2286 (2009)
64. M. Watanabe et al., Application of ionic liquids to energy storage and conversion materials and devices. Chem. Rev. **117**(10), 7190 (2017)
65. J.P. Hallett, T. Welton, Room-temperature ionic liquids: solvents for synthesis and catalysis. 2. Chem. Rev. **111**, 3508 (2011)
66. A. Pinkert, K.N. Marsh, S.S. Pang, M.P. Staiger, Ionic liquids and their interaction with cellulose. Chem. Rev. **109**, 6712 (2009)
67. P. Hapiot, C. Lagrost, Electrochemical reactivity in room-temperature ionic liquids. Chem. Rev. **108**, 2238 (2008)
68. J.S. Wilkes, M.J. Zaworotko, Air and water stable 1-Ethyl-3-Methylimidazolium Based Ionic Liquids. J. Chem. Soc.—Chem. Commun. **965** (1992)
69. M.V. Fedorov, A.A. Kornyshev, Ionic liquids at electrified interfaces. Chem. Rev. **114**, 2978 (2014)
70. A.M. Smith, K.R.J. Lovelock, N.N. Gosvami, T. Welton, S. Perkin, Quantized friction across ionic liquid thin films. Phys. Chem. Chem. Phys. **15**, 15317 (2013)
71. M.V. Fedorov, A.A. Kornyshev, Ionic liquid near a charged wall: structure and capacitance of electrical double layer **112**, 11868 (2008)

72. C. Merlet, B. Rotenberg, P.A. Madden, M. Salanne, Computer simulations of ionic liquids at electrochemical interfaces. Phys. Chem. Chem. Phys. **15**, 15781 (2013)
73. M.Z. Bazant, B.D. Storey, A.A. Kornyshev, Double layer in ionic liquids: overscreening versus crowding. Phys. Rev. Lett. **106**, 046102 (2011)
74. J. Sweeney et al., Control of nanoscale friction on gold in an ionic liquid by a potential-dependent ionic lubricant layer. Phys. Rev. Lett. **109**, 155502 (2012)
75. S. Baldelli, Surface structure at the ionic liquid-electrified metal interface. Acc. Chem. Res. **41**, 421–431 (2008)
76. S. Baldelli, Interfacial structure of room-temperature ionic liquids at the solid-liquid interface as probed by sum frequency generation spectroscopy. J. Phys. Chem. Lett. **4**, 244 (2013)
77. S. Watanabe, M. Nakano, K. Miyake, R. Tsuboi, S. Sasaki, Effect of molecular orientation angle of imidazolium ring on frictional properties of imidazolium-based ionic liquid. Langmuir **30**, 8078 (2014)
78. H. Li, R.J. Wood, M.W. Rutland, R. Atkin, An ionic liquid lubricant enables superlubricity to be 'switched on' in situ using an electrical potential. Chem. Commun. **50**, 4368 (2014)
79. H. Li, M.W. Rutland, R. Atkin, Ionic liquid lubrication: influence of ion structure, surface potential and sliding velocity. Phys. Chem. Chem. Phys. **15**, 14616 (2013)
80. F. Federici Canova, H. Matsubara, M. Mizukami, K. Kurihara, A.L. Shluger, Shear dynamics of nanoconfined ionic liquids. Phys. Chem. Chem. Phys. (2014). https://doi.org/10.1039/c4cp00005f
81. O.Y. Fajardo, F. Bresme, A.A. Kornyshev, M. Urbakh, Electrotunable lubricity with ionic liquid nanoscale films. Sci. Rep. **5**, 7698 (2015)
82. A.M. Smith, M.A. Parkes, S. Perkin, Molecular friction mechanisms across nano fi lms of a bilayer- forming ionic liquid. J. Phys. Chem. Lett. **5**, 4032 (2014)
83. R.M. Espinosa-Marzal, A. Arcifa, A. Rossi, N.D. Spencer, Ionic liquids confined in hydrophilic nanocontacts: structure and lubricity in the presence of water. J. Phys. Chem. C **118**, 6491 (2014)
84. A. Kailer et al., Influence of electric potentials on the tribological behaviour of silicon carbide. Wear **271**, 1922 (2011)

Chapter 10
In Situ Friction Tests in a Transmission Electron Microscope

Fabrice Dassenoy

Abstract Post mortem characterization techniques are essential for the understanding of the lubrication mechanisms of complex tribological systems. The information obtained using these techniques can be used to propose hypothesis of mechanisms that have then to be definitively validated by probing the interfacial material in real time during the friction test. To go further in the understanding of the action modes of some tribological systems, it is important to set up an analytical methodology using in situ experimental techniques in order to (i) probe directly the behavior of the interfacial material in the contact zone and (ii) dissociate, for a better understanding, the different components of the tribological stress (pressure and shear). There are several techniques that combine mechanical stress and in situ analysis (Raman tribology, Infra-Red tribology, EXAFS under pressure …). However, the most interesting techniques are certainly those that permit a visualization of the contact area in real time area during the friction test, down to a nanometer scale. The objective of this chapter is to present through some examples the potential of in situ friction tests done inside a transmission electron microscope.

10.1 Introduction

The dream of any tribologist is certainly to observe in real time and down to the smallest detail the behavior of the interfacial material in the tribological contact during mechanical stress. When the observation takes place at the atomic scale, this dream becomes a grail. The only technique for this is the Transmission Electron Microscopy (TEM). Recently, the development of new sample holders for TEM including either a nanoindenter or an atomic force microscope (AFM) made possible compression and friction tests in a TEM, thus allowing the simulation of the tribological contact.

F. Dassenoy (✉)
Laboratoire de Tribologie et Dynamique des Systèmes, CNRS, UMR 5512, Ecole Centrale de Lyon, 69134 Ecully, France
e-mail: fabrice.dassenoy@ec-lyon.fr

M. Dienwiebel and M.-I. De Barros Bouchet (eds.), *Advanced Analytical Methods in Tribology*, Microtechnology and MEMS,
https://doi.org/10.1007/978-3-319-99897-8_10

These techniques have a very high potential because they allow combining friction and in situ observation/analysis, thus permitting to go further in the understanding of the reaction mechanisms operating in a tribological contact.

In situ TEM techniques were used in many works to determine nanoscale mechanical properties concurrently with real-time microstructure characterization of various types of samples [1] such as thin film [2–4], carbon nanotube [5], nanowire [6, 7] and various types of nanoparticles [8–13]. Minor et al. [2, 3] used an in situ TEM nanoindenter holder to investigate dislocation nucleation and motion in silicon and aluminium thin films. Deneen et al. [8] used the same device to study the mechanical behaviour of individual silicon nanospheres. The experiments were carried out on some 200 nm diameter particles. Both elastic and plastic deformation as well as fracture was observed. In situ TEM Nano indentation experiments were also performed on polycrystalline CdS hollow spheres ranging in diameter from 200 to 450 nm [7] and on some clusters of silicon particles around 50 nm in size [10]. In the first work, it was shown that the particles can achieve both a high compression to failure and withstand very high shear stresses with respect to their ideal strength, while in the second one a rotation of the nanoparticles was observed, followed by cracking at the interface between two nanoparticles. More recently Carlton et al. [11] investigated the deformation of silver nanoparticles under compression using diffraction contrast and phase contrast TEM. The authors observed the disappearance of dislocation in single crystal silver nanoparticles after nanoindentation during real time experimental observations. The use of in situ TEM techniques for solving complex tribological problems has emerged recently. The objective of this chapter is to illustrate through few examples of in situ compression and sliding tests performed on single nanoparticles the potential of this technique for tribological applications.

10.2 Example 1: In Situ Compression Test on a Single MoS_2 Nanoparticle

In Lahouij et al. [13], performed compression tests of individual IF-MoS_2 nanoparticles in a High Resolution Transmission Electron Microscope using an in situ TEM nanoindenter. Thanks to this technique, in situ observations of the behavior of single nanoparticles under pressure were performed for the first time. In addition, the mechanical response of the particle under compression was followed and force-time and force-displacement curves were recorded. The compression experiments were carried out using a Nanofactory Instruments HN200 single tilt Nanoindenter Micro-Force probing holder (Fig. 10.1) mounted on the specimen holder of a High resolution JEOL 2010 FEG microscope operating at 200 kV accelerating voltage. The nanoindentation sensor used for the experiments was equipped with a diamond tip. The force range of the sensor was between 0 and 3 mN and its stiffness constant was 3500 N/m with a force resolution of 100 nN. Thanks to this system, in situ force measurement recording was possible. This TEM nanoindenter was used for (i) the characterization

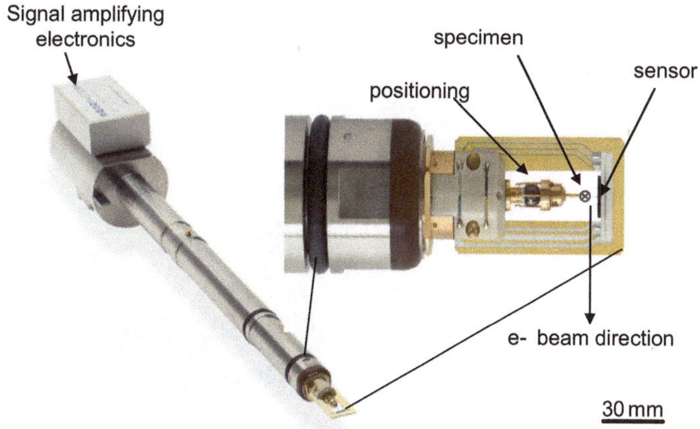

Fig. 10.1 Schematic of the TEM nanoindenter sample holder produced by Nanofactory company [13]

of the mechanical properties of the fullerenes, (ii) their nanomanipulation at the sub-nanometer scale, (iii) the observation of their elastic and plastic deformation during the test. The nanoindenter was also used to carry out sliding experiments by moving sideways the substrate on which the particles were deposited by nanomanipulation experiment. Bright Field imaging and the observation of the stress/strain-induced structural changes in the particle were possible in addition of TEM nanoindentation experiments. The TEM-Nanoindentor permits to image the nanocompression process with real time imaging, to record movies, as well as to acquire the force-displacement and force-time data. For compression and sliding experiments, substrates on which the nanoparticles have to be previously deposited must be nanomachined to be thin enough for electron transparency in one dimension, long enough in second dimension and moderately short in the third one, in order to avoid crashing onto the sensor. A short wedge sample made in silicon was used. FIB-SEM was used to prepare the Si-wedge on which the nanoparticles were deposited. Figure 10.2 shows the Si wedge after machining. The presence of 10 μm long "terraces" used for the nanoparticles deposition can be observed. A thin gold wire was used to fix the Si wedge to the positioning part of the TEM sample holder.

Figure 10.3 shows a series of images captured from a video recording during a nanocompression test carried out on a single IF-MoS_2 particle of about 100 nm diameter. The particle does not present any empty core and presents a highly crystalline and faceted structure. Compression tests were carried out with a truncated diamond tip of 1 μm width. Nanoscale alignment of the particles and tip along the microscope electron beam axis was performed by monitoring the relative focus between the tip and the particles at high TEM magnification. This operation is not straightforward and can take a long time. Compression was performed by forward actuation of the support wedge along the z-axis using the fine piezoelectric tube. The exact 3D shape of the entire particle is difficult to know here because the TEM image is a pro-

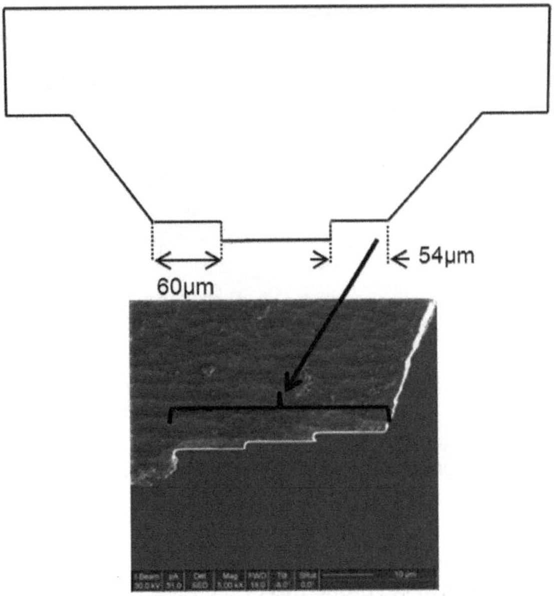

Fig. 10.2 Si wedge after FIB machining. The micromachining enables the creation of some 10 μm steps used for the particles deposition

jection in two-dimension. Each stage consisted of 100 nm displacement controlled compression of the tip at a loading rate of approximately 13 nm s^{-1}.

Figure 10.4a, b show the corresponding force/time and displacement/time curves for this test.

Figure 10.3a shows the nanoparticle on the Si substrate just before compression. It corresponds to $t = 0$ s on the force/time curve. From this stage, the nanocompression test starts and the Si substrate moves towards to the tip. At $t = 2$ s, the tip is brought into real contact with the fullerene surface (Fig. 10.3b). The particle is further loaded and the evolution of the force can be followed on the force/time curve (Fig. 10.4a). On Fig. 10.3c, the particle starts to change of shape in the TEM image. This change is more and more pronounced with the load increase (Fig. 10.3d–g). The facetted shape of the particle changes progressively to an ellipsoidal shape. To be sure that the thickness of the particle has increased in the direction of the electron beam to form the oblate, the intensity of the electron signal in the center of the particle was measured. The signal is proportional to the grey level because the images have been recorded practically in the same illumination conditions. The results indicate that the particle is oblate after compression and accommodates the contact pressure and progressively rotates and lies parallel to the two contacting surfaces (Fig. 10.3g). When the maximum force is reached (10 μN at $t = 7.5$ s), then the particle is ejected from the contact zone and slides outside the compression zone certainly due to low adhesion to the substrate and the fact that the two surfaces are not strictly parallel (Fig. 10.3h, i). The sudden sliding of the oblate particle causes an exfoliation of its outer sheets where the curvature of the lamellar is maximum. These isolated sheets stick to the diamond tip (Fig. 10.3i—zoomed part). Afterwards, when the particle is

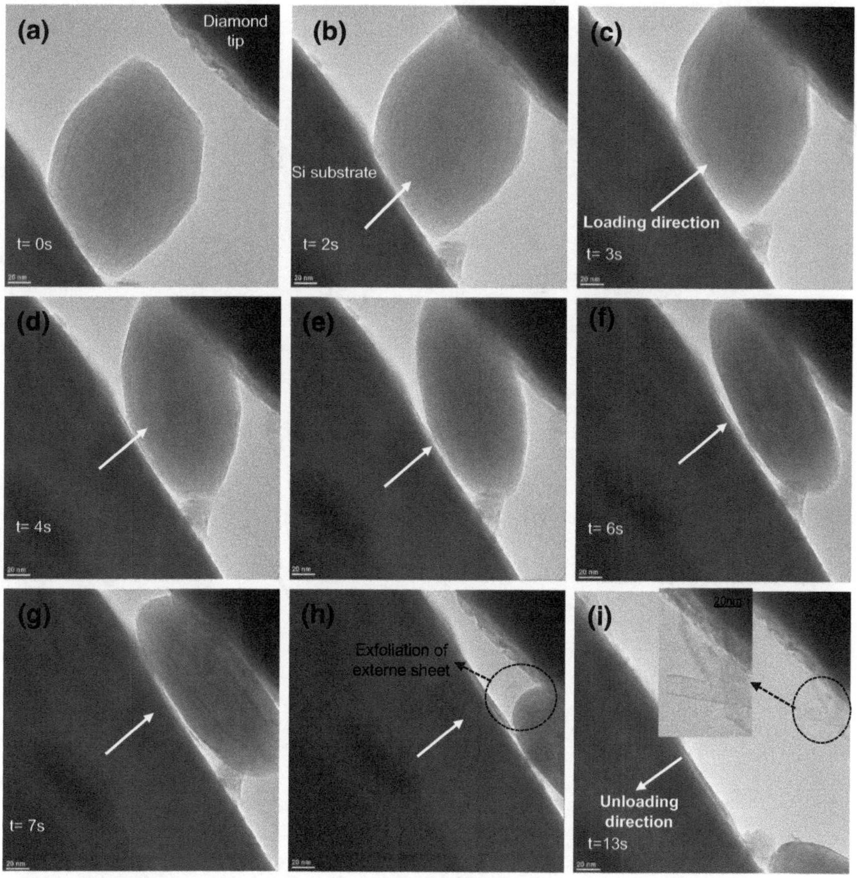

Fig. 10.3 Image captures obtained from a video recorded during a compression experiment carried out with a single MoS_2 particle [13]

unloaded, it remains deformed (Fig. 10.3i). Unfortunately, due to intrinsic motion of the sample, the resolution in the images does not permit to observe the basal planes of MoS_2 in the structure of the particle during solicitation. The question of either the particle elastically rotates or plastically deforms cannot be answered properly here, because of the lack of 3-D imaging in the TEM. However, the observation of the force/time curve (Fig. 10.4a) does not show any sliding effect, suggesting that the particle is partially plastically deformed.

It is possible to estimate the contact area between the tip and the sample during loading by assuming that (i) the contact area of the faceted MoS_2 nanoparticle is a square and (ii) the contact between the tip and the faceted MoS_2 nanoparticle is a flat-on-flat configuration. The contact area between the fullerene and the tip was measured in the selected video images. Thus, on Fig. 10.3g, the estimated contact

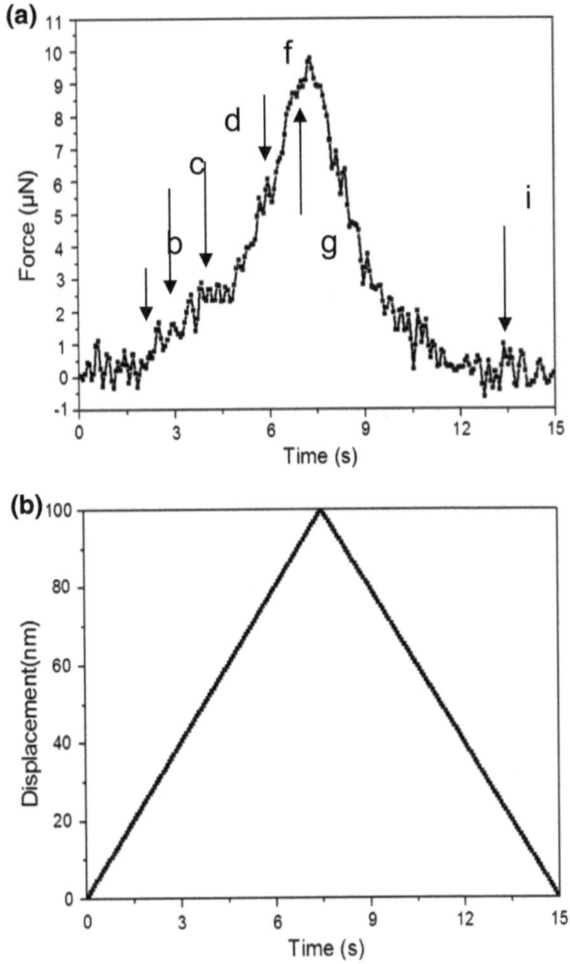

Fig. 10.4 Image captures obtained from a video recorded during the compression experiment shown in Fig. 10.3 carried out with a single MoS_2 particle [13]

area is ~10,000 nm^2. If we consider that the maximum force applied on the particle is 10 µN, this gives an estimated average contact pressure of 1 GPa.

In this work, Lahouij et al. [13] investigated in real time the deformation and degradation behavior of spherical and well crystallized single inorganic fullerene nanoparticles of MoS_2 under compression using a HRTEM equipped with a nanoindentation holder. For the first time, the exfoliation of the outer sheets of a fullerene nested structure was imaged. However, it was found that well crystallized and round shape nanoparticles were difficult to exfoliate. Under 1–1.5 GPa uniaxial pressure, it was found that the shape of single IF-MoS_2 nanoparticles was preserved and only external layers were exfoliated. A uniaxial pressure higher than 1.5 GPa was found to be necessary to crush the particles. The resistance of the IF-MoS_2 to exfoliation was attributed to their good mechanical properties due to their well crystalized structure and the round shape of the particle. However, the behavior of nanoparticles during mechanical stress strongly depends on the size, the morphology and the crystallinity of the particle.

10.3 Example 2: In Situ Compression Test on a Single WS_2 Nanoparticle

Figure 10.5 shows a compression test carried out on a single WS_2 nanoparticle. The particle present a hollow core compared to the MoS_2 nanoparticle studied earlier. Test was done with a displacement of 45 nm leading to an applied maximum force of 3.5 μN and a maximum contact pressure estimated at 0.45 GPa. The load/time and displacement/time curves are shown in Fig. 10.6. The behavior under pressure of this WS_2 nanoparticle is completely different to that of the MoS_2 nanoparticle. As soon as the particle is loaded, the sheets are deformed. The particle seems to lengthen ($t = 8$ s). The central cavity loses volume and is no longer visible ($t = 11$ s). Despite the loss of image resolution due to vibration, we can see that the particle starts to be strongly damaged ($t = 15$ s) giving rise to a large number of WS_2 sheets (from $t =$ 15 s to $t = 45$ s). At the end of the test, the particle is completely damaged.

10.4 Example 3: In Situ Sliding Test on a Single MoS_2 Nanoparticle

In [15], Lahouij et al. gave evidence that the lubricating properties of poorly crystallized MoS_2 nanoparticles (with many defects) were better than those of perfectly crystallized spherical ones. On the basis of the *post-mortem* analyses of the worn surfaces, the authors attributed the good lubricating properties of the poorly crystallized IF-MoS_2 to their ability to exfoliate immediately and to form rapidly a homogeneous tribofilm on the surface made of MoS_2 layers aligned in the sliding direction. At the opposite, it was observed that the tribofilm obtained with well-crystallized IF-MoS_2 nanoparticles was heterogeneous and composed of a mixture of MoS_2 layers, iron oxide nanoparticles and intact particles embedded in the tribofilm. It was also observed that the shape of most of the well crystallized particles was preserved after friction. The authors ascribed this finding to the perfect crystallinity of the particles together with the large number of closed layers that confer them with a higher mechanical resistance. It was also suggested that these perfect particles could be considered to behave as genuine nano-ball bearings, at least temporarily, until they gradually deform and start to exfoliate their outer layers. However, only a direct observation of the sliding contact allows accessing the real behavior of the fullerenes during a mechanical stress and to answer the question of the influence of the structure/morphology of the particles on their lubrication mechanisms (exfoliation, sliding

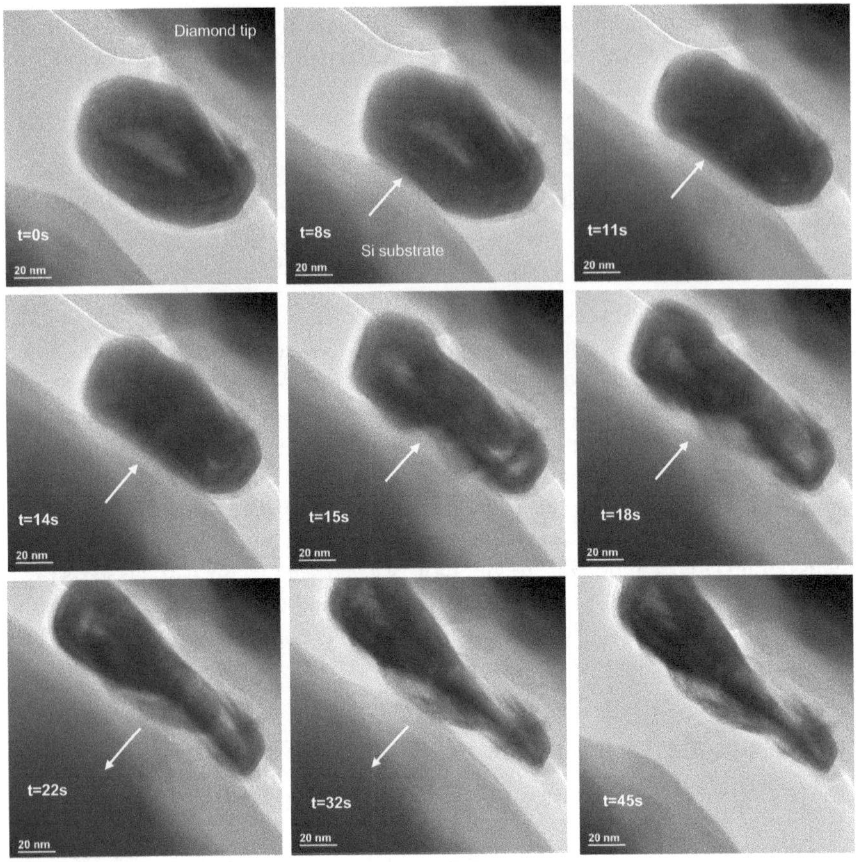

Fig. 10.5 Image captures obtained from a video recorded during a compression experiment carried out with a single WS_2 particle [14]

and rolling). In [16], the authors focused on the influence of the crystal structure of two types of IF-MoS_2 nanoparticles (namely perfectly crystallized IF-MoS_2 and poorly crystallized IF-MoS_2) on their behaviors during sliding tests performed inside the TEM. Figure 10.6 shows a series of image captures obtained from a video recorded during a sliding experiment carried out with a single perfectly crystallized IF-MoS_2 nanoparticle. Figure 10.6a corresponds to the particle compressed between the tip and the Si substrate before starting the sliding test at $t = t_0$. Figure 10.6h shows the particle at the end of the sliding test at $t = t_{\text{final}}$. A white point was arbitrary placed on the particle in order to easily follow its movement during the sliding test. The white arrow marks the starting point on the diamond tip. When the sliding test starts (by moving the Si substrate parallel to the tip—the white arrow on the Si substrate indicates the direction of the movement of the Si substrate), it can be observed from the image captures that the IF-MoS_2 particle rolls in the contact. The distance trav-

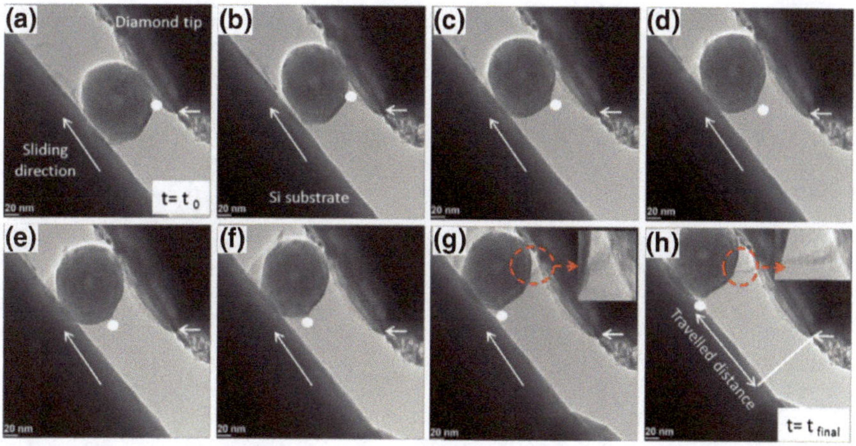

Fig. 10.6 Image captures obtained from a video recorded during a sliding experiment carried out with a single perfectly crystallized IF-MoS_2 nanoparticle. **a–h** shows the rolling of the particles and **g–h** shows the exfoliation of an outer layer. A white point was arbitrary placed on the particle in order to easily follow the movement of the particle during the test. The white arrow marks the starting point on the diamond tip. The distance travelled by the particle from $t = t_0$ to $t = t$ final is estimated to be 160 nm [16]

elled by the particle from Fig. 10.6a–h was estimated to be of ~160 nm. This means that the particle has travelled a little more than half of its circumference calculated to be of ~250 nm. It can be also observed from these images that the structure and the shape of the particle are preserved during sliding and that only an outer layer is delaminated as shown on the zoomed parts of Fig. 10.6g, h).

In the same way, the behavior of a single poorly crystallized IF-MoS_2 particle was observed during a sliding test. Figures 10.7 and 10.8 show respectively two sequences recorded during a sliding test performed from a single isolated IF-MoS_2 of about 30 nm of diameter. The two sequences of the sliding test were performed with an average normal force of, respectively, 1 and 4 μN and a constant velocity of 1 nm/s. Figure 10.7a shows the single particle on the Si substrate just before starting the sliding test. The surface of the diamond tip is covered (contaminated) by some MoS_2 layers deposited during previous tests. Figure 10.7b–f show the behavior of the particle during the sliding test. Figure 10.7b shows the particle when the normal force of 1 μN is applied to the particle (without shear stress) corresponding to an estimated contact pressure of 110 MPa. No important structural change of the particle is observed, except for a small compression. When the shear stress is applied, a huge compression with ensuing exfoliation and material transfer is observed. This leads to a delivery of exfoliated MoS_2 layers in the contact as shown in Fig. 10.7c–e. At the end of this first sequence, the normal force is unloaded, and it can be seen in Fig. 10.7f that both the surface of the wedge and the tip are now covered by some MoS_2 layers aligned along the sliding direction. The second sequence consists in the observation of the deformation of the exfoliated MoS_2 layers during the sliding

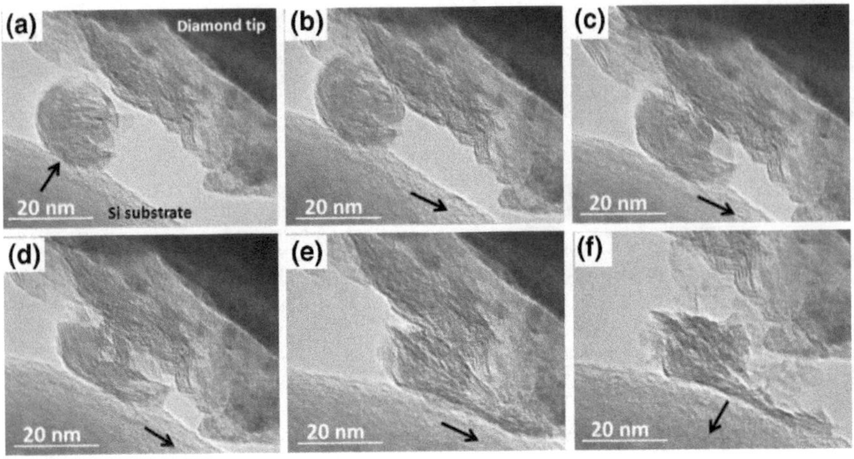

Fig. 10.7 Image captures obtained from a video recorded during a sliding experiment carried out with a single poorly crystallized IF-MoS_2 nanoparticle demonstrating exfoliation. The black arrow on the Si substrate indicates the direction of the movement of the Si substrate [16]

process (Fig. 10.8). Figure 10.8a corresponds to the final state of the first sequence (Fig. 10.7f) and is also the initial state of the second sliding sequence. The substrate covered by exfoliated MoS_2 layers was placed in contact with the MoS_2 layers deposited on the diamond tip. A normal load of 4 μN was applied. Figure 10.8b shows the MoS_2 layers compressed between the wedge and the tip. When the wedge starts to move sideways, a transfer of material from the wedge to the tip can be observed (Fig. 10.8c). The displacement of the wedge induces the shearing of MoS_2 layers (Fig. 10.8d, e). At the end of the test, we can see that a few sheared MoS_2 layers were detached from the tip and adhere now on the wedge (Fig. 10.8f). This sequence shows the easy shearing properties of the MoS_2 film under the combined effect of pressure and shear stress.

10.5 Example 4: Friction Coefficient Measured on a Single WS_2 Nanoparticle

Through an in situ series of experiments using a special TEM-AFM holder that permitted the simulation of a tribological contact, Lahouij et al. in [13, 16] described the lubrication mechanism of Inorganic Fullerene Like nanoparticles made of metal disulfide. The experiment provided valuable qualitative results. Rolling of MoS_2 nanoparticles was observed at estimated contact pressure up to 100 MPa, while exfoliation at approximately 1 GPa. It was observed that the nanoparticles can behave differently under seemingly similar experimental conditions: rolling or sliding/exfoliating. In their experiments, the sample holder used had increased sensibility for measuring

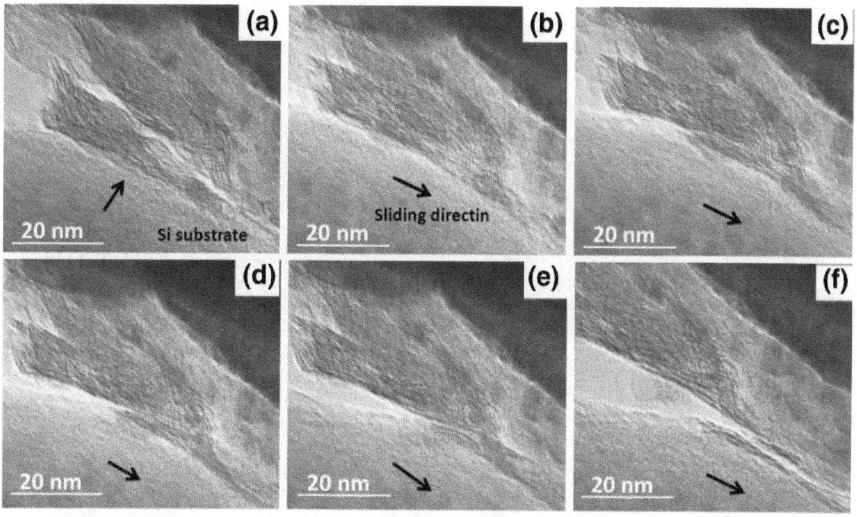

Fig. 10.8 Image captures obtained from a video recorded during a sliding experiment carried out with a single poorly crystallized IF-MoS_2 nanoparticle demonstrating the easy shearing between the exfoliated MoS_2 layers. The white arrow on the Si substrate indicates the direction of the movement of the Si substrate [16]

normal forces, in the direction of the main axis of the device. It was able to move in normal and lateral direction, but it was not able to measure lateral force, hence the friction coefficient. Recently, Jenei et al. [17] went further in the investigation of the lubrication mechanisms of metal disulfide nanoparticle. An in situ friction test was carried out on a single WS_2 nanoparticle in a transmission electron microscope (TEM) using a nano indentation device able to measure normal and lateral forces independently. For the first time, quantitative results were reported: friction forces were recorded during the test, and thus friction coefficient values were calculated. The exfoliation of the particle was observed during the experiment, and was linked to the friction coefficient modification. The in situ tests were conducted in an FEI Titan environmental transmission electron microscope (ETEM) operated on 300 kV accelerating voltage, equipped with a Cs corrector, with the help a Pi-95 Picoindenter from Hysitron, Inc. The sensor (transducer) of the picoindenter is a MEMS based device, which has a diamond tip attached to it. The transducer is capable of controlling the tip with a 0.02 nm displacement resolution and it is capable of measuring forces with a 3 nN resolution. The use of a 2D transducer makes it possible to measure normal and lateral forces at the same time, independently from each other. The tip of the transducer used in this experiment is a so called "at punch", it has a flat surface section with a length of approximately 160 nm (see Fig. 10.9).

A tribological contact was mimicked with a single WS_2 nanoparticle being held between the silicon wedge and the diamond tip of the transducer. A "nanofriction test" was designed in the following way: the diamond tip was aligned to the selected WS_2 nanoparticle, a constant load was applied on the nanoparticle after which the tip

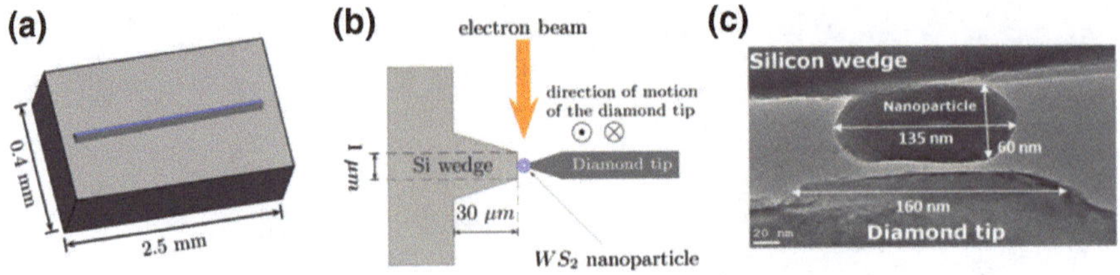

Fig. 10.9 **a** Schematic representation of the silicon wedge, the plateau is depicted in light blue. **b** Schematic representation of a nanoparticle deposited on the plateau, the diamond tip slides along the length of the wedge. **c** A TEM image of the WS_2 nanoparticle on the silicon wedge before the experiment [17]

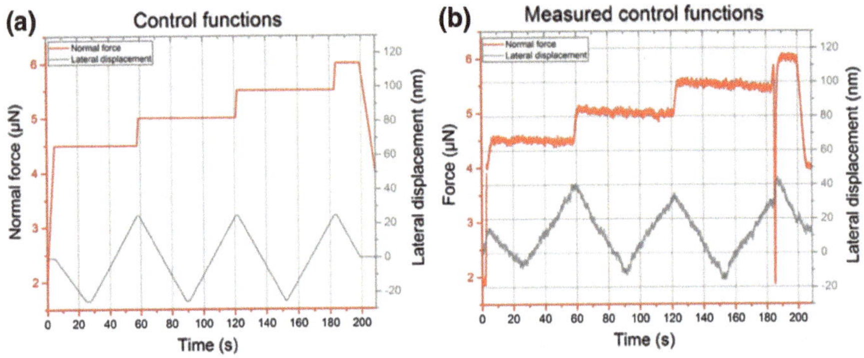

Fig. 10.10 **a** The control function for the in situ friction test and **b** the measured normal force and lateral displacement during the experiment [17]

was moved laterally to the right and then to the left (right and left strokes). This was considered as a cycle or a segment. At the end of the segment the load was increased. Between each stroke there was a short pause. In total 4 segments were recorded. The test was performed in the so called "load control" mode which means that during the test the instrument made sure that the predefined load was being applied to the nanoparticle. During the test the normal and lateral forces were recorded, as well as the normal and lateral displacement of the tip. A TEM image in Fig. 10.9c shows the WS_2 nanoparticle on the silicon wedge with the aligned diamond tip. The normal load and lateral displacement during such a test was predefined. Figure 10.10a shows the control function for the experiment. The length of one stroke is 50 nm, exceptions are the first and last strokes where the displacements are only 25 nm. The normal load in the first segment was 4.5 μN, then it is set to increase with a 0.5 μN step in each segment, reaching a maximum of 6.5 μN at the end of the experiment. The length of the friction test was set to be a little longer than three minutes. The actual normal force and lateral displacement values measured during the experiment can be seen in Fig. 10.10b.

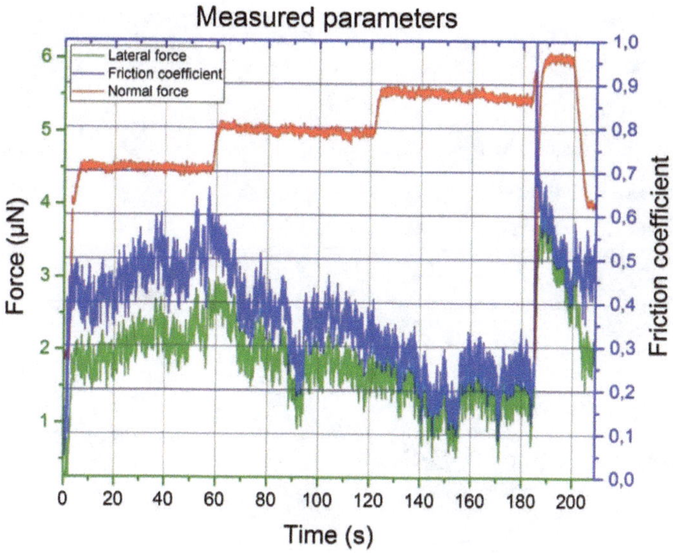

Fig. 10.11 The normal and lateral forces and the friction coefficient values recorded during the experiment [17]

Figure 10.11 shows the normal and lateral force and the calculated friction coefficient (COF) during the whole experiment. The COF is calculated as the ratio of the lateral and normal force. During the first couple of strokes, the friction force is relatively high (compared to the rest of the experiment), after which there is a significant decrease of the COF. The friction coefficient reduces from a 0.55 level to 0.15. At the beginning of the last stroke, where the normal load is increased to 6.5 μN, the nanoparticle slips out from between the diamond tip and the silicon wedge. The COF increases instantly. It is interesting to see that as the stroke progresses, the friction coefficient follows a decreasing trend, although here the diamond tip is sliding on the silicon wedge. Figure 10.12 shows some image captures from the video of the "nano" friction test. In Fig. 10.12a the exfoliation during the right stroke in segment 1 can be observed. In Fig. 10.12b the adhesion of the previously exfoliated layer is visible; in (c): exfoliation during the right stroke in segment 1; in (d) exfoliation at the left stroke in segment 1; in (e) exfoliation at the end of the left stroke in segment 1, this layer is not completely detached from the nanoparticle; in (f) exfoliation in the left stroke of segment 2; in (g) an exfoliation can be observed at the right cap of the particle just before is slips out from the contact; in (h) several exfoliated layers can be seen between the tip and the silicon wedge after the particle got out from the contact.

This experiment permitted to follow in real time the evolution of the friction coefficient during the tribotest done in situ inside a TEM and to establish a correlation with the results obtained at the micro scale.

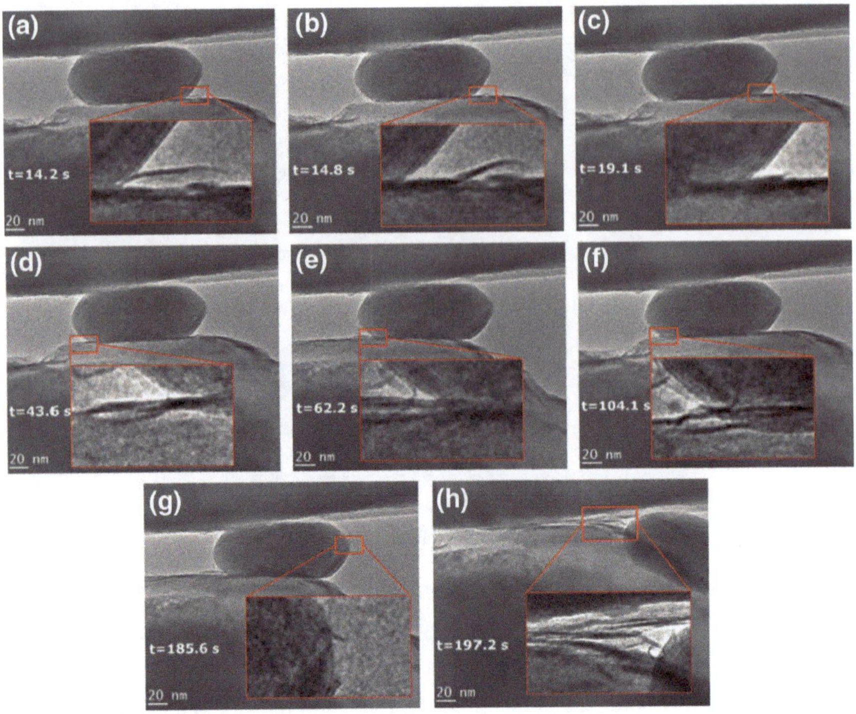

Fig. 10.12 Frames captured from the video of the "nano" friction test performed on a WS_2 nanoparticle [17]

10.6 Conclusion

Friction experiments carried out in a Transmission Electron Microscope (TEM) constitute a way to go further in the understanding of the reaction mechanisms operating in a tribological contact. They enable a direct visualization of the behavior of the interfacial material during the tribological test, thus making possible the follow in real time of any structural and morphological change in the stressed material. With the emergence of new microscopes that allow carrying out experiments in a controlled environment, it will be soon possible to perform similar tests in presence of gas, thus making possible the in situ monitoring of chemical reactions between the stressed materials and this gas. Complementary techniques such as Electron Energy Loss Spectroscopy (EELS) will be in this respect very useful.

References

1. J.M. Howe, H. Mori, Z.L. Wang, In situ high-resolution transmission electron microscopy in the study of nanomaterials and properties. MRS Bull. **33**(2), 115–121 (2008)
2. A.M. Minor, J.W. Morris Jr., E.A. Stach, Quantitative in situ nanoindentation in an electron microscope. App. Phys. Lett. **79**, 1625–1627 (2001)
3. E.A. Stach, T. Freeman, A.M. Minor, D.K. Owen, J. Cumings, M.A. Wall, T. Chraska, R. Hull, J.W. Morris Jr., A. Zettl, U. Dahmen, Development of a nanoindenter for in situ transmission electron microscopy. Microsc. Microanal. **7**, 507–517 (2001)
4. N. Li, J. Wang, J.Y. Huang, A. Misra, X. Zhang, In situ TEM observations of room temperature dislocation climb at interfaces in nanolayered Al/Nb composites. Scr. Mater. **63**, 363–366 (2010)
5. Z.L. Wang, P. Poncharal, W.A. de Heer, Measuring physical and mechanical properties of individual carbon nanotubes by in situ TEM. J. Phys. Chem. **61**, 1025–1030 (2000)
6. X. Han, K. Zheng, Y.F. Zhang, X. Zhang, Z. Zhang, Z.L. Wang, Low-temperature in situ large-strain plasticity of silicon nanowires. Adv. Mater. **19**, 2112–2118 (2007)
7. A. Asthana, K. Momeni, A. Prasad, Y.K. Yap, R.S. Yassar, In situ observation of size scale effects on the mechanical properties of ZnO nanowires. Nanotechnology **22**, 265712 (2011)
8. J. Deneen, W.M. Mook, A.M. Minor, W.W. Gerberich, C.B. Carter, In situ deformation of silicon nanospheres. J. Mater. Sci. **41**, 4477–4483 (2006)
9. Z.W. Shan, G. Adesso, A. Cabot, M.P. Sherburne, S.A. Syed Asif, O.L. Warren, D.C. Chrzan, A.M. Minor, A.P. Alivisatos, Ultrahigh stress and strain in hierarchically structured hollow nanoparticles. Nat. Mat. **7**, 947–952 (2008)
10. A.J. Lockwood, B.J. Inkson, In situ TEM nanoindentation and deformation of Si nanoparticle clusters. J. Phys. D Appl. Phys. **42**, 035410 (2009)
11. C.E. Carlton, P.J. Ferreira, In situ TEM nanoindentation of nanoparticles. Micron **43**, 1134–1139 (2012)
12. I. Lahouij, F. Dassenoy, L. De Knoop, J.M. Martin, B. Vacher, In situ TEM observation of the behavior of an individual fullerene-like MoS_2 nanoparticle in a dynamic contact. Tribol. Lett. **42**, 133–140 (2011)
13. I. Lahouij, F. Dassenoy, B. Vacher, J.M. Martin, Real time imaging of compression and shear of single fullerene-like MoS2 nanoparticle. Tribol. Lett. **45**, 131–141 (2012)
14. I. Lahouij, Ph.D. thesis, Ecole Centrale de Lyon, France (2013)
15. I. Lahouij, B. Vacher, J.M. Martin, F. Dassenoy, IF-MoS_2 based lubricants: influence of size, shape and crystal structure. Wear **296**, 558–567 (2012)
16. I. Lahouij, B. Vacher, F. Dassenoy, Direct observation by in situ TEM of the behavior of IF-MoS_2 nanoparticles during sliding tests. Lubr. Sci. **36**(3), 163–173 (2014)
17. I. Jenei, F. Dassenoy, Friction coefficient measured on a single WS_2 nanoparticle: an in situ transmission electron microscope experiment. Tribol. Lett. **65**, 86 (2017)

Chapter 11
In Situ Digital Holography for 3D Topography Analysis of Tribological Experiments

Martin Dienwiebel and Pantcho Stoyanov

Abstract This chapter reviews topography measurement techniques that allow performing the in situ quantification of roughness and the analysis of the surface morphology during sliding experiments. The method of digital holographic microscopy is then introduced in detail and examples of topography evolution are presented.

11.1 Introduction

In situ tribometry has significantly advanced the field of Tribology over the last 2 or 3 decades because temporal information of the state of the sliding contact is very beneficial in obtaining in-depth understanding of the processes that lead to differences in the friction and wear performance of a tribosystem. However the term is widely used to describe very different methods and experimental situations. It is mostly used when it is possible to look into the tribological by using a transparent counter surface. This setup allows to measure the real area of contact between surfaces (see e.g. chap. 9 by Dassenoy), changes in chemistry (see chaps. 4, 6 and 7) or the thickness of a lubricant layer [1]. When it is not possible or desirable to use a transparent sliding surface it is possible to monitor changes in the wear track close to the tribological contact as described in a recent review by Wahl and Sawyer [2].

The term in situ is also used when the tribometer is located inside another instrument such as a TEM or when it is operating in a special environment such as an ultra-high vacuum chamber, as beautifully described in chap. 5 by Tysoe. This chapter will

M. Dienwiebel (✉) · P. Stoyanov
Institute for Applied Materials, Microtribology Center μTC, Karlsruhe Institute of Technology KIT, Strasse am Forum 7, 76131 Karlsruhe, Germany
e-mail: martin.dienwiebel@kit.edu; martin.dienwiebel@iwm.fraunhofer.de

P. Stoyanov
e-mail: pantcho.stoyanov@gmail.com

M. Dienwiebel · P. Stoyanov
Fraunhofer-Institute for Mechanics of Materials, Wöhlerstrasse 11, 79108 Freiburg, Germany

M. Dienwiebel and M.-I. De Barros Bouchet (eds.), *Advanced Analytical Methods in Tribology*, Microtechnology and MEMS,
https://doi.org/10.1007/978-3-319-99897-8_11

give an overview of in situ topography analysis during a tribological experiment and will describe the method of digital holographic microscopy (DHM) with a focus on tribology.

11.1.1 Optical Roughness Measurements

Optical roughness measurement techniques or optical profilers are well established and are being used in tribology research as well as in industrial quality control. Optical profilers can be divided into instruments that are interferometry based and confocal instruments. All optical techniques have in common that the lateral resolution is typically an order of magnitude lower that the resolution in the normal direction and that it depends on the used objective lens of the microscope. The optics also leads to differences in the maximum slope that can be observed with either confocal microscopy, phase shift or white light interferometry (Fig. 11.1).

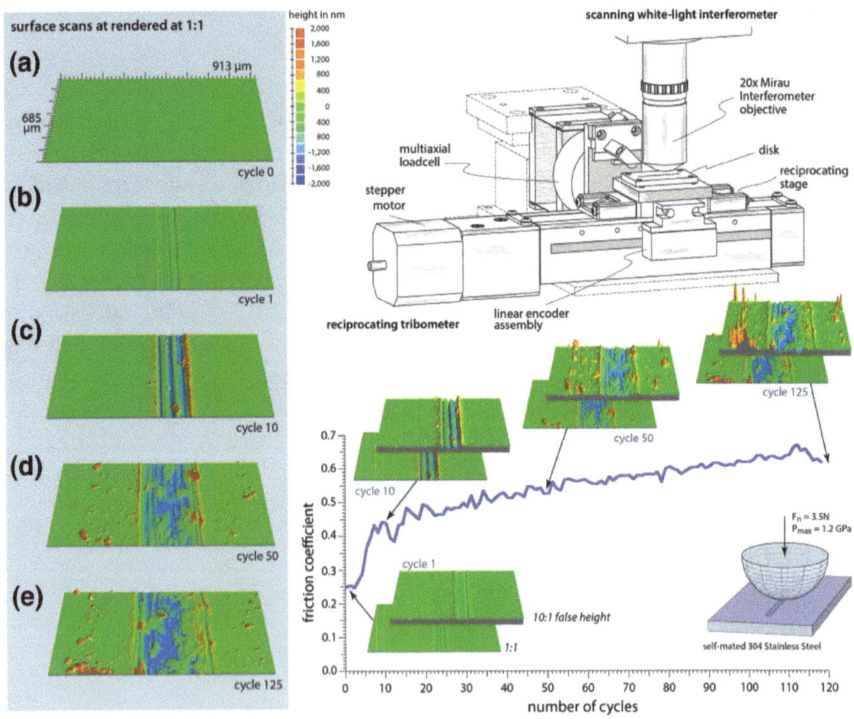

Fig. 11.1 Top: Schematic of a combined White Light Interferometer and Tribometer. Bottom: Friction as function of sliding cycles and snapshots of the wear track topography. Adapted from [3] with permission from *W*. Gregory Sawyer and Kathryn *J*. Wahl

With the introduction of the new standard EN ISO 25178 the measurement of 3D surface parameters using different instruments and their calibration are nowadays available. In this standard, parameters such as R_a, R_z and R_t are substituted by their corresponding 3D parameters S_a, S_z and S_t. As will be shown in this chapter the analysis of sliding surfaces should not be limited to the measurement of roughness parameters. Especially the observation of changes in time at the same location of the sliding track can be very valuable.

The implementation of an optical profiler into a tribological experiment is possible when the wear track can be accessed optically, i.e. when the length of the wear track is larger than the diameter of the counter sample and the objective lens. To overcome ambiguities in height due to phase jumps either the objective lens or the surface has to be scanned vertically and later the topography data is constructed from a large number of single images. The acquisition time thus depends on the total roughness of the surface and is typically in the order of several seconds.

11.1.2 Digital Holographic Microscopy (DHM)

A digital holographic microscope uses a coherent object wave that is brought into superposition with a reference wave. Figure 11.2 shows an optical configuration and is described in detail elsewhere [4, 5]. Instead of a photographic film or plate that is used to capture a classic hologram in digital holography the hologram is recorded with a CCD camera [6, 7] and therefore the acquisition is not limited by the exposure of the film and thus depends on illumination and frame rate of the CCD. The reconstruction of the object wave is obtained in DH by numerical computation with a computer using diffraction theory. Recent progress in computational speed and in the development of CCD cameras has made DHM a quasi "real-time" method. However, similar to the case of White Light Interferometry the reconstructed topography will suffer from phase jumps when the roughness of the sample is larger than the wavelength λ of the coherent light source. In such a case numerical phase unwrapping algorithms can be used. However, these algorithms can fail under certain circumstances such as steep wear tracks that extent over the whole field of view. Another way to circumnavigate phase ambiguities is to use two or more lasers [8]. Also Kühn et al. show that by using a dual-wavelength approach the a new synthetic wavelength Λ can be expressed as [9]:

$$\Lambda = \frac{\lambda_1 \lambda_2}{\lambda_2 - \lambda_1} \tag{11.1}$$

For two laser diodes with $\lambda_1 = 680$ nm and $\lambda_2 = 760$ nm the resulting synthetic wavelength, where phase jumps will occur is $\Lambda = 6.46$ μm. Therefore dual-wavelength DHM allows measuring surfaces with larger roughness without the numerical phase unwrapping.

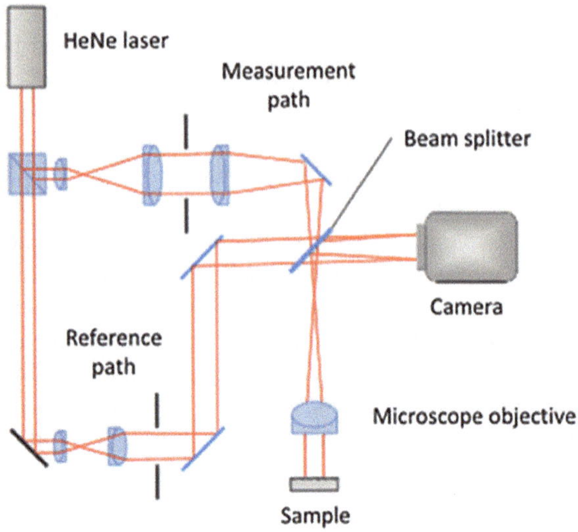

Fig. 11.2 Digital holographic microscope for reflection imaging. Reprinted with permission from [4] OSA

11.2 Integration of DHM into a Tribometer

In order to create a sequence of topographic images in a sliding experiment the topography acquisition has to be performed at the same position after each pass of the counter surface (Fig. 11.3). The precision of the positioning thus has to be better than the lateral resolution of the microscope. In the homebuilt tribometer that was constructed to integrate a DHM, this is accomplished by using a high precision *x*-*y* table that is steered by electromagnetic fields and that is used to move the lower sliding surface [10]. The position is monitored and adjusted optically with a resolution of 20 nm. The table can achieve a speed of 0.5 m/s and has a total travel of 200 mm in each direction. The position repeatability is better than 1 μm. The *x*-*y* table is mounted on a granite block and bridge in order to reduce mechanical vibrations that also deteriorate the DHM image acquisition. In order to further reduce vibrations the granite block is mounted on air dampers.

In a different example, a DHM was used to provide a better understanding of the cavitation phenomenon in tribological contacts [F]. In particular, the authors used DHM to measure cavitation bubble formation and thickness. The schematic representation of this measurement technique is shown in Fig. 11.4.

In the following sections of this chapter, several examples are given in order to demonstrate where DHM can help to obtain a better understanding of tribological phenomena by investigating the topography evolution of sliding surfaces.

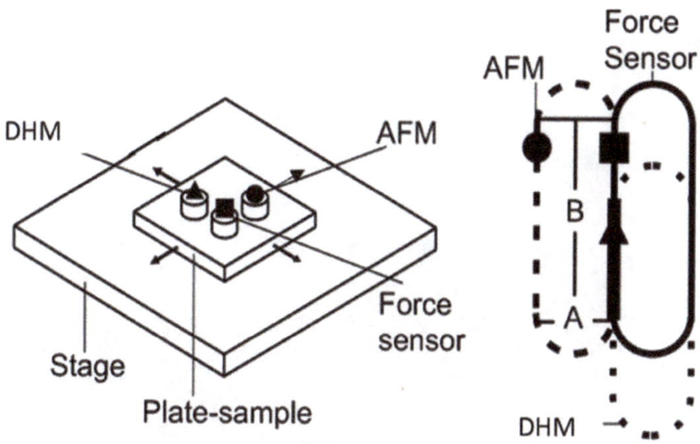

Fig. 11.3 The concept of integrating DHM, pin (square), and an atomic force microscopy (AFM) (circle). On the right, the travel path for the three components is indicated as AFM (dashed line), DHM (dotted line) and pin (solid line). Reprinted with permission from [11]

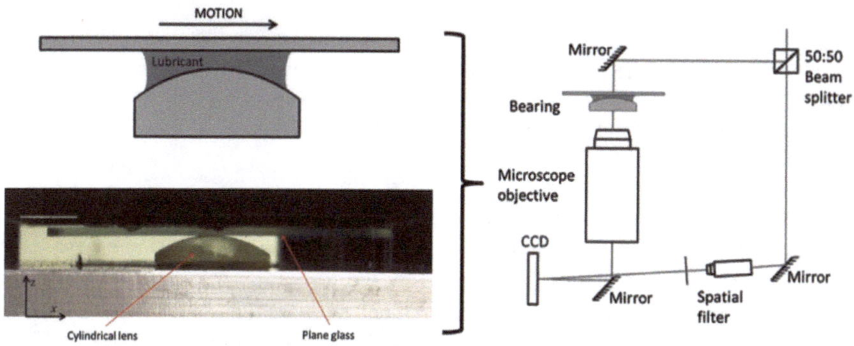

Fig. 11.4 Schematic representation of using DHM to improves the understanding of the cavitation phenomenon in tribological contacts. Reprinted under creative commons licence from [12]

11.2.1 Analysis of Plowing Friction

Plowing generally describes a situation where a sharp asperity indents into and plastically deforms a softer surface. The friction that is generated in the case of plowing is described as the sum of a shear contribution F_S and a plowing contribution F_P. According to Bowden Moore and Tabor the two terms can be written as [13, 14]:

$$F_F = F_S + F_P = \tau A_R + p' A' \tag{11.2}$$

Here τ is the shear strength of the softer material, A_R the real area of contact, p' is the flow pressure and A' the projected area of the asperity in direction of sliding. Thus the plowing friction depends mainly on the shape of the asperity and the normal

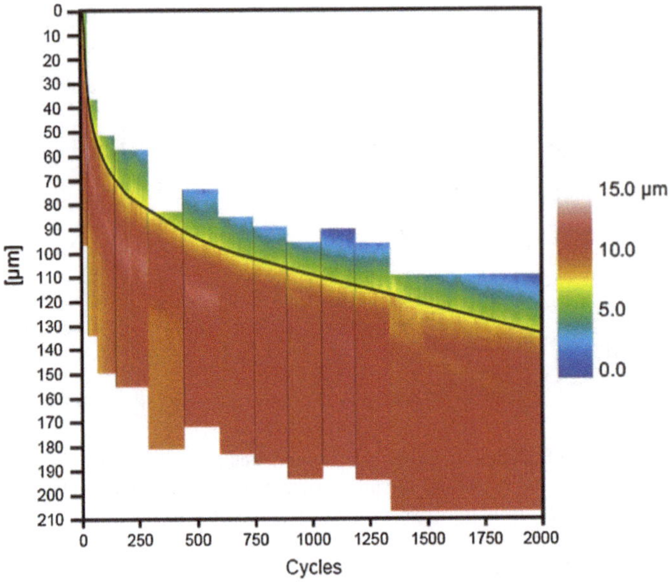

Fig. 11.5 Extracted profile lines from individual DHM frames. The profile lines are stitched together to compensate for the repositioning of the DHM microscope. The black solid line indicates the edge of the sliding track. Reprinted with permission from [15]

force. It is assumed in (11.2) that the asperity creates a single scratch and indeed plowing often dominates the initial phase of the running-in phase, whereas in a later stage the running-in of metals is governed by the formation of a third body. In the case that the plowing track subsequently widens and deepens, the model has to be modified. Therefore, in order to model friction in the case of multiple reciprocating sliding, the sliding track created by a spherical indenter was monitored as function of sliding cycles and correlated to the change in friction. The plowing experiments were conducted on a polished high purity copper sample. As indenter a ruby sphere with a diameter of 1 mm was chosen, mainly because of the high precision of the shape and the low roughness of the sphere. The surface was lubricated using base oil (PAO 8) in order to transport wear particles out of the sliding track and to prevent oxidation. The experiments were performed at room temperature and a humidity of approx 45% RH.

During the experiment the sliding track widened very rapidly and thus the DHM had to be repositioned perpendicular to the sliding track in order to keep one side of the track in the field-of-view of the microscope. At each sliding cycle one profile line across the sliding track was evaluated. All profile lines were added to a profile versus sliding cycle map. To correct for the repositioning all profile lines had to be stitched together after the experiment (Fig. 11.5).

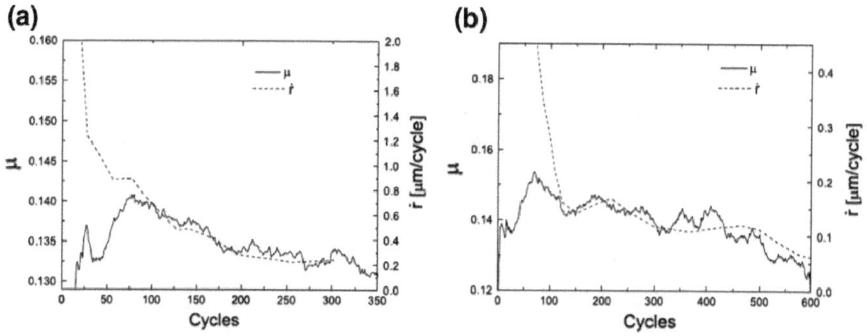

Fig. 11.6 Friction coefficient μ (solid line) and widening rate (dotted line) for 8 N (**a**) and 4 N (**b**) normal load and for 20 mm/s (**a**) and 15 mm/s (**b**) speed. Reprinted from [15]

From the position of the track edge the radius r of indentation can be extracted assuming that the widening of the track is symmetric. According to (11.2) the plowing contribution to friction should be proportional to r. However, in our experiments we find a better correlation between the friction coefficient and the widening rate per sliding cycle $\dot{r}$. The reason for this apparent discrepancy is the fact that (11.2) describes the situation of a single scratch or pass whereas the experiments here are performed by sliding multiple times in the same track. In this case a plowing contribution to the friction force is only present if the sphere sinks in between two subsequent cycles, since the effective cross section that contributes to the friction is $A' = K_i - K_{i-1}$, and the plowing cross section

$$K_i = \frac{1}{2}\left[r_s^2 \cos^{-1}\frac{R_i}{r_s} - R_i\sqrt{r_s^2 - \left(\frac{R_i}{2}\right)^2}\right] \tag{11.3}$$

Here r_s is the radius of the sphere and R_i is half of the width of the plowing track [16]. If the increase of R_i with every cycle is small, then it can be shown that A′ scales with the widening rate. Since the track widens very fast during the first cycles the above correlation can only be seen in the case of a ruby sphere sliding against copper from cycles 75–100 onwards depending on the loading conditions (Fig. 11.6).

11.2.1.1 The Influence of Running-in on the Topography of Brass

In a very similar fashion as presented in the first example, different running-in experiments were conducted on CuZn5 brass plates. The plates were annealed at a temperature of 650 °C for 45 min. The average grain size after the heat treatment was 30 ± 3 μm. Counter surfaces were produced from100Cr6 spheres with a diameter of 3 mm which were first hand-grinded and then hand-polished on one side in order to obtain flat-on-flat geometry. The tribocontact was lubricated by spraying poly-alpha

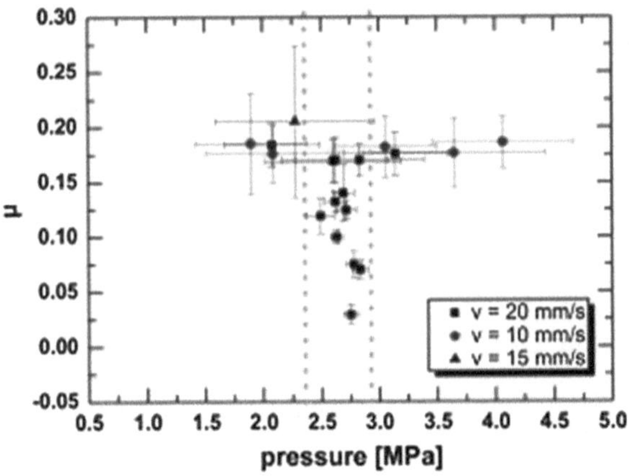

Fig. 11.7 Average friction coefficient μ as a function of speed and nominal contact pressure. (Reprinted with permission from [17], Elsevier)

olefin (PAO-8 with a viscosity of 45.8 mm^2/s at 40 °C). The oil temperature throughout the sliding experiments was kept at 35 °C. For the experiments we use a linear reciprocating path with a length of 120 mm. After each cycle the topography was recorded at the same location by DHM. In order to influence the running-in behavior the speed was varied from 10 to 20 mm/s and the nominal contact pressure from 1 to 4 MPa respectively.

Figure 11.7 shows the average friction coefficient recorded over approximately 5000 sliding cycles. Only few experiments located in a narrow range from 2.2 to 2.9 MPa (dashed lines) led to a significant friction reduction. The experiments that were conducted in this pressure range show a markedly decrease in friction (from 0.1 to below 0.05) with considerably smaller scatter of the friction force from one to another cycle. Figure 11.8 shows the evolution of the COF and for comparison the average roughness as a function of sliding cycles for one experiment that was conducted in the low-friction corridor. The roughness was computed from DHM topography maps. In this case the roughness follows roughly the trend in the friction. Another example of an experiment that is in terms of normal pressure located outside the low-friction corridor is presented in Fig. 11.9. Here the friction force and also the roughness show strong fluctuations although and possess a similar behavior although there is no strict correlation between the two signals.

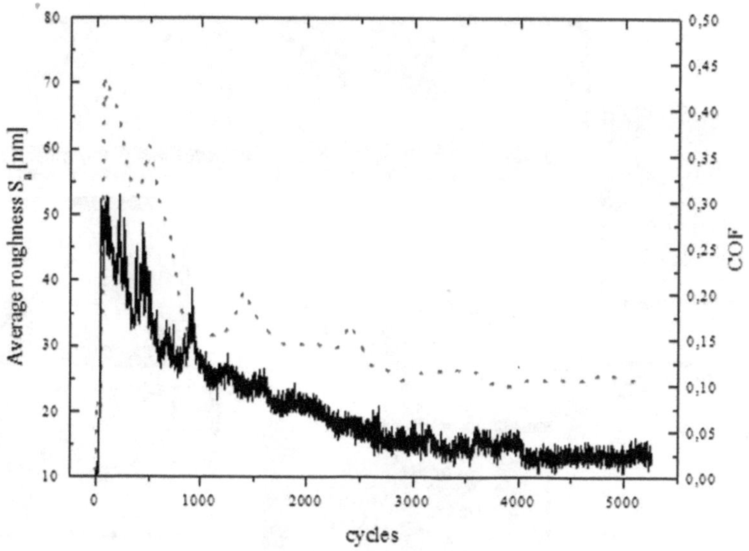

Fig. 11.8 Coefficient of friction (solid line) and average roughness S_a (dotted line) for a constant sliding speed of 20 mm/s and a nominal contact pressure of 2.7 MPa. The experiment was performed under lubrication using a synthetic PAO-8 base oil. (Adapted from [17])

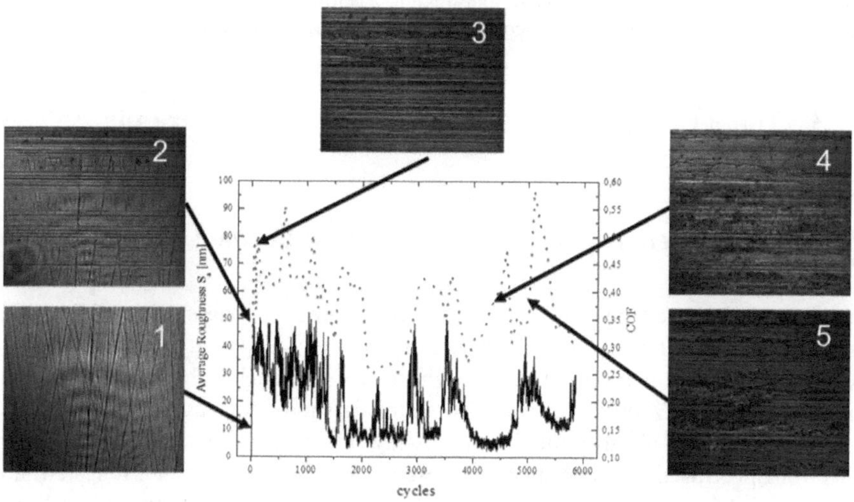

Fig. 11.9 Coefficient of friction (solid line) and average roughness S_a (dotted line) for a constant sliding speed of 20 mm/s and a nominal contact pressure of 3.6 MPa. The insets show topography maps of the wear track. The experiment was performed under lubrication using a synthetic PAO-8 base oil. (Adapted from [17])

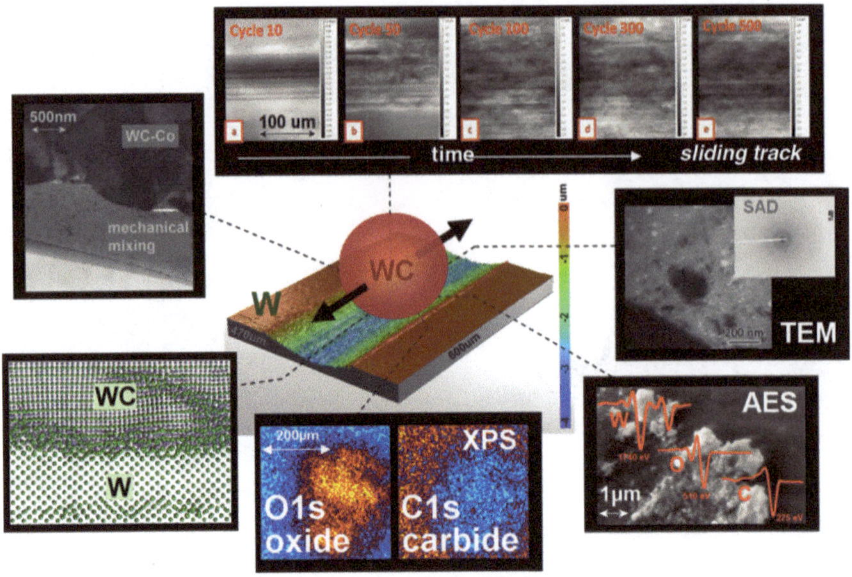

Fig. 11.10 An approach of studying dynamic interfacial processes leading to the variations in the friction and wear. This concept involves linking on line experiments to atomistic simulations with realistic bond order potentials. The structural/chemical changes observed in the simulations are compared to ex situ analysis using transmission electron microscopy, X-ray photoelectron spectroscopy, Auger Electron Spectroscopy, and Raman spectroscopy (Adapted from [21])

11.3 Topography and Wear Evolution in Metal/Ceramic Tribosystems

11.3.1 Tungsten Sliding Against Tungsten Carbide

In one study, we used the DHM in order to evaluate the roughness in dry sliding of a metal versus a ceramic in order to capture the dynamic interfacial processes leading to the variations in the friction and wear. The overall approach is shown in Fig. 11.10, linking in situ DHM characterization to atomistic simulations and ex situ analysis using transmission electron microscopy, X-ray photoelectron spectroscopy, Auger Electron Spectroscopy, and Raman spectroscopy. For these experiments tungsten and tungsten carbide were chosen in order to perform molecular dynamics simulations with realistic potentials. The simulations are presented elsewhere [18–20]. In the present example a 20× objective lens is used that is not operating in immersion, therefore allowing a larger field of view and measurement of the complete wear track.

Upon initial sliding the friction is approximately 0.1 and subsequently levels off to 0.6 where it remains nearly constant, as shown in Fig. 11.11a. Similarly, the scatter in the coefficient of friction increases after 40 cycles. The average roughness, measured by DHM, followed the trend of the friction coefficient closely. A similar behavior is observed from the roughness values parallel and perpendicular to the sliding direction. However, there are evident differences between the two directions, as shown in Fig. 11.11b; while the roughness in the perpendicular direction increases instantaneously (i.e. within 10 cycles) to nearly 80 nm, the roughness in the parallel direction took more than four times as long to reach the same value. This can be explained by the formation of grooves and scratches along the sliding direction and has been previously referred to as an adaptation of the surface topographies between the sliding interfaces [22]. These processes typically consist of plowing or cutting events caused by the asperities of the harder material onto the softer one (see [14]).

The wear depth values, obtained using the DHM data, are used to calculate the wear rate (i.e. depth/sliding distance) and analysed in terms of cycle number, Fig. 11.11c. The wear depth is calculated only up to the 80th cycle due to the width of the wear track being larger than the image size for the remaining cycles. The wear rate of the last cycle is also included in the analysis; however it is calculated using the depth obtained from the confocal microscope at the end of the test. The values measured by DHM and ex situ suggest that the wear rate remains constant after the first 50 sliding cycles. Interestingly, while the coefficient of friction and the average roughness are increasing, the wear rate is decreasing in the same phase of the experiment.

The in situ digital holography approach revealed a different behavior in lubricated ceramic/metal tribocouples (i.e. WC/W). The DHM images and corresponding friction coefficients are shown in Fig. 11.12. As expected, the friction is approximately 0.1 throughout steady state. Similarly to the unlubricated WC/W sliding contacts, the DHM images upon initial sliding show grooves parallel to the sliding direction indicating adaptation of the two surface topographies, as shown with the DHM images in Fig. 11.12. However, unlike the lubricated sliding case, the grooves here remained visible for the remainder of the test and the roughness values parallel to the sliding was low throughout the duration of the test. This correlated well with the low and steady friction values in the lubricated conditions.

An added benefit of integrating DHM into a tribometer is the observations of individual particles from consecutive images within the lubricant. An example of the consecutive images within the wear track is shown in Fig. 11.13. The observations clearly indicate that the particles are mobile and float within the hexadecane for the WC/W tribocouple with hexadecane. More details on this analysis can be found elsewhere [23].

11.3.2 Tungsten Sliding Against Diamond-like Carbon

In situ digital holography was also used more recently on diamond-like carbon coatings in order to provide a better understanding of the interfacial phenomena in lubricated and unlubricated sliding conditions. Figure 11.14 shows the average friction coefficient as a function of the cycle number for the two conditions. Similarly to the

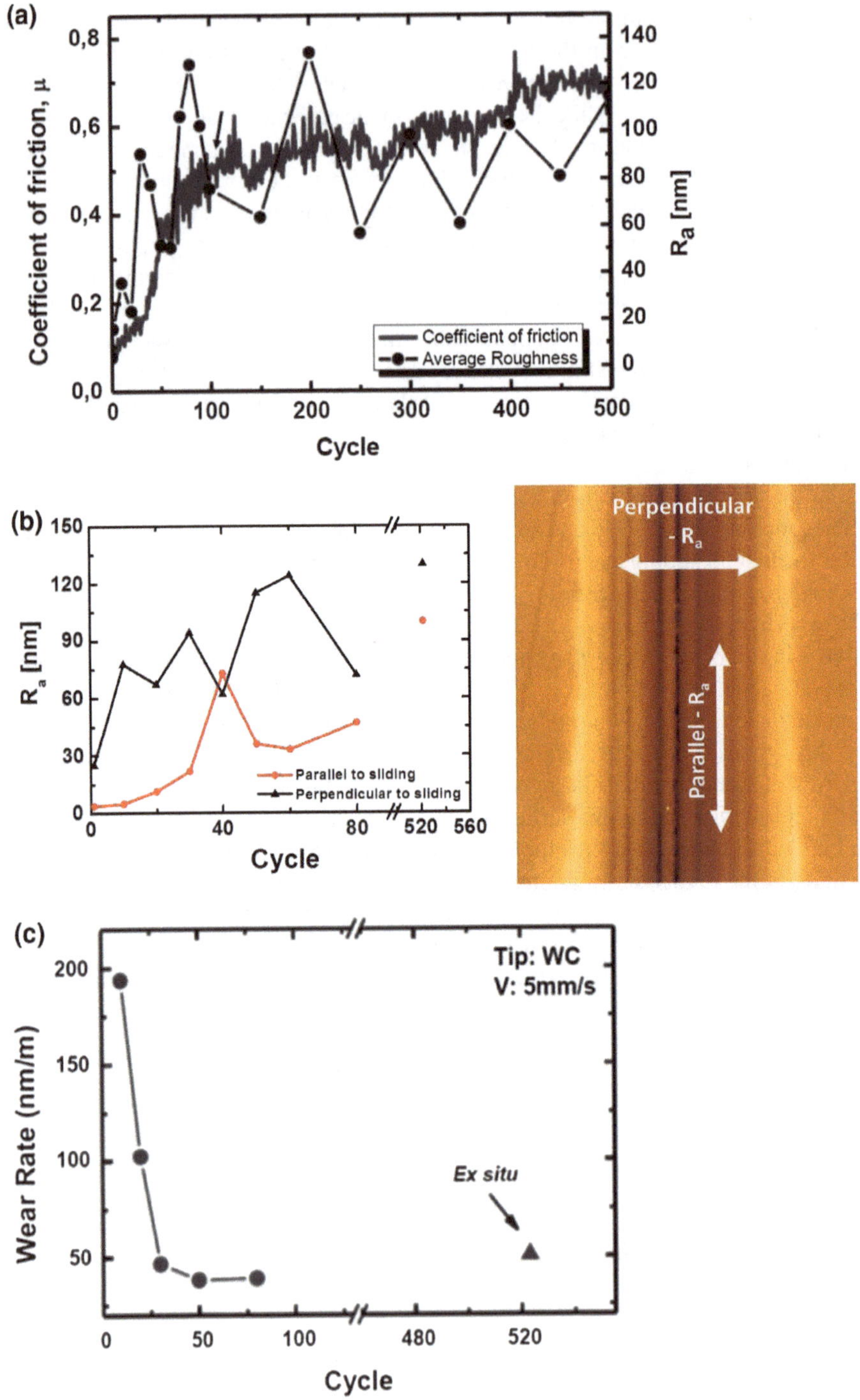
(a)
Coefficient of friction, μ
Rₐ [nm]
Cycle
Coefficient of friction
Average Roughness
(b)
Rₐ [nm]
Cycle
Parallel to sliding
Perpendicular to sliding
Perpendicular - Rₐ
Parallel - Rₐ
(c)
Wear Rate (nm/m)
Cycle
Tip: WC
V: 5mm/s
Ex situ

◀**Fig. 11.11** **a** Evolution of friction coefficients and average roughness. **b** Roughness values obtained using a thin rectangle parallel to the sliding direction and a thin rectangle perpendicular to the sliding direction (right panel). The roughness values obtained from the two directions versus the cycles are shown in left panel. **c** Wear rate for dry sliding W against WC. Roughness and wear rate are obtained from in situ holographic microscopy. Due to the large wear track created with the WC tip, wear rates are only possible to be obtained up to the 80th cycle. The last data point of the wear rate is obtained using ex situ confocal microscopy. It represent the average wear over ~520 cycles (Reprinted with permission from [23])

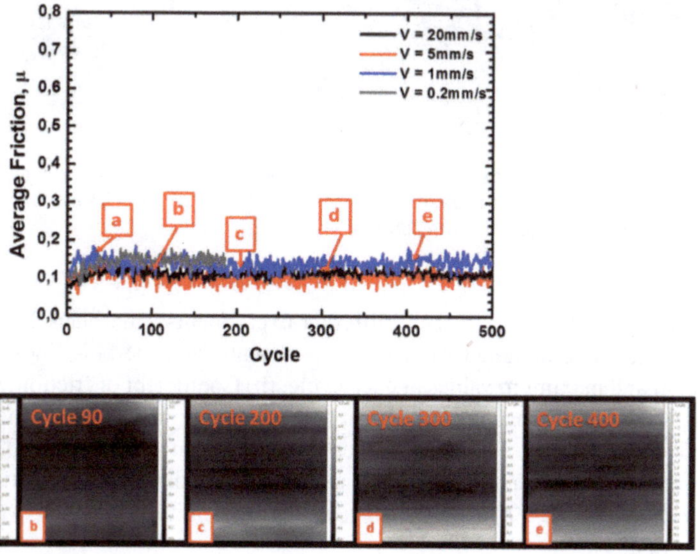

Fig. 11.12 Top: Evolution of the coefficient of friction as function of the sliding speed of a tungsten carbide sphere sliding against tungsten lubricated by PAO for sliding speeds ranging from 0.2 to 20 mm/s. Bottom: Topography of the wear track at different stages of the experiment (Reprinted from [23])

lubricated tungsten carbide/tungsten tribocouple. the friction coefficient of a-C:H sliding against tungsten with hexadecane is approximately 0.1 and remains nearly constant throughout the test (Fig. 11.14a). As a comparison, the dry sliding of a-C:H against tungsten resulted in slightly higher steady state friction values (Fig. 11.14b) however increased to above 0.4 between 2000 and 3000 cycles (Fig. 11.14c). The DHM images of the different conditions are shown within the figures and the observations followed the friction behavior closely; the lubricated test showed smooth wear tracks with lower roughness parallel to the sliding direction, while the dry sliding resulted in higher roughness parallel to the sliding. The behavior is also observed with the holographic images within the figures, where the lubricated conditions showed smoother wear tracks compared to the dry sliding case. Consistently with the increase in friction between 2000 and 3000 cycles for the unlubricated condition, the holographic images showed an increase in surface roughness and increased presence of

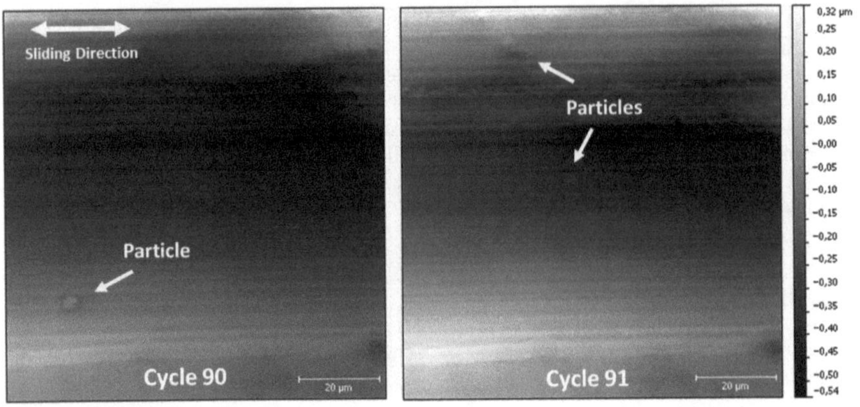

Fig. 11.13 DHM images of two consecutive cycles (i.e. cycles 90 and 91) for a sliding velocity of 5 mm/s. Observations of individual particles indicate that the particles are mobile and float within the hexadecane (Reprinted with permission from [23])

debris particles (Fig. 11.14c) different experiments. The black line represents the average value obtained from two experiments, and the blue lines show the minimum and maximum values at each cycle. (b) Coefficient of friction vs. cycle as well as holographic images of the wear track under dry sliding conditions for the first 1000 cycles and (c) long experiments (i.e. between 1000 and 4000 cycles). Experiments with higher cycle numbers show that the coefficient of friction increases up to 0.5 between 2500 and 3000 cycles, after which it fluctuates between 0.2 and 0.5. From [19].

11.3.3 Aluminum Sliding Against Sapphire

In a more recent study, in situ DHM was combined with optical in situ observation of the contact in order to study the interfacial processes in lubricated metallic (i.e., aluminum based) sliding conditions, as shown in Fig. 11.15 (see [24]). The DHM analysis in this study showed that the roughness evolution followed the coefficient of friction trend closely, with initially low values followed by higher roughness during steady state. Consistently, the optical in situ observation of the contact revealed that the transfer film behavior correlated well with the roughness of the worn surfaces and the subsurface microstructure of the worn surfaces [24].

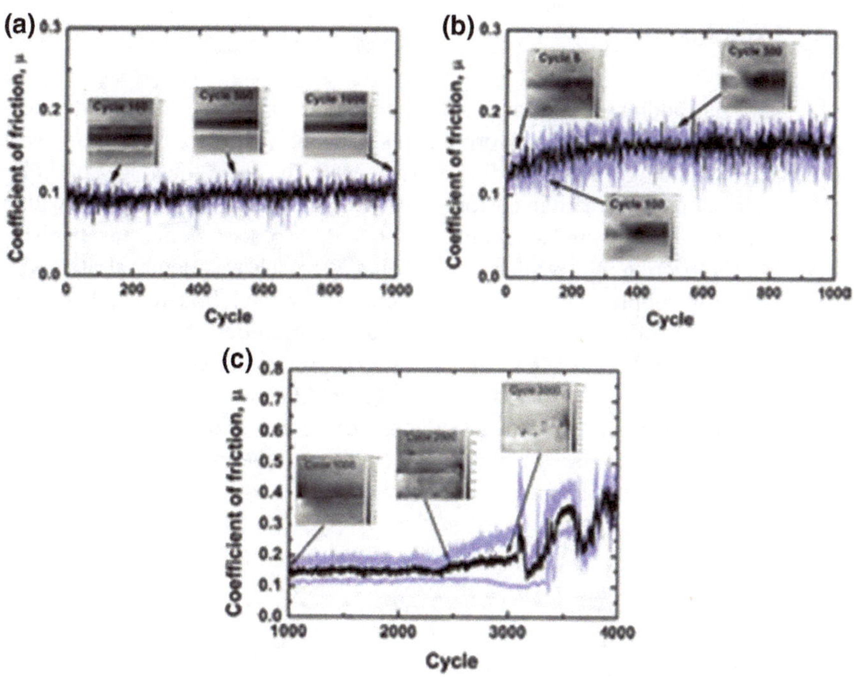

Fig. 11.14 **a** Friction coefficient as a function of the number of sliding cycles of a a-C:H coated sphere sliding against a tungsten plate lubricated by hexadecane at room temperature. The friction values are averages of two consecutive experiments. The insets show snapshots of the topography recorded by DHM. **b** Friction coefficient vs. cycles for an unlubricated a-C:H sphere against tungsten with 1000 cycles (b) and 4000 cycles (c). Instabilities in friction appear after approximately 3000 cycles (Reprinted with permission from [19])

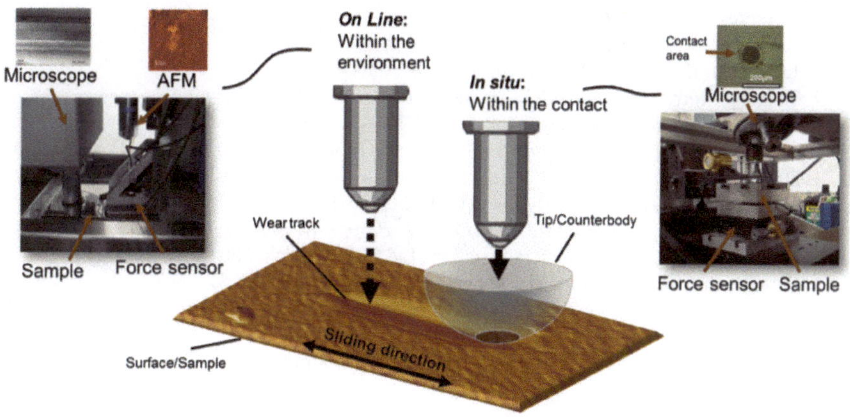

Fig. 11.15 Schematic representation of combining in situ and on-line methods for studying the interfacial processes in sliding couples. (Reprinted with permission from [24])

11.4 Conclusion

This chapter has addressed and outlined fundamental aspects of topography measurement techniques that allow performing in situ quantification of roughness, wear and surface morphology during sliding experiments. The main focus was on holographic microscopy integrated in custom-build tribometers. The use of DHM within tribometers broadens our understanding of the sliding processes and mechanisms that govern the friction and wear behavior. This approach was demonstrated with different material couples including metallic contacts in the lubricated and unlubricated conditions. In addition, there exists several opportunities of combining DHM with other in situ techniques (e.g. optical observation of the contact) in order to study the interfacial processes in lubricated metallic. As a future outlook, combining in situ techniques with ex situ transmission electron microscopy or scanning electron microscopy will provide a better understanding on the interfacial phenomena and velocity accommodation modes of tribosystems.

References

1. R. Gohar, A. Cameron, Nature **200**, 458 (1963)
2. W.G. Sawyer, K.J. Wahl, MRS Bull. **33**, 1145 (2008)
3. K.J. Wahl, W.G. Sawyer, MRS Bull. **33**, 1159 (2008)
4. E. Cuche, P. Marquet, C. Depeursinge, Appl. Opt. **38**, 6994 (1999)
5. P. de Groot, Adv. Opt. Photonics **7**, 1 (2015)
6. M.K. Kim, SPIE Rev. **1**, 18005 (2010)
7. M. Matrecano, M. Paturzo, P. Ferraro, Opt. Eng. **53**, 112317 (2014)
8. J. Gass, A. Dakoff, M.K. Kim, Opt. Lett. **28**, 1141 (2003)
9. J. Kühn, T. Colomb, F. Montfort, F. Charrière, Y. Emery, E. Cuche, P. Marquet, C. Depeursinge, Opt. Express **15**, 7231 (2007)
10. S. Korres, M. Dienwiebel, (n.d.)
11. S. Korres, M. Dienwiebel, Rev. Sci. Instrum. **81**, 63904 (2010)
12. T. Tang, N. Morris, J. Coupland, L. Arevalo, Tribol. Lett. **58**, 5 (2015)
13. F.P. Bowden, A.J.W. Moore, D. Tabor, J. Appl. Phys. **14**, 80 (1943)
14. F.P. Bowden, D. Tabor, *The Friction and Lubrication of Solids, Part I* (Clarendon Press, Oxford, 1950)
15. S. Korres, T. Feser, M. Dienwiebel, Wear **303**, 202 (2013)
16. Y.S. Kim, J.S. Ha, W. J. Kim, in *Des. Process. Prop. Adv. Eng. Mater. Pts 1 2*, ed. by S.G. Kang, T. Kobayashi (Trans Tech Publications Ltd, Stafa-Zurich, 2004), pp. 597–600
17. T. Feser, P. Stoyanov, F. Mohr, M. Dienwiebel, Wear **303**, 465 (2013)
18. P. Stoyanov, P.A. Romero, T.T. Järvi, L. Pastewka, M. Scherge, P. Stemmer, A. Fischer, M. Dienwiebel, M. Moseler, Tribol. Lett. (2013)
19. P. Stoyanov, P.A. Romero, R. Merz, M. Kopnarski, M. Stricker, P. Stemmer, M. Dienwiebel, M. Moseler, Acta Mater. **67**, 395 (2014)
20. P. Stoyanov, R. Merz, P. Romero, F.C. Wählisch, O.T. Abad, R. Gralla, P. Stemmer, M. Kopnarski, M. Moseler, R. Bennewitz, M. Dienwiebel, ACS Nano **9**, 1478 (2014)
21. P. Stoyanov, P. Stemmer, T.T. Järvi, R. Merz, P.A. Romero, M. Scherge, M. Kopnarski, M. Moseler, A. Fischer, M. Dienwiebel, A.C.S. Appl, Mater. Interfaces **5**, 6123 (2013)
22. S. Hanke, I. Samerski, J. Schöfer, A. Fischer, Wear **267**, 1319 (2009)

23. P. Stoyanov, P. Stemmer, T.T. Järvi, R. Merz, P.A. Romero, M. Scherge, M. Kopnarski, M. Moseler, A. Fischer, M. Dienwiebel, ACS Appl. Mater. Interfaces **5**, (2013)
24. P. Stoyanov, J.M. Shockley, M. Dienwiebel, R.R. Chromik, MRS Commun. **6**, 301 (2016)

Part IV
Computer Simulations

Chapter 12
Understanding the Tribochemistry of Lubricant Additives by *Ab initio* Calculations: The Case of Phosphites

M. Clelia Righi

Abstract The search for novel lubricants to improve the energy efficiency of engines and help mitigate the environmental effects of carbon dioxide emissions has gained increasingly importance in recent years. Commercial lubricants contain a wide range of compounds including those that undergo tribochemical reactions and form friction- or wear-reducing films. A microscopic understanding of tribochemical reactions is of paramount importance for designing new, environmental-friendly lubricants. However, many aspects of tribochemistry remain elusive due to the difficulties in experimentally probing the sliding buried interface. Simulations can play a decisive role in this context, in particular ab initio molecular dynamics (AIMD), where both the ionic and electronics degrees of freedom are fully taken in into account. This is essential for an accurate description of reactions in situations of enhanced reactivity imposed by the tribological conditions. This chapter offers an example of application of ab initio methods in tribochemistry. A twofold analysis of the reaction mechanisms and effects of tribochemical reactions involving organophosphorus additives at iron interfaces allowed to understand the mechanisms of function of phosphorus-based additives in boundary lubrication.

12.1 Introduction

Many tribological phenomena can ultimately be described as chemical reactions occurring in the presence of mechanical forces, i.e., **tribochemical reactions (TRs)**. One important example is represented by boundary lubrication, where tribologically-induced modifications of the surfaces interacting with lubricant-additives (or with other molecules present in the environment surrounding the sliding media) can dramatically change the adhesion and friction of the materials in contact. Therefore,

M. Clelia Righi (✉)
Department of Physics, Informatics and Mathematics,
University of Modena and Reggio Emilia,
Via Campi 213/a, Modena, Italy
e-mail: mariaclelia.righi@unimore.it

M. Dienwiebel and M.-I. De Barros Bouchet (eds.), *Advanced Analytical Methods in Tribology*, Microtechnology and MEMS,
https://doi.org/10.1007/978-3-319-99897-8_12

it is highly important understand their activation mechanisms. Our knowledge of the reaction kinetics and thermodynamics driving forces at the open surface is not sufficient for this aim, since the tribological conditions alter the reaction kinetics. Tribological conditions include **frictional heating, molecular confinement, shear stress, and high pressure**. Usually, the reaction rate is observed to increase, a phenomenon referred to as "mechanical catalysis" in the field of **mechanochemistry**, where chemical and materials syntheses are achieved with the application of mechanical forces. Many aspects of mechanochemistry as well as tribochemistry remain elusive due to the difficulties in directly probing the interface: From the analysis of the reaction products, typically by means of *post mortem* spectroscopic analysis, it is possible to acquire information on the type of chemical reactions occurred, but very little can be inferred on the activation mechanisms, reaction pathways and rates. **Simulations** can play a decisive role here, in particular ab initio **molecular dynamics (AIMD)**, where both the ionic and electronics degrees of freedom are fully taken in into account. This is essential for an accurate description of **bond-breaking and bond-forming reactions in situations of enhanced reactivity imposed by the tribological conditions**. However, the use of AIMD in tribology has been very scarce and most of the existing simulations of tribochemistry are based on force fields, i.e., on a parameterization of the atomic interactions. While many force fields are available for selected chemical environments, their transferability is often poor, especially to describe the non-ordinary conditions present at the tribological interface. In this chapter we show that ab initio methods can succesfully be applied to understand the functionality of lubricant additives.

The design of lubricants to decrease friction and wear in machine components is an important way to increase the energy efficiency of mechanical systems while taking into account restrictive environmental requirements and technological advances [1]. Lubricants are formulated products composed of base oil and a package of additives designed for specific performance needs. The additives can be classified as chemically active, i.e., designed to chemically interact with the surface and form protective layers, or chemically inert, i.e., with the function of improving the physical properties of the bulk materials. Most of lubricant additives have a non-polar part, usually consisting of a hydrocarbon chain, the functionality of which is to solubilize the molecule into the base oil and a functional polar group that reacts with the surface. Here we consider extreme-pressure (EP) and anti-wear (AW) additives where the key elements of the functional group are sulfur and phosphorus. EP, AW additives are typically used in synchronizer systems, [2] which reduce the speed difference between the shaft and the idler gear [3]. The synchronizing system is a key component of manually operated vehicle transmissions, as fluent transmission operations have become increasingly important to drivers in the recent years. Gear oils must perform in conditions and applications that can vary drastically, therefore they are enriched by additives that protect seals, improve thermal, oxidative and viscosity stability, provide micropitting resistance, bearing corrosion protection, foam resistance, enhanced demulsibility and load-carrying capacity. The concentration of EP and AW Additives that work in severe conditions is higher in gear oils, which are enclosed in gear boxes, than in engine oils [4, 5].

S- and P-based compounds are largely used as EP and AW additives for gear oils [6–8]. EP additives typically adsorb onto the metal surface either by physical or chemical attraction [9, 10]. Under severe tribological conditions, they react with the surfaces forming surface films that prevent the welding of opposing asperities and avoid scuffing that is destructive to sliding surfaces under high loads [11].

The action mechanism of the sulfur-based EP/AW additives was first proposed by Davey et al. [12] and then refined by Forbes et al. [13]. The sulfur compounds adsorb on an iron surface and under tribological conditions form inorganic iron sulfide film, which can reduce frictional wear and prevent seizure. The Fe–S formation under boundary lubrication conditions has been identified in different studies [14–16]. Recently, Li et al. considered three novel S-containing alkyl phenylboric esters [17, 18]. After the tribotest, the worn surface was investigated by X-ray absorption near edge structure spectroscopy (XANES) and Fourier transform infrared (FT-IR) spectroscopy. The surface analysis confirmed the presence of iron sulfide.

The chemical structure of P-containing additives influences the nature and the tribological performances of the tribofilm as highlighted in [19–22], where phosphite and phosphate have been compared. Typically, organic phosphites function as friction-modifiers whereas phosphates as anti-wear additives. Post mortem analysis of the boundary lubrication film formed on frictional surfaces showed that an iron phosphate compound is often obtained from phosphates [15, 22–24]. First, the organic phosphate molecule adsorbs on the iron substrate and then, iron phosphate is formed due to thermal/mechanical decomposition, [26–32] but also due to hydrolysis [23, 25]. The formation of iron phosphate as a boundary lubrication film that prevents wear-out has been observed with steel ball-on-disc tribotest followed by X-ray photoelectron spectroscopy (XPS) [33]. Philippon et al. have proposed a decomposition mechanism of a model EP additive, trimethylphosphite (TMPi), on nascent iron surface [34] that leads to the formation of iron phosphide film under tribological condition [35–37].

Ab initio calculations have been applied to investigate the functionality of P-containing additives. A twofold analysis have been performed: (i) the tribochemical reactions that lead to the formation of a tribofilm starting from P-containing additive molecules have been identified by combining chemisorption studies and ab initio molecular dynamics simulations; (ii) the functionality of the P-containing tribofilms has been elucidated by first principles calculations of intrinsic tribological properties of iron interfaces. The main stages of this analysis are described in the following to provide an example of a computational protocol that can be applied to the research in lubricant additives.

12.2 Combining First Principles Calculations and XPS Experiments to Study the Adsorption of Lubricant Additives

Heavy molecular weight organophosphorous additives like triphenylphosphite are used as in commercial lubricants. The functional group containing phosphorous affects the frictional properties while the long hydrocarbon chains have the function to improve the additive solubility in oil. Since it is difficult to evaporate such heavy molecular weight additives in UHV, 10^{-9} hPa, smaller molecules, having similar active gropus, are used as model compounds for experimental research in lubricant additives. Among all types of phosphite additives, trimethylphosphite (TMPi), with chemical formulae $P(CH_3O)_3$, is probably the simplest compound containing the phosphite ion, thus it was chosen as a model to provide a basic understanding on the functionality of organophosphorous additives. First principles calculations based on density functional theory (DFT) were applied to elucidate the mechanism of P release from TMPi decomposition at the iron surface. The numerical results were discussed in relation to the in situ XPS analysis performed on metallic iron after TMPi thermally-controlled adsorption [38].

Spin-polarized DFT calculations were performed within the generalized gradient approximation proposed by Perdew–Burke–Ernzerhof (PBE) to describe the exchange correlation functional [39]. The electronic wave-functions were expanded on a plane waves basis and the ionic species were described by pseudopotentials. The Fe(110) surface was considered since it is the most stable among the densely packed iron surfaces [40]. The surface is modeled by means of periodic supercells containing an iron slab of (4×4) in-plane size, e.g. 16 atoms per layer, and a vacuum region 20 Å thick. It has been verified that the choice of the (4×4) in-plane size is sufficient to avoid lateral interaction of TMPi with its periodic replicas. To identify the microscopic mechanisms that govern the release of P and its adsorption onto the Fe(110) surface by TMPi dissociation, the adsorption of TMPi was first studied and then the reaction paths for molecular dissociation were calculated by means of the Nudged elastic band (NEB) method [41]. The NEB method allows us to obtain the minimum energy path (MEP), which is the path with the greatest statistical weight, followed by the system in a transition from an initial to a final state. In the adsorption study, the initial and final states corresponded to the undissociated and dissociated adsorption configurations identified as the most stable ones.

Temperature-programmed reaction spectroscopy (TPRS), X-ray photoelectron spectroscopy (XPS) and low-energy electron diffraction (LEED) [42] experiments have shown that TMPi decomposes into adsorbed phosphorous and gaseous CO and H_2 via the methoxy (CH_3O) intermediate on clean Fe(110), leaving behind a phosphide film. In agreement with these observations, the calculated reaction energies indicate that the molecular dissociation via methoxy detachment is a highly exothermic process [38]. Furthermore, the calculations revealed that the adsorption of molecular fragments is more favorable when the fragments are isolated on the surface, i.e., they do not share metal atoms with neighboring adsorbed fragments.

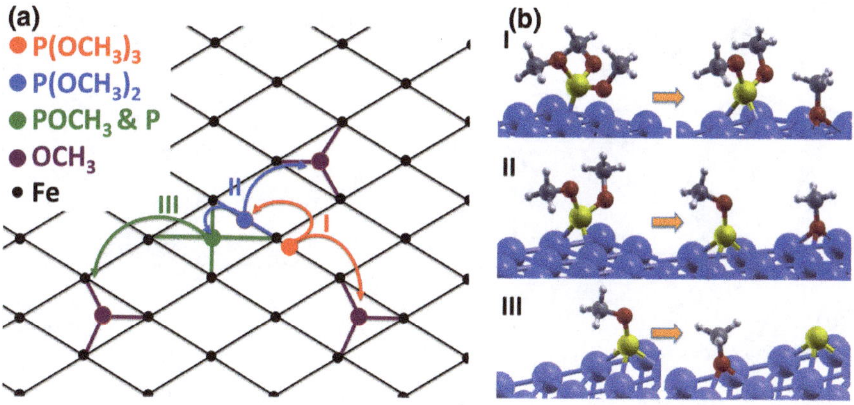

Fig. 12.1 Schematic representation of the TMPi dissociation into atomic phosphorus and adsorbed methoxy groups on the Fe(110) surface (**a**). Initial and final states for each methoxy detachment process (**b**) [31]. Fe is colored in blue, P in yellow, O in red, C in grey, and H in white

This result indicates that TMPi dissociative adsorption is favored in the presence of open Fe sites, as evidenced by the experiments that highlighted the key role played by the nascent metallic surfaces on the tribochemistry of phosphites [43, 44].

A schematic representation of TMPi dissociation in atomic phosphorous and methoxy groups on the Fe(110) surface is offered in Fig. 12.1a, while the initial and final states of each reaction of methoxy detachment are reported in Fig. 12.1b in a ball-and-stick representation. The TMPi molecule binds to the surface forming a P–Fe bond. Via methoxy detachment, the number of P–Fe bonds increases until the P atom is adsorbed into a long bridge site, where it is bonded to four Fe atoms. The energy of the dissociated adsorption configurations decreases with the number of detached methoxy groups, and the most stable adsorption configuration is obtained when all the methoxy groups are detached and atomic P is released at the surface. The passivation of the Fe surface with phosphorus through TMPi dissociation is thus an energetically favorable process.

The reaction paths and corresponding energy barriers obtained by means of the NEB method are reported in Fig. 12.2. The results indicate that an activation energy of $E_A = 0.75$ eV is necessary for the first methoxy detachment, and produces an energy gain of $E_R = -0.73$ eV (Fig. 12.2a). The remaining $P(CH_3O_2)$ is less stable than the entire molecule, thus the second methoxy detachment requires a lower activation energy ($E_A = 0.23$) and produced an higher gain ($E_R = -1.20$) than the first methoxy dissociation and (Fig. 12.2b). Such energy gain is sufficient to promote the decomposition of the residual $P(CH_3O)$ fragment in elemental P and adsorbed methoxy, which is still an exothermic process ($E_R = -0.78$).

The formation of iron phosphide from the TMPi adsorption onto metallic iron at different temperatures has been investigated thanks to in situ surface analyses. Figure 12.3 shows the P_{2p} XPS spectra of phosphorous obtained in situ after adsorption of TMPi onto an etched steel surface at various temperatures. The chemisorption

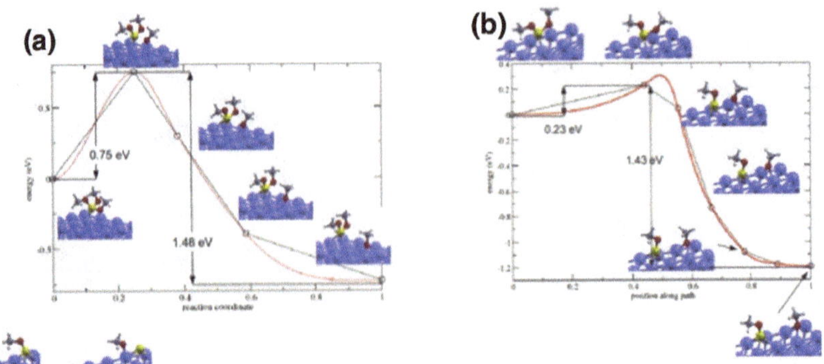

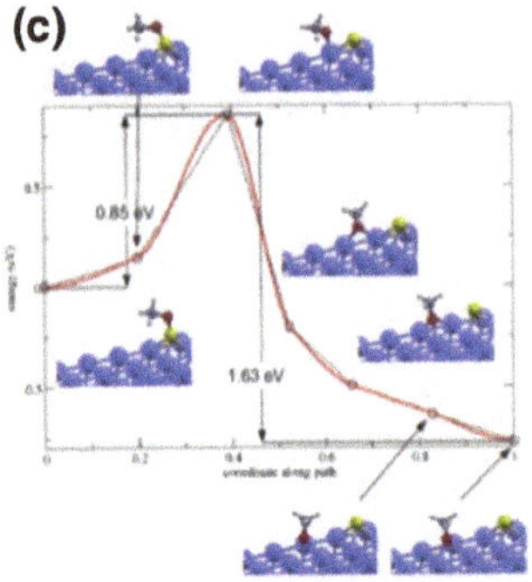

	Initial state	Final state	E_R (eV)	E_A (eV)
a)	$P(OCH_3)_3$	$P(OCH_3)_2$ + OCH_3	- 0.73	0.75
b)	$P(OCH_3)_2$	$POCH_3$ + OCH_3	- 1.20	0.23
c)	$P(OCH_3)$	P + OCH_3	- 0.78	0.85

Fig. 12.2 Reaction paths, energies, E_R, and barriers, E_A, for subsequent methoxy detachment from the the TMPi molecule [31]

of TMPi molecules is evidenced by the appearance of the P_{2p} XPS peak at 133.2 eV at 100 °C. The characteristic peak of iron phosphide appears clearly at a temperature of 300 °C with a distinct characteristic binding energy of 130 eV. These results strongly suggest that the formation of iron phosphide is thermally-activated. In fact, temperatures above 200 °C increase the formation of phosphide to a large extent, providing sufficient energy to dissociate the TMPi molecule. Although iron phosphide formation is evidenced, undissociated TMPi molecules are also observed on the surface, as shown by the characteristic binding energy at 133.5 eV.

The combined computational and experimental analysis above described revealed that the full TMPi dissociation, leading to phosphorus release, can occur once the activation energy is provided, e.g. by heating the sample. In the next section the effects of mechanical stresses (load and shear) in the activation of the reaction will be considered.

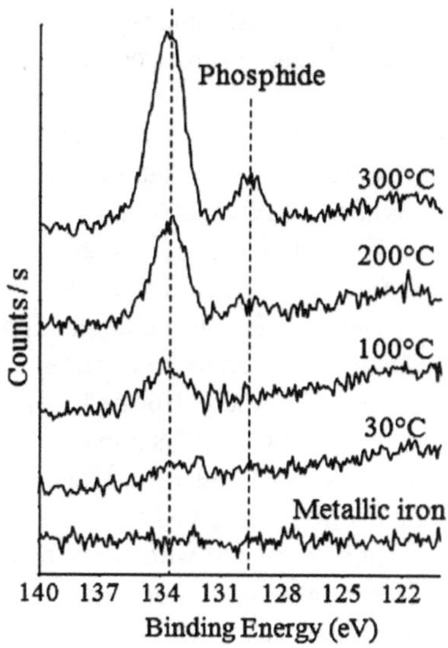

Fig. 12.3 In situ X-ray photoelectron spectroscopy P_{2p} lines recorded after TMPi thermally controlled adsorption onto etched steel, e.g. metallic iron [31]

12.3 Effects of Mechanical Stresses on Additive Decomposition at the Tribological Interface

The tribological conditions, which include high temperatures, molecular confinement, load and shear stresses can highly enhance the rate of chemical reactions. The first proposed mechanisms [45] attributed the increase of the reaction rates to extremely high temperatures, "flashes temperatures," that develop at few discrete "hot spots," where the contact between the two surfaces occurs [46]. Such temperature increase has been also proposed to produce local excitations, "magama plasma," that decay rapidly into a local heating of the surface [47]. However, recent works have shown that tribochemical reactions can take place even though the temperature raise during sliding is negligible or limited and highlighted the primary role of mechanical stresses in the activation of chemical reactions [48–57]. AIMD simulations of tribochemical reactions involving water molecules confined at diamond interfaces have shown that confinement under pressure is the driving force for molecular dissociation and reaction rates are functions of load. In particular, it was found that the higher the load, the more effective the surface passivation by water dissociation is, which in turn leads to a reduction of friction [58, 59].

Applying AIMD to study the tribochemistry of additives at iron interfaces represents a great computational challenge due to difficulty imposed by the presence of a metallic system where the ionic and electronic degrees of freedom can exchange energy due to the absence of a band gap. In addition, the magnetic character of iron imposes the use of spin polarization. This challenge has been successfully faced [60] by adopting the Born Oppenheimer scheme, where the ions are moved according to the Hellmann–Feynman forces obtained from the total electronic energy, which is minimized at fixed ionic position every MD step. In this way the ionic and electronic degrees of freedom are decoupled and the dynamics of metallic systems, even including magnetization, can be described in accurate way. Iron interfaces, composed by two self-mated Fe(110) surfaces, were modeled by means of two slabs with (4×4) in-plain size. The Brillouin zone sampling was realized by means of a $(2 \times 2 \times 1)$ Monkhorst Pack grid [61]. The bottom layer of the lower slab (the substrate) was held rigid, while the upper slab (the counter-surface) was moved at constant velocity of 200 m/s. A vertical force was applied to the atoms belonging to the top layer of the counter-surface to model an applied pressure the value of which was varied to evaluate the dependence of the reaction rates on load. In the initial configuration, common to all the performed simulations, a TMPi molecule per cell (of ~100 Å^2 area) was adsorbed in its most stable configuration on the iron substrate and the system was relaxed under the effects of the applied pressure. After the relaxation, the system was equilibrated at a constant temperature of 300 K and the countersurface was laterally moved with constant velocity.

In the first simulation, the pressure was fixed to 0.5 GPa. During a simulated time interval of 7 ps the system dynamics consisted in molecular vibrations, mainly rotations of each methyl group around the axis along the CO bond, and we did not observe any dissociative reaction. By increasing the pressure to 2 GPa, molecular dissociation is, instead observed after 1 ps. As can be seen in Fig. 12.4a, b, as soon as the iron counter-surface gets closer to the molecule, the oxygen atoms become attracted by the iron countersurface, the PO bonds get stretched and finally dissociatevily adsorb on the countersurface while elemental phosphorus remains adsorbed on the substrate (Fig. 12.4c).

The reaction path just described is perfectly consistent with that predicted by the NEB method and observed experimentally at the open surface (Sect. 12.2). Molecular dissociation occurs, in fact, through methoxy detachment and the importance of clean Fe sites for favoring methoxy adsorption appears evident in the dynamic simulations, where the clean Fe counter-surface attracts methoxy groups. We also observed that the released elemental phosphorus at the end of the MD simulation was located at a long bridge site, which was indicated as the most favorable site for P adsorption at the Fe(110) surface (Fig. 12.1a). It is interesting to notice that the molecular dissociation observed during the dynamic simulation occurred as expected on the basis of thermodinamical driving forces. However, the activation time for TMPi dissociation observed in the dynamic simulation under tribological conditions turned out to be much shorter than the activation time obtained by applying the Arrhenius law at room temperature and considering a dissociation barrier of 0.75 eV, as reported in Sect. 12.2. This result, along with the different trajectories that we obtained for the two simulations at 0.5 and 2 GPa, indicate that the mechanical stresses play a crucial role in accelerating the reaction rates.

In the second part of the dynamic simulation at 2 GPa pressure, it was observed the tendency of phosphorus at low concentration on the substrate to interact with the counter-surface. Once Fe–P bonds were established across the interface the surface separation was reduced (Fig. 12.4d) and a direct contact between few metal atoms occurred (Fig. 12.4e). The direct metal-metal contact was sufficient to promote the

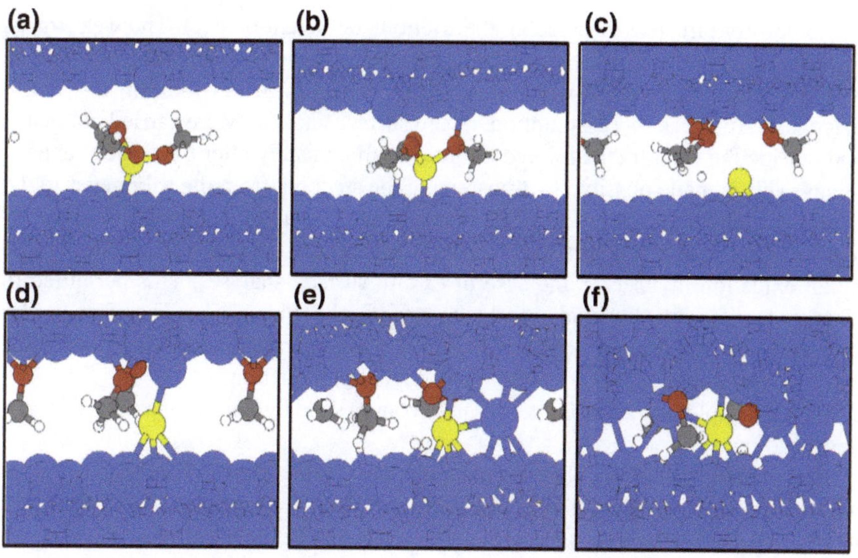

Fig. 12.4 Snapshots acquired during AIMD simulation of TMPi tribochemistry at iron interfaces under 2 GPa pressure and room temperature. The simulation time increases from panel (**a**) to panel (**f**) [60]

cold sealing of the two iron surfaces and the molecular fragments remained embedded in the iron matrix (Fig. 12.4f). This event in real life would be detrimental, corresponding e.g. to engine seizure. An increased phosphorous coverage is crucial to prevent direct metal-metal contact [62].

A closer inspection of the molecular fragments embedded in the iron matrix revealed that the methoxy group got dissociated into H atoms on CO molecules. This result, in agreement with the observation that the dissociation of a CO molecule on the Fe(110) surface is an endothermic process, [63] reveals that O adsorption does not compete with P adsorption from TMPi decomposition. This can explain the higher efficiency of phosphites than phosphates observed in the experiments [62].

12.4 Functionality of Phosphorous in Reducing Adhesion and Friction

Gas phase lubrication (GPL) experiments are ideal experiments to provide understanding on the functionality of extreme pressure additives because the chemical compounds are tested without the presence of the base oil, thus mimicking boundary lubrication conditions. In a gas phase lubrication experiment the friction coefficient of a steel-on-steel sliding contact was first measured in ultra high vacuum (UHV) and then in the presence of 1 hPa of TMPi. As can be seen in Fig. 12.5a, TMPi turned out to be very effective in reducing the friction coefficient of steel. The wear tracks formed on the surface of the steel flat at the end of the experiments in the presence of TMPi were observed by optical microscopy (Fig. 12.5b). A tribofilm was clearly evidenced. To check this assumption, in situ surface analyses were carried out inside and outside the wear tracks formed on the flat immediately after the friction experiments. This is made possible by transferring the steel flat from the tribometer to the analysis chamber thanks to the UHV intermediate preparation chamber. The significant advantage of in situ analyses is that the signals are not changed and/or masked by an oxidation in air as in the case of ex situ surface analyses. This permitted to clearly identify iron phosphide inside the wear tracks (Fig. 12.5b). Therefore, a film composed by iron and elemental phosphorus was generated during the tribological test and it was responsible for the observed friction reduction.

First principles calculations were carried out to shed light into capability of phosphorus to reduce iron friction.

A computational protocol to calculate from first principles two intrinsic tribological properties of solid interfaces, namely the *work of separation* W and the *ideal interfacial shear strength*, τ, was proposed in [64] These properties, which correspond to the energy required to separate two surfaces from contact and to the static friction force per unit area, are ruled by the physical/chemical interactions between the surfaces in contact. First principles calculation based on Density Functional Theory (DFT) can accurately describe surface-surface interactions, offering the possibility to characterize in silico the adhesive and shear strength of materials and their change due to surface chemical modification. The protocol was applied to identify the effects of surface passivation in diamond [65], to compare the friction reduction properties of phosphorus and sulfur [66], and to understand the lubricating properties of graphene [67]. The protocol has been recently implemented as a workflow to perform high throughput calculations of the intrinsic tribological properties of solid interfaces [68]. Here we show how it can be applied to quantify the effects of phosphorus on the adhesion and resistance to sliding og iron.

It is well known that phosphorus segregates in iron and forms 2D chemisorbed overlayers, [69] but there is presently no experimental information on their structure in the literature. The most favorable adsorption site for P at the Fe(110) surface, was indentified by means of first principles calculations by comparing the adatom energy at high symmetry locations [20]. The long bridge site, having the highest coordination, turned out to be the most stable adsorption location. Then, the effects of coverage were evaluated by comparing the P adsorption energy in the configurations represented in Fig. 12.6. It turned out that a coverage increase with respect to 0.25 ML (Fig. 12.6b) destabilizes the overlayer. In particular, the adsorption energy per atom increases by 6% for a coverage increase to 0.5 ML (Fig. 12.6c) and by 21% at full coverage (Fig. 12.6d) [70].

The interfaces were modeled by self-mating the optimized surfaces represented in Fig. 12.6. A ball-stick representation of the optimized interface structures obtained for the PES minima is presented in Fig. 12.7, the corresponding work of separation, W, and the distance between interfacial Fe layers, z_{eq}, are reported in Table 12.1.

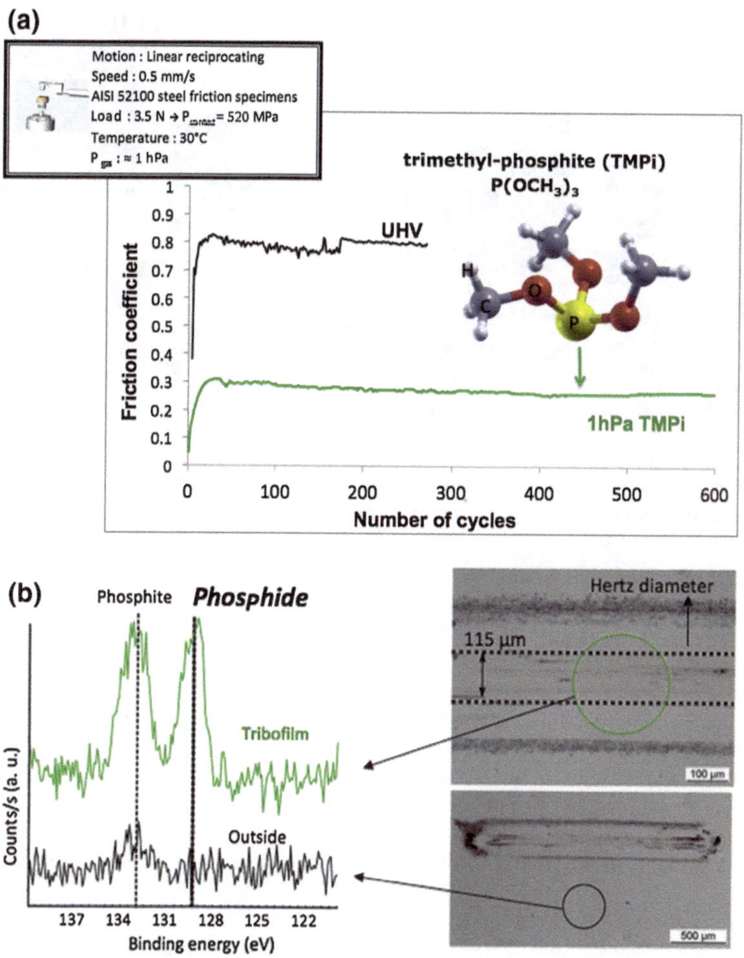

Fig. 12.5 Steady-state friction coefficient obtained under GPL at room temperature for steel/steel friction pair in vacuum and in the presence of TMPi (**a**). High resolution XPS spectra recorded in situ after the tribo-test in the presence of TMPi with and without the wear tracks (**b**)

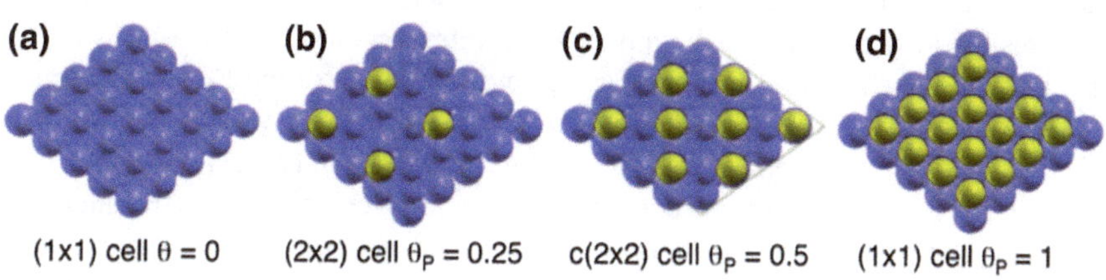

Fig. 12.6 Top view representation of the Fe(110) surface with different P coverages θ [20]

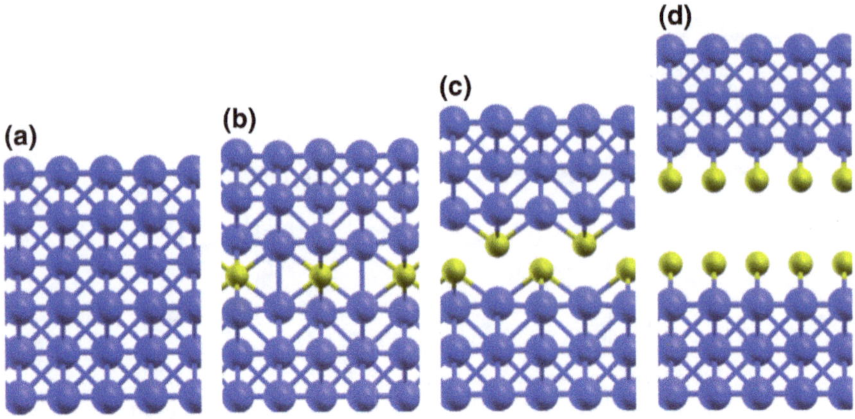

Fig. 12.7 Lateral view representation of the iron interfaces obtained by self-mating the surfaces with different P coverages represented in Fig. 12.6. The lateral and vertical relative position of the self-mated surfaces correspond to the optimal ones [62]

Table 12.1 The work of separation, W, and the equilibrium distance, z_{eq}, separating the innermost Fe layers at the interface are reported for the PES absolute minimum

Interfacial coverage	$W_{sep(J/m^2)}$	Z_{eq} (Å)	τ_x (GPa)	τ_y (GPa)
$\theta = 0$	4.8	2.0	10.1	10.1
$\theta_P = 0.25$	1.71	3.2	4.0	4.0
$\theta_P = 0.5$	0.6	4.3	3.3	2.1
$\theta_P = 1.0$	0.04	7.3	0.1	0.1

The shear strengths, τ, are calculated for the MEPs along the [−110] and [001] sliding directions [20]

The adhesion energy was calculated for different relative lateral positions, constructing in this way the PES for the sliding interface. In Fig. 12.8 the PESes of the bare interface and that obtained including interstitial phosphorus at an interfacial coverage of 0.25 are compared. A common energy scale is used, the blue color indicates the PES minimum, which is taken as reference and the red color is for the PES maximum. As can be seen in Fig. 12.8a, the minima of the PES for the clean interface are located at LB positions, i.e., where a new layer would be positioned in the bcc structure of iron bulk. Two clean iron surfaces in contact undergo a cold sealing (Fig. 12.7a): the optimized interfacial separation corresponds to the interlayer distance in Fe bulk, the work of separation W_{sep} is equal to the energy required to form two Fe(110) surfaces and the shear strength $\tau_{x,y} = 10$ GPa is typical of bulk iron.

The situation is completely altered if the mating surfaces are P-terminated. At interfacial coverage $\theta_P = 0.25$ (Fig. 12.7b) the distance between the interfacial Fe layers increases by 60% and the adhesion decreases by one order of magnitude with

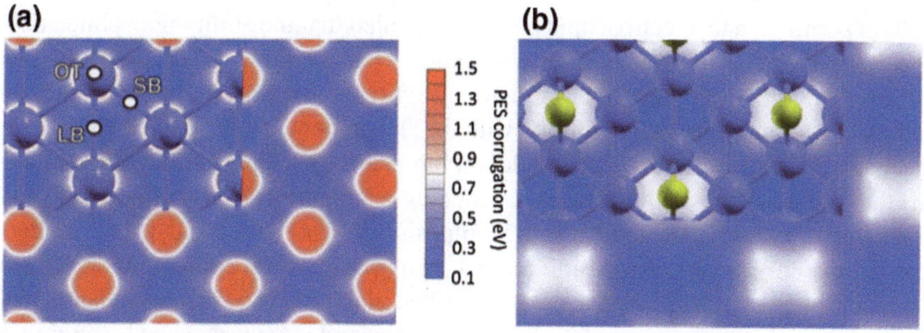

Fig. 12.8 Two-dimensional representation of the PESes obtained for the clean (**a**) and the P-containing iron interface (P atoms are in yellow) at relative concentration of 0.25 (**b**). A common energy scale is used for the potential corrugation [62]

respect to the clean iron interface. A decreased adhesion gives rise to a smoother PES, as can be seen in Fig. 12.8b, and lower frictional forces: the shear stress decreases by 60%.

In Fig. 12.7c, d it can be seen that by increasing the concentration of interfacial phosphorus to $\theta_P = 0.\ 5$ and $\theta_P = 1$, the distance between the surfaces increases and chemical bonds are no longer present across the interface, the surfaces are hold together by physical interactions. The disappearance of chemical bonds produces a considerable decrease in the adhesion and shear strength (last two columns of Table 12.1). In particular, in the case of two fully passivated surfaces in contact the short-range repulsion between interfacial P layers causes a decrease of adhesion and shear strength by two orders magnitude with respect to clean iron. The magnitude of the physical interactions are most likely underestimated by DFT-PBE. However, such consistent changes give a clear indication of the dramatic effect of phosphorus in reducing friction and adhesion at iron interfaces, in agreement with the tribological experiments and with the well known effect of iron embrittlement caused by phosphorous.

The above described results clearly indicate that atomic phosphorsus is able to effectively reduce the adhesion and shear strength of iron, in excellent agreement with the experimental observation of a friction reduction promoted by the formation of an iron phosphide tribofilm. This clearly exemplify the power of first principles calculation in the research on lubricant material, as chemical surface modifications, accurately described by first principles methods, can impact macroscale properties as the friction coefficient.

12.5 Conclusions

The power of ab initio methods in the research in lubricant materials has been highlighted through an example study on the functionality of organophosphorus additives.

GPL experiment and spectroscopic analysis revealed that the ability phosphite additives to reduce friction relies on the formation of an iron phosphide tribofilm. Ab initio calculations were applied to provide insights into these experimental findings and a twofold analysis was performed with the aim at identifying: (i) the microscopic mechanisms that lead to iron-phosphide formation through phosphite decomposition, (ii) the functionality of phosphorus in reducing friction of iron. The first issue was addressed by studying the molecular decomposition at the surface and at the tribological interface by means of ab initio molecular dynamics simulations. The atomistic mechanisms for P release were clearly identified and the role of mechanical stresses in promoting the process were highlighted. The second issue was faced by applying a computational protocol that allows one to calculate intrinsic interfacial properties as adhesion and shear strength, which provided results in excellent agreement with the experimental observations.

References

1. H. Spikes, Tribol. Lett. **60**(5), 1–26 (2015)
2. X. Wanli, Z. Wai, S. Bin, X. Ximeng, Wear **328–329**, 475–479 (2015)
3. A. Bouffet, SAE Paper 01-2022 (2004)
4. J. Vinci, US20060276352, Lubrizol Corporation (2006)
5. J. Zakarian, M. Haire, US20060252657, Chevron Texaco Corporation (2006)
6. S. Matsui, H. Komatsubara, US20070287643, Nippon Oil Corporation (2007)
7. T. Craig, T. Thomas, WO2007111741, Lubrizol Corporation (2007)
8. O. Furlong, F. Gao, Tribol. Int., 2008, 40, 699
9. Z. Tang, S. Li, Curr. Opin. Solid State Mater. Sci., **18**, 119 (2014)
10. H. Spikes, Lubr. Sci. **20**, 103 (2008)
11. J. Van Rensselar, Power Transm. Eng. (2003)
12. W. Davey, et al., Wear **1**, 291 (1958)
13. E.S. Forbes, A.J.D. Reid, ASLE Trans. **16**, 50 (1973)
14. R.C. Coy, T.F.J. Quinn, ASLE Trans. **18**, 163 (1975)
15. S. Mori, K. Hori, Y. Tamai, Junkatsu **27**, 505 (1982)
16. J. Tannous, B.M.I. de Bouchet, T. Le Mogne, P. Charles, J.M. Martin, Tribol.-Mater. Surf. Interfaces **1**, 98 (2007)
17. Z. Li, Y. Li, Y. Zhang, T. Ren, Y. Zhao, RSC Adv. **4**, 25118 (2014)
18. J. Li, B. Fan, T. Ren, Y. Zhao, Tribol. Int. **88**, 1 (2015)
19. A. Riga, J. Cahoon, W.R. Pistillo, Tribol. Lett. **9**, 219 (2000)
20. M.I. De Barros-Bouchet, M.C. Righi, D. Philippon, S. Mambingo-Doumbe, T. Le-Mogne, J.M. Martin, A. Bouffet, RSC Adv. **5**, 49270 (2015)
21. F. Gao, P.V. Kotvis, D. Stacchiola, W.T. Tysoe, Tribol. Lett. **18**, 377–384 (2005)
22. F. Gao, O. Furlong, P.V. Kotvis, W.T. Tysoe, Langmuir **20**, 7557–7568 (2004)
23. O. Beek, J.W. Givens, E.C. Williams, Proc. R. Soc. London Ser. A **177**, 103 (1940)
24. D. Godfrey, ASLE Trans. **8**, 1 (1965)
25. F.T. Barcroft, S.G. Daniel, Trans. ASME Ser. D **87** (1965)
26. Y. Yamamoto et al., Wear **50**, 343 (1978)
27. D. Sung, A.J. Gellman, Tribol. Lett. **13**, 1 (2002)
28. A.W. Holbert, J.D. Batteas, A. Wong-Foy, T. S. Rufael, C.M. Friend, Surf. Sci. **401**, l437 (1998)
29. D. Ren, G. Zhou, A.J. Gellman, Surf. Sci. **61**, 475 (2001)
30. D. Ren, A.J. Gellman, Tribol. Int. **34**, 353 (2001)

31. M.C. Righi, S. Loehlé, M.I. De Barros-Bouchet, D. Philippon, J.M. Martin, RSC Adv. **5**, 101162 (2015)
32. A. Rossi, F.M. Piras, D. Kim, A.J. Gellman, N.D. Spencer, Tribol. Lett. **23**, 197 (2006)
33. K. K. Mistry, A. Morina, A. Erdemir and A. Neville, Tribol. Trans., 2013, 56, 4
34. D. Philippon, et al., Tribol. Int. **44**, 113 (2011)
35. F.T. Barcro, Wear **3**, 440 (1960)
36. D. Philippon, M.I. De Barros Bouchet, Th. Le Mogne, E. Gresser, J.M. Martin, Tribol.-Mater. Surf. Interfaces **1**, 113 (2007)
37. D. Philippon, M.I. De Barros-Bouchet, Th. Le-Mogne, B. Vacher, O. Lerasle, J.-M. Martin, Thin Solid Films **524**, 191 (2012)
38. M.C. Righi, S. Loehlé, M.I. de Barros Bouchet, D. Philippon, J.M. Martin, RSC Adv. **5**, 101162–101168 (2015)
39. J.P. Perdew, K. Burke, M. Ernzerhof, Phys. Rev. Lett. **77**, 3865 (1996)
40. P. Blonski, A. Kiejna, Surf. Sci. **601**, 123 (2007)
41. G. Henkelman, B.P. Uberuaga, H.J. Jonsson, J. Chem. Phys. **113**, 9978 (2000)
42. A.W. Holbert, A. Batteas, A. Wong-Foy, T.S. Rufael, C.M. Friend, Surf. Sci. **401**, L437–L443 (1998)
43. D. Philippon, M.-I. De Barros-Bouchet, T. Le Mogne, O. Lerasle, A. Bouffet, J.-M. Martin, Tribol. Int. **44**, 684 (2011)
44. D. Philippon, M.-I. De Barros-Bouchet, T. Le Mogne, E. Gresser, J.-M. Martin, Tribol.-Mater. Surf. Interfaces **1**, 113 (2007)
45. T.E. Fischer, Ann. Rev. Mater. Sci. **18**, 303 (1988)
46. G. Heinicke, *Tribochemistry* (Carl Hanser Verlag, Munchen, 1984)
47. P.A. Thiessen, K. Meyer, G. Heinicke, *Grundlagen der Tribochemie* (Akademie Verlag, Berlin, 1967)
48. N.N. Gosvami, et al., Science **348**, 102 (2015)
49. B. Gotsmann, M.A. Lantz, Phys. Rev. Lett. **101**, 125501 (2008)
50. C.R. Hickenboth et al., Nature **446**, 423 (2007)
51. O.J. Furlong et al., ACS Appl. Mater. Interfaces **3**, 795 (2011)
52. O.J. Furlong et al., RSC Adv. **4**, 24059 (2014)
53. A. Ghanbarzadeh et al., Tribol. Lett. **1**, 61 (2016)
54. J.R. Felts et al., Nat. Commun. **6**, 6467 (2015)
55. D.B. Jacobs, R.W. Carpick, Nat. Nanotechnol. **8**, 108 (2013)
56. H.L. Adams, M.T. Garvey, S.U. Ramasamy, Z. Ye, A. Martini, W.T., Tysoe, J. Phys. Chem. C **119**, 7115 (2015)
57. J. Zhang, H. Spikes, Tribol. Lett. **63**, 24 (2016)
58. S. Kajita, M.C. Righi, Carbon **103**, 193 (2016)
59. G. Zilibotti, S. Corni, M.C. Righi, Phys. Rev. Lett. **111**, 146101 (2013)
60. S. Lohelé, M. C. Righi, Lubricants **6**, 31 (2018)
61. H.J. Monkhorst, J.D. Pack, Phys. Rev. B **13**, 5188 (1976)
62. M.I. De Barros-Bouchet et al., RSC Adv. **5**, 49270 (2015)
63. D.E. Jiang, et al., Surf. Sci. **570**, 167 (2004)
64. G. Zilibotti, M.C. Righi, Langmuir **27**, 6862 (2011)
65. M.I. De Barros Bouchet, G. Zilibotti, C. Matta, M.C. Righi, J. Phys. Chem. C **116**(12), 6966–6972 (2012)
66. M.C. Righi et al., RSC Adv. **6**, 47753 (2016)
67. P. Restuccia, M.C. Righi, Carbon **106**, 118–124 (2016)
68. P. Restuccia et al. Comp. Mat. Sci. **154**, 517 (2018)
69. W. Arabczyk, T. Baumann, F. Storbeck, H.J. Mussig, A. Meisel, Interaction between phosphorus and oxygen on the Fe(111) surface. Surf. Sci. **189**, 190–198 (1987)
70. G. Fatti, P. Restuccia, C. Calandra, M.C. Righi (to be published)

Index

A
Ab Initio Molecular Dynamics (AIMD), 310
Abrasion, 31
Additives, 111
Adhesion, 31
Ag/AgCl – Electrode, 251
AISI 52100, 113
Alkyl cyanobiphenyl (ACB), 225
Alkyl cyanocyclohexylphenyl (ACC), 225
Aluminum, 302
Amorphous carbon (a-C), 83
Analytical Electron Microscopes (AEM), 16
Anti-Wear (AW) additives, 310
Antimony trioxide, 25
Antiwear mechanism, 201
Atomic Force Microscopy (AFM), 248
ATR crystal, 136
ATR-FTIR, 177
Attenuated Total internal Reflection Infrared spectroscopy (ATR-IR), 136
Attenuated Total Reflection (ATR) IR spectroscopy, 227
Au(111), 254
Auger stectroscopy, 138
Austenitic materials, 34

B
Beer-Lambert's Law, 65
Bis(trifluoromethylsulfonyl)imide (Tf_2N), 263
Blocking dislocations, 34
Borate, 111
Boundary lubrication, 108
Brass, 295
Bremsstrahlung, 17
Bright-field (BF), 15
Brilliance, 238
Butylated triphenyl phosphorothionate (b-TPPT), 178
1-butyl-3-methyl-imidazolium ([BMIM]), 265

C
CoCr29Mo6, 34
17CrNiMo7-6, 48
18CrNiMo7-6, 36
CoCrMo, 40
Collision cascade, 6
CrN, 238
Cross sections, 7
CuZn5, 295
Cyclic Strain Gradient, 33
Cyclic Voltammetry (CV), 251
Cyclic Voltammogram, 252

D
Dark-field (DF), 15
Debris, 302
Delamination, 34
Density Functional Theory (DFT), 312
Dialkyldithiophosphate (DDP), 159
Diamond Anvil Cell (DAC), 224
Diamond-like carbon coating, 20
Digital Holographic Microscopy (DHM), 290, 291
Dimethyl disulphide (DMDS), 141, 146
Dimple grinding, 52
DLC, 176

E
EDS, 124
Electrochemical AFM (EC-AFM), 253

M. Dienwiebel and M.-I. De Barros Bouchet (eds.), *Advanced Analytical Methods in Tribology*, Microtechnology and MEMS,
https://doi.org/10.1007/978-3-319-99897-8

Electrochemical Double Layer (EDL), 249
Electrochemical Friction Force Microscopy (EC-FFM), 247, 254
Electrochemically modified Surface Force Apparatus, 253
Electron Backscattering Diffraction (EBSD), 3, 10
Electron Energy Loss Spectroscopy (EELS), 44
Electron microscopy, 3
Emulsions, 228
EN ISO 25178, 291
Entrance Grid Bias (EGB), 71, 73
Environmentally Controlled Analytical Tribometer (ECAT), 109
Ethyl-3-methyl-imidazolium (EMIM), 265
Extended X-ray Absorption Fine Structure (EXAFS), 64
Extreme-Pressure (EP), 310

F
Fe(110), 316
Fe–O–S ternary system, 174
FeOOH, 261
Fermi's Golden Rule, 69
Ferritic steel, 41
Fluorescence Yield (FY), 68, 71, 189
Focused Ion Beam (FIB), 3, 5, 52, 121
Fourier-Transform Infrared spectroscopy (FTIR), 216
Frictogram, 259

G
Gas Phase Lubrication (GPL), 108, 130, 146, 317
Glassy polyphosphate, 188
Glycol, 224
Gold oxide, 261
Grain contrast images, 6
Grease, 229

H
Hard and Soft Acids and Bases (HSAB), 165
Heat-Assisted Magnetic Recording (HAMR), 83
Hexadecane, 301
1-hexanethiol(C_6H_{13}-SH), 113, 116
1-hexanethiol 1-hexyl-3-methyl-imidazolium (HMIM), 265
Hip joints, 48
Hydrogen-free tetrahedral amorphous carbon (ta-C), 88
Hydrogenated amorphous carbon (a-C:H), 75, 177
Hydrolytic decomposition, 169

I
IF-MoS_2, 274, 275
Inner Helmholtz Plane (IHP), 249
In situ AES, 117
In-situ AFM, 182
In situ IR analysis, 217
In-situ lift-out, 8
In situ TEM, 274
In-situ XANES, 182
In-situ XAS, 183
Interfacial Force Microscope (IFM), 192
Inverse pole figure maps, 13
Ion damage, 6
Ionic Liquids (ILs), 263
Ion milling, 52
IR spectrum, 221
ISE imaging, 7

K
Kikuchi bands, 10

L
Lateral force map, 262
LI-ZDDP, 168
LIGA, 20
Ligand exchange reaction, 163
Liquid Metal Ion Source (LMIS), 5
Low-Energy Electron Diffraction (LEED), 143, 312
Lowest Unoccupied Molecular Orbital (LUMO), 67

M
Magnetically-guided imaging NEXAFS spectroscopy, 91
Main wear mechanisms, 32
Mechanical mixing, 32, 49
Metallic alloys, 3
Methane, 132
Mica, 256
Microanalysis, 16
Microelectromechanical systems (MEMS), 20
Micro-FTIR, 216
Molecular Dynamics (MD) simulations, 88
Molybdenum disulfide (MoS_2), 25
Multivariate Statistical Analysis (MSA), 18

N
N-oleoylsulcosine, 231
N-alkanethiols, 112
$NaClO_4$, 256
Nanocompression, 275
Nanocrystalline metals, 23
Nanoindentation, 192

Nanoindenter, 274
NaOH, 261
Near edge X-ray absorption fine structure (NEXAFS), 63
NEXAFS spectroscopy, 63
Nudged Elastic Band (NEB) method, 312

O
Oiliness agent, 220
Oleic acid, 221
Outer Helmholtz Plane (OHP), 249

P
$P(CH_3O)_2$, 313
P-based additives, 163
Partial Electron Yield (PEY), 71
Perfluoropolyether (ZDOL), 87
Phosphate, 111
Phosphite, 111
Photoabsorption edge, 66
Photoelectric absorption coefficient, 65
Photoemission Electron Microscopy (PEEM), 71, 87
Piston ring, 43
Plasma source ion column, 5
Plowing friction, 293
Point of Zero Charge (PZC), 249
Polarized IR spectroscopy, 225
poly α-olefin (PAO), 147, 294
Polybutenyl saccimide, 231
Polybutylene glycol, 222
Polypropylene glycol (PPG), 222
Precursor gas, 7
Pressure-viscosity coefficient, 222
Principal Components Analysis (PCA), 18
Protective layer, 7
PTWA Coating, 43

Q
Quasi-Reference Electrode (QRE), 251

R
Ratio of sliding to rolling, 187
"Redhead" equation, 149
Reference Electrode (RE), 251
Reflection-Absorption Infrared Spectroscopy (RAIRS), 135
Roughness, 291
Running-in, 294

S
Sandia's Automated eXpert Spectral Image Analysis (AXSIA), 18
Saturated Calomel Electrode (SCE), 251
Schwoebel-Ehrlich barrier, 257
Selected-Area Diffraction (SAD), 15
Self-Assembled Monolayers (SAMs), 112
SEM cross section, 36
Shake-up structures, 68
Shear-induced surface-to-bulk transport, 155
Silicon-and oxygen-containing hydrogenated amorphous carbon (a-C:H:Si:O), 77
Silicon oxide-doped diamond-like carbon (SiO_x-DLC), 85
Slide-to-Roll Ratio (SRR), 187
Spacer Layer Imaging Method (SLIM), 190
Spectral image analysis, 17
Standard Hydrogen Electrode (SHE), 251
Sum Frequency Generation (SFG), 227, 265
Surface Fatigue, 31
Surface-Enhanced Raman Spectroscopy (SERS), 135
Surface-Force Apparatus (SFA), 189
Synchrotron, 63
Synchrotron beamline, 70
Synchrotron radiation, 242

T
TEM-AFM holder, 282
TEM-Nanoindentor, 275
Temperature-Programmed Desorption (TPD), 147
Temperature-Programmed Reaction Spectroscopy (TPRS), 312
Tenacity, 196
Tetrafluorobarate (BF_4), 263
Thermal decomposition, 164
Thermal films, 164
Thermal liquid cell, 183
Thermo-oxidative decomposition, 166
Thickener, 229
Ti-doped (Ti-C:H) DLC coating, 177
TiAl6V4, 48
TiAlN, 241
Time-of-Flight Secondary-Ion Mass Spectrometry (ToF-SIMS), 121
Topography, 296
Total Electron Yield (TEY), 71, 189
Transition Dipole Moment (TDM), 69
Transition Metal Dichalcogenides (TMD), 25
Transition zone, 32
Transmission Electron Microscopy (TEM), 5, 12, 29, 30, 121
Transmission Kikuchi Diffraction (TKD), 8, 10
Tribochemical reactions, 31, 32, 111, 309
Tribochemistry, 130
Tribocorrosion, 249
Tribofilm, 116

Tribomaterial, 39
Tricresylphosphate (TCP), 204
Trimethylphosphite (TMPi), 311
Tris(pentafluoroethyl)trifluorophosphate (FAP), 263, 265
Tungsten, 298
Tungsten disulfide (WS_2), 25

U
Ultra-High Vacuum (UHV), 108, 140, 160
Ultra-mild sliding wear, 29, 30
Ultrananocrystalline diamond (UNCD), 75
Urea-type thickener, 230

V
Viscosity Index Improver (VII), 222

W
Water-based lubricants, 229
Wavy slip, 34
Wear particle analyses, 47
Wear particle preparation, 54
Wear particles, 49
Wear-induced recrystallization, 20
Wire drawing, 231
Working Electrode (WE), 251
WS_2 nanoparticle, 279

X
X13CrMnMoN18-14-3, 34
X5CrNiMo17-13-2, 34
X6CrNiMoN22-10-4-3, 34
X-ray Photoelectron Spectroscopy (XPS), 110, 312
X-ray Absorption Near Edge Structure (XANES), 64, 238
X-ray Photoemission Electron Microscopy (XPEEM), 190

Z
ZDDP tribofilm, 166
Zincdialkyldithiophosphate (ZDDP), 159

FSC
www.fsc.org
MIX
Papier | Fördert gute Waldnutzung
FSC® C083411

Kolibri 360 GmbH
Ferdinand-Jühlke-Straße 7
99095 Erfurt, Deutschland
produktsicherheit@kolibri360.de